Green Energy and Technology

Climate change, environmental impact and the limited natural resources urge scientific research and novel technical solutions. The monograph series Green Energy and Technology serves as a publishing platform for scientific and technological approaches to "green"—i.e. environmentally friendly and sustainable—technologies. While a focus lies on energy and power supply, it also covers "green" solutions in industrial engineering and engineering design. Green Energy and Technology addresses researchers, advanced students, technical consultants as well as decision makers in industries and politics. Hence, the level of presentation spans from instructional to highly technical.

Indexed in Scopus.

More information about this series at http://www.springer.com/series/8059

Caroline Hachem-Vermette

Solar Buildings and Neighborhoods

Design Considerations for High Energy Performance

Springer

Caroline Hachem-Vermette
School of Architecture
Planning and Landscape
University of Calgary
Calgary, AB, Canada

ISSN 1865-3529 ISSN 1865-3537 (electronic)
Green Energy and Technology
ISBN 978-3-030-47015-9 ISBN 978-3-030-47016-6 (eBook)
https://doi.org/10.1007/978-3-030-47016-6

All figures and illustrations are prepared by Dr. C. Hachem-Vermette and M. R. Verma

This Springer imprint is published by the registered company Springer Nature Switzerland AG
The registered company address is: Gewerbestrasse 11, 6330 Cham, Switzerland

I dedicate the Design Considerations for Solar Buildings and Neighborhoods book to Alyss-Andra and Marc

Preface

The risks of climate change are becoming more tangible, as the number of worldwide climate attributed disasters is increasing. In response, a worldwide consensus is reached to reduce the negative impact of the built environment, since it is responsible for a significant portion of energy consumption and greenhouse gas emissions. This entails increasing energy efficiency of buildings and increasing the implementation of renewable energy resources.

Solar energy is considered as one of the most desirable and affordable renewable energy sources. A rapid growth in the production of solar technologies and their implementation, worldwide, highlights the role of solar energy as a significant future energy source.

One of the benefits of solar energy is related to the fact that it can be implemented in various applications and harnessed in different methods. Solar energy can be utilized in passive applications, such as in passive space heating, daylighting as well as in passive cooling. In such passive applications, solar energy is directly utilized by buildings, without the need for mechanical devices. In addition, solar radiation can be converted by various types of active technologies into electrical and thermal energy.

Notwithstanding the general interest in exploiting solar energy as renewable energy source, in buildings and urban areas, there are still obstacles that hinder the full implementation of solar technologies, as well as the application of passive solar design principles, especially in urban planning. For instance, the effects of design parameters of buildings and neighborhoods on solar capture and utilization are still not well defined. Such design parameters include architectural, engineering, and planning considerations, which need to be systematically integrated into a holistic approach, to allow a practical and functional solar, energy-efficient neighborhood. Such approach should consider the interaction between individual buildings, and methods of assemblage of these buildings in varying density configurations and site layouts.

This study presents the main principles of the design of buildings and neighborhoods with increased potential of capturing and utilization of solar energy. It discusses practical issues in the design of the built environment and their impact on energy performance. These principles span from building component level, and specifically building envelope design, its overall impact on energy performance and solar energy capture, to the general aspect of neighborhood design, including buildings' density and streets' layout.

The information presented in this book will allow the reader to understand the impact of early stages of architectural, planning as well as engineering design considerations on the energy performance of buildings and communities. Such decisions can be easily modified in the planning stages, as compared to later stages of design, and can significantly affect the performance during building operations.

This book is primarily addressed at building professionals, including architects, engineers, and urban planners. It can also be used as background source in courses on energy efficiency in buildings and urban planning. The information provided can be divided into two main parts: solar buildings and solar neighborhoods. Following is a brief outline of the main chapters.

Solar Buildings

This part, consisting of Chaps. 1–5, is concerned mainly with building design to increase the potential of solar capture and utilization. The main focus is on the envelope and layout of buildings.

Chapter 1: Principles of Passive Solar Design
The main principles to capture and utilize solar energy in buildings are presented in Chap. 1. It includes a historical overview of solar energy application, including ancient examples related to specific civilizations, as well as some more modern examples of solar design applications. The chapter outlines basic principles of solar radiation and of solar energy utilization in the built environment, highlighting the difference between passive and active design applications. Passive solar design process is then discussed in more detail, including principles and main applications. Passive solar energy applications include passive heating and cooling, and daylighting. The role of building design in capturing and utilization of passive solar energy is highlighted.

Chapter 2: Impact of Building Envelope
Chapter 2 focuses on the building envelope, presenting its role and functional aspect, as well as its main components. This chapter discusses the impact of building envelope on energy performance and methods to enhance it. A summary of heat transfer mechanisms through the building envelope is presented together with the basic design considerations that need to be implemented in order to reduce energy loss through it, while not compromising the admission of solar radiation into the indoor space. The chapter discusses building-integrated passive systems, including advanced building envelope and components that aim at enhancing solar energy capture and utilization.

Chapter 3: Selected High-Performance Envelopes
Chapter 3 focuses on selected high-performance building envelope systems, presenting such envelope systems as key components in the design of solar buildings and communities. The chapter divides building envelope systems into two main categories, those associated with low-rise (up to 3 floors) and multistory buildings. Discussed low-rise building envelopes include double-stud walls, structural insulated panels, and insulated concrete forms. Multistory building envelope systems presented in this chapter include double-skin facades and climate-adaptive facade systems.

Chapter 4: Active Solar Technologies
This chapter presents a summary of active devices, which can convert solar radiation into thermal and electric energy to be utilized in various building applications. These include space heating, domestic hot water, and various electric loads. Solar technologies include different photovoltaic (PV) technologies ranging from the traditional PV panels to transparent and semi-transparent PV modules, hybrid PV thermal collectors, and solar thermal collectors. The chapter discusses methods of integration of these technologies in buildings.

Chapter 5: Advanced Active Solar Design
This chapter presents advanced geometrical roof designs for low-rise buildings, and facades, mostly applied to multistory buildings, to allow increased solar electricity production. The chapter discusses a number of roof and facade designs, ranging from simple commonly applied to more sophisticated designs, based on multifaceted geometries. The impact of these designs on the thermal performance of specific building examples is discussed. In addition, the chapter explores the impact of buildings' layout, on the design of roof and facade surfaces to integrate solar collectors as well as on the overall energy performance. This chapter relies on hypothetical examples that are designed, and systematically simulated and analyzed, to demonstrate various conceptual designs of PV-integrated roof and facade systems.

Solar Communities

This part presents design principles for building clusters and neighborhoods to increase solar access and energy efficiency. Issues related to density, street layouts, and building layouts are discussed. Various concepts of sustainable neighborhoods are presented, including net-zero energy and carbon-neutral neighborhoods. In addition, the impact of neighborhood spatial design on issues, such as transportation and associated GHG emissions and the overall resilience, is discussed.

Chapter 6: Introduction to Solar Neighborhoods
The general concept of solar neighborhoods and its main characteristics are presented. This includes an overview of urban-scale energy performance, comprising

energy consumption and potential of renewable energy generation. The chapter underlines specific design parameters that affect the neighborhood solar energy access and utilization, such as building design, density, and community layout. In addition, advanced neighborhood design trends such as *Net-zero Energy Neighborhoods* and *Zero-Carbon Emissions Neighborhoods* are briefly discussed, and the impact of reducing the dependence on fossil fuel and utilization of solar energy in such neighborhoods is highlighted. The role of modeling and simulation methods in studying energy efficiency and solar potential strategies, applied on neighborhood scale, is highlighted.

Chapter 7: Residential, Low-Density Neighborhoods
The main principles underlying the design of solar residential, low-density neighborhoods are outlined. The parameters influencing the capture of solar energy by buildings are discussed. These parameters consist mainly of density and street layout, as well as building shapes. A number of hypothetical case studies of residential neighborhoods are employed to highlight design principles and their impact. The selected examples aim at presenting flexibility of design, while maximizing solar capture and utilization.

Chapter 8: Mixed–Use Neighborhoods
Chapter 8 consists of an overview of the main design issues, as well as opportunities, in planning mixed-use solar neighborhoods. The chapter delineates the impact of neighborhood's shape, layout of streets, density of buildings, and combination of buildings' types, on the overall neighborhood performance. The application of energy generation and storage technologies is discussed, including the design of PV, solar thermal collectors, and borehole thermal storage. The chapter highlights some design considerations to improve energy and environmental performance, as well as the overall resilience of neighborhoods. These design considerations include optimal mixture of building types, the impact of neighborhood design on transportation, and its related energy consumption and GHG emissions, as well as the impact of street layout on the overall resilience of the neighborhood for disasters.

Calgary, Canada Caroline Hachem-Vermette

Contents

Abbreviations

ACH	Air Change per Hour
AES	Alternative energy solutions
AP	Apartment buildings
AR	Aspect Ratio
A-Si	Amorphous silicon
BIPV	Building integrated PV systems
BIPV/T	Building integrated PV /thermal system
BIST	Building integrated solar thermal collectors
BPV	Biophotovoltaic Panels
BSI	Building System Integrator
BTES	Borehole Thermal Energy Storage
CBD	Central business district
CdTe	Cadmium Telluride
CFARi	Commercial floor area ratio index
CHP	Combined heat and power generation
CIS	Copper Indium Diselenide
CLF	Commercial land fraction
DG	Distributed energy
DHW	Domestic hot water
DSF	Double skin facade
DSW	Double stud wall
EPS	Expanded polystyrene foam insulation
ETFE	Ethylene Tetrafluoroethylene
EVA	Ethylene Vinyl Acetate
FAR	Floor area ratio
FF	Flat facade
FTO	Fluorine-Doped Tin Oxide
GHG	Greenhouse gas emissions
GHP	Geothermal heat pump
GSHPs	Ground source heat pumps

HDPE	High Density Polyethylene
HP	Heat pump
HRV	Heat recovery ventilator
HST	Horizontal saw-tooth
HVAC	Heating, Ventilation, and air conditioning
IAQ	Indoor air quality
IBE	Intelligent Building Envelope
ICF	Insulated Concrete Forms
Low-E	Low emissivity
LSC	Luminescent solar concentrator
MURBs	Multi-Unit Residential Building
NEC	Net energy consumption
PCMs	Phase change materials
PV	Photovoltaic
RES	Renewable energy sources
RH	Relative humidity
ROP	Ratio of performance
RP	Rectangular pyramids
SD	Single detached houses
SDD	Summer design day
SDR	Shading depth ratio
SHGC	Solar heat gain coefficients
SIP	Structural Insulated Panels
STC	Solar thermal collectors
STPV	Semi-transparent PV
TH	Townhouses
TND	Traditional Neighborhood Development
TOD	Transit Oriented Development
TP	Triangular pyramids
u/a	Units per Acres
VIP	Vacuum Insulated Panels
VLT	Visible Light Transmittance
VST	Vertical saw-tooth
VT	Visible Transmittance
WDD	Winter design day
WtE	Waste to Energy
WWR	Window to wall area ratio
XPS	Extruded Polystyrene

List of Figures

Chapter 1
Principles of Solar Design

This chapter summarizes the main solar principles to allow capture and utilization of solar energy in buildings. A brief historical view of solar energy application is first presented, followed by a brief outline of solar energy utilization in the built environment. The chapter highlights the difference between passive and active solar design, and focuses on passive solar design principles and main applications. Utilization of passive solar energy in buildings, summarized in this chapter, includes passive heating and cooling, and daylighting. This chapter highlights the general role of building design in capturing and utilization of passive solar energy.

1.1 Solar Energy: Brief Historical Overview

The capture and utilization of solar energy in buildings and the built environment is not a new trend in the history of architecture. Passive solar energy was exploited by numerous civilizations, including Egyptian, Greek, Roman, Persian, Hindu, and the Native American civilizations. The utilization of solar energy for passive heating, daylighting, and other innovative applications was referred to in the writings of philosophers, such as Aristotle, Xenophon, and Vitruvius [1, 2]. In fact, it is only in the last 100 years that building design started to depend significantly on mechanical systems and on artificial lighting, steering away from relying on natural solar energy in buildings.

1.1.1 Examples from Ancient History

Designing homes to exploit solar heat during winters and avoid it during summers was part of the Chinese traditional architecture [3]. For instance, Neolithic Chinese homes were designed with the sole opening facing south. On a larger scale, Chinese

C. Hachem-Vermette, *Solar Buildings and Neighborhoods*, Green Energy and Technology, https://doi.org/10.1007/978-3-030-47016-6_1

urban planners constructed the main streets of towns to run east–west, so all houses can benefit from solar energy [4].

A striking example of ancient teachings on methods of designing houses to adapt solar radiation capture to domestic human needs is attributed to the Greek philosopher Socrates. He stated: "in houses that look toward the south, the sun penetrates the portico in winter, while in summer the path of the sun is right over our heads and above the roof, so that there is shade" [2]. Aristotle also states that houses should be sunny in winter and ventilated in summer.

Ancient Greeks designed their houses to take advantage of passive solar radiation even on the scale of whole cities. Houses, for example, were constructed in block rows running east–west. The north wall was built of adobe with few or no openings, to reduce heat loss during winter. Main living areas were designed on the south orientation. An interesting feature of these houses was the portico, constructed on the south facade, which shade the house in summer, and allows the penetration of solar radiation during winter. North Hill in the city of Olynthus was the first planned solar city [5]. The city consists of orthogonal street network with main streets running east–west, allowing thus all houses to be built with south exposure.

The Roman civilization also applied solar architecture principles. Vitruvius wrote in his book, *The Ten Books of Architecture*, that houses should be shut in on the north orientation and exposed on the south [6]. Vitruvius also recommended the location of some specific rooms, so as to take advantage of solar energy. Romans were early to realize that increasing the potential of buildings to capture solar heat allows to reduce the cost for heating the buildings, by designing smaller furnaces that require less fuel.

Romans were the first to explore trapping passive heat in indoor spaces, employing transparent glass [7, 8]. The roman baths present a prominent application of this principle, where passive heat gain is employed to keep the water warm during cold weather [8]. The Romans also employed glazing for horticultural purposes to enhance cultivation of fruits and vegetables [8].

Another Roman invention consisted of employing thermal storage to keep the floor of sun-heated rooms warm during the winter nights. They constructed shallow pits under the floor, filled with rubble and topped with a black cover (e.g., Dark sand). This allows absorption of solar heat, which is transferred to the pebbles where it is stored. The stored heat is then released when the ambient temperature drops below the temperature of the rubble [9].

In North America, there are few pre-colonial examples of the application of solar design at the scale of communities. The cliff dwellings at Mesa Verde (twelfth century) in southwest Colorado are a striking example of passive solar buildings [10, 11]. These dwellings have openings on the south, with overhangs to control solar radiation and a massive construction that allows the regulation of diurnal temperature.

1.1.2 Examples in Modern History

Solar energy applications persisted in South Europe through the nineteenth century. When such interest in solar design reached Northern Europe, it was poorly applied. It was common to see porticos and window openings applied at inappropriate orientations, resulting in opposite impact to that originally intended [9].

A novel approach to building design took place in Germany in the nineteenth century, taking into account solar trajectory over the seasons, thus facilitating thermal comfort year-round. Such principles had impact on architectural teaching and practice, resulting in initiation and dissemination of solar homes throughout Europe.

Despite the emergence of interest in solar energy utilization, a systematic application of passive solar design in architecture started only around the mid-twentieth century, principally due to rising cost of energy from conventional sources, and due to potential energy shortage issues.

In the United States, a number of architects were active since the 1930s in designing and building solar houses [12]. Prefabricated buildings implementing passive design strategies started to appear during the same period (1940s). The main strategies that these buildings implemented include large south-facing windows and large overhangs to control and regulate solar radiation [2]. Figure 1.1 represents an example of the application of passive solar design principles in new built houses.

Experimentation with passive solar design was conducted in different areas of the world, with the objective of finding innovative methods to collect, control, and store heat gain. For instance, during the 1950s, the Trombe wall was developed (by

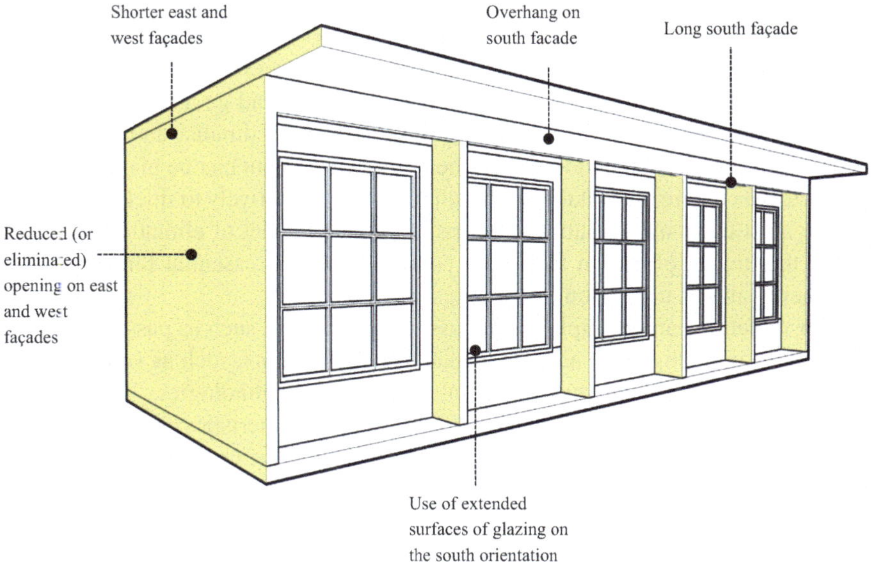

Fig. 1.1 Illustration based on a solar house designed by George Keck [9]

Felix Trombe, France), becoming one of the best-known examples of indirect heat gain collection (see Sect. 1.3). Trombe constructed the entire south wall of a house of dark, thick masonry with few or no windows and added a layer of glass on the outside of this wall, to allow solar energy to be transmitted to the masonry layer [13]. The purpose of this design is to collect heat during the day within the masonry (or the thermal mass), to be released gradually to the interior space during the colder hours of the evening and night. This method of design is presented in detail below (Sect. 1.3.1).

The recent increase in the application of and interest in passive solar energy, worldwide, is partly attributed to the awareness of the pressing risk of climate change and the need to tackle it. A variety of methods of employing solar energy for heating within buildings have been introduced in relatively cold climate. Passive solar design is becoming more applicable in cold climate due to the advancement in technologies and building materials, and their availability at reasonable cost. The development of high-performance windows plays a major role in the implementation of larger south-oriented window areas, which enhances solar potential capture for passive utilization in buildings.

1.2 Solar Energy Utilization

1.2.1 Solar Energy and the Built Environment

The availability of solar energy and its potential utilization in buildings and the built environment depends on a large number of factors that can be divided into two categories: those that can be controlled through planning and architectural design of buildings, and those that are given depending on climate and geographic locations. While solar radiation intensity depends on atmospheric, climatic, and geographic conditions, design of the buildings and the built environment can be manipulated to maximize the capture of available solar radiation, or, alternatively to shield the indoor space against excessive radiation. Understanding the impact of climatic factors, as well as the impact of design factors on solar radiation, is essential for harnessing solar energy and its utilization in buildings.

Solar radiation can be employed in passive applications such as passive heating, cooling, and in daylighting, as well as in active applications, such as generation of electricity, space, and water heating, employing diverse technologies.

This section presents a brief summary of the main characteristics of solar radiation as it reaches the earth, and the main climatic and geographic conditions that control its intensity.

Solar radiation

The radiation reaching the surface of the earth can be divided into shortwave radiation with wavelengths between about 0.29 and 4 μm, and thermal longwave radiation with

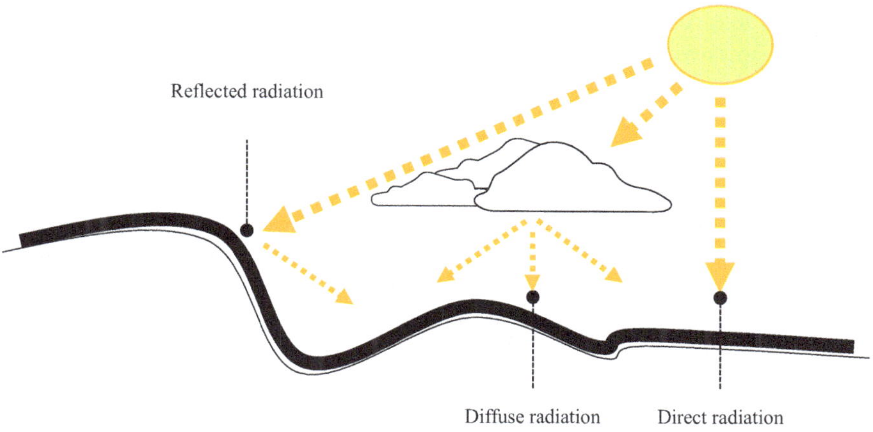

Fig. 1.2 Illustration of solar radiation components: direct, diffuse, and reflected

wavelengths between 4 and 100 μm [14, 15]. Shortwave radiation can be generally absorbed by various bodies including clouds, and re-emitted as longwave radiation (thermal radiation).

The shortwave radiation may be direct, diffuse, or reflected (Fig. 1.2). Direct radiation occurs mainly on clear days, when there are no obstacles intercepting the path of the solar radiation. Diffuse radiation is the portion scattered by gases and aerosols including dust particles, sulfate particles, sea salt particles, pollen, and others. On a clear day, the diffuse component of the solar radiation constitutes some 10–30% of the total incident radiation. On a cloudy day, solar radiation can be totally diffused.

Reflected radiation is the portion of the solar radiation reflected from the terrain, its components (such as snow and grass), and from surrounding bodies and surfaces.

In designing buildings and the built environment for solar energy capture, these three components—direct, diffuse, and reflected radiation—should be taken into consideration, as they can affect significantly the design factors.

Determining factors

The quantity of solar radiation reaching a surface is determined by various climatic and geographic conditions such as the time of year, atmospheric diffusion, cloud cover, and characteristics of the surface including its shape, position, and reflectivity.

Atmospheric and climatic conditions: The location of the sun in the sky for a given geographic location depends on two main factors: the season and the time of day. In addition, the atmosphere surrounding the earth modifies the amount of solar radiation that reaches earth's surface. Cloudy conditions or polluted atmosphere will interfere with the intensity of solar radiation and its type (direct or diffuse). Diffuse sky condition is most beneficial for daylighting in buildings.

Location: The geographic location of a surface is mainly expressed by its latitude—representing a measurement of the location with respect to the equator (north or south of the equator). Seasonal variations increase gradually with higher latitude (towards north or south of the equator). In higher latitude locations, summer days are longer than they are in areas located close to the equator, while winter days are shorter. In addition, higher latitude locations experience lower solar radiation intensity per unit area than near-equatorial locations, and larger seasonal swing of sunrises and sunsets.

Topography and landscaping: The amount of solar energy incident is significantly affected by the topography of a specific location on the Earth's surface. Topographic characteristics such as variability in elevation, slope, slope orientation, and shadowing, can create strong local gradients in solar radiation [16–19]. The distribution of slopes in hilly and mountainous terrains, for instance, has major effect on solar radiation incident on the surface. In a rough terrain, some areas although south facing may not receive any direct radiation throughout the year because of the surroundings (e.g., high hills) [20].
Landscaping plays an important role in passive solar design. Deciduous trees should be considered on south orientation, to avoid blocking solar radiation in the winter.

1.2.2 Passive Solar Energy Design

Solar energy can be utilized in the built environment in two main ways: Passively and actively.

Passive design consists mainly of capturing and applying solar energy in buildings without the need of mechanical means for moving heat and its distribution. In contrast, active solar design employs specific means to convert solar energy to usable thermal or electrical energy by means of solar collectors. These collectors include thermal collectors that generate thermal energy, for domestic hot water (DHW) and space heating, as well as photovoltaics (PV) for generation of electricity. Additional types of hybrid photovoltaic/thermal (PV/T) systems are also employed to generate both electricity and thermal energy. Active solar technologies for electric and thermal energy generation are presented in detail in Chap. 4.

Passive solar design involves a number of strategies including the following:

- Employing a holistic approach that relies on the integration of a building's architecture, its envelope design, and construction materials, together with the mechanical systems for heating and cooling, in both design and operation stages.
- The collection, storage, and redistribution of solar energy.
- Cutting heat loss, maximizing solar heat gains in winter and passive cooling in summer, and providing daylighting, thus reducing overall energy consumption.

Advantages and risks

Advantages

There are a number of benefits and advantages associated with the exploitation of passive solar energy in buildings. Those are summarized below:

- Implementing passive methods for passive heating or cooling in buildings reduces the dependence on mechanical systems, with associated reduction of fuel consumption to operate those systems. Similarly, maximizing daylighting results in reduced artificial lighting and associated electricity consumption. The combined effect leads to reduction of the overall energy cost for building operations.
- Potential environmental benefits of reduced GHG emissions, associated with reduced energy consumption.
- Overall human comfort, when the design is properly implemented, including thermal and visual comfort. This is in addition to reducing negative effects associated with mechanical systems, such as space and sound insulation requirements.
- Psychological benefits and increased productivity due to abundant but controlled solar radiation admitted to the interior space.
- Passive design can be aesthetically pleasing, if properly implemented, as it involves large windows, sunny interiors, and open floor plans.

Risks

A number of negative effects may arise if passive design is not properly implemented. These include the following:

- Overheating: The issue of overheating arises when solar radiation is admitted into the space in large quantities, without control or tempering of the indoor peak temperature (such as use of shading devices).
- Material deterioration: Excess solar radiation can damage artwork or furnishings inside buildings if these experience long-term exposure, unmitigated by shading devices.
- Glare: This phenomenon is caused by non-uniform luminance distribution and high contrasts of luminance level between the glare source (e.g., window) and its surroundings. Glare can lead to discomfort and reduced visual performance.

Passive Solar Design Principles

A successful passive solar design application entails a thorough, simultaneous implementation of a number of principles and building design considerations. These principles can be modified in relation to the climatic conditions in which a building is planned.

These considerations are summarized in five main principles, presented below [3].

- Solar Gain: The design should allow getting sufficient amount of solar radiation inside the building, when it is most useful throughout the year (e.g., during the cold months).

- Thermal Storage: This consists of storing some of the solar energy heat gain captured within the building, to keep the indoor space warm during colder hours. Designing passive storage assists as well in avoiding overheating during the day by reducing the temperature swing (see below).
- Conservation of energy: Heat transfer (especially heat loss) through the building envelope should be avoided. This can be achieved by designing a high-performance building envelope (e.g., highly insulated and airtight; see Chap. 2).
- Distribution: Distributing solar gain collected in a specific location in the building to other locations that do not have direct access to solar radiation allows various areas of the interior space to benefit from solar heat gain.
- Control of heat gain: Use of specific design elements such as shades (natural or architectural) to control solar heat gain when it is not needed (for instance in summer), to reduce potential overheating.

Due to the impact of the architectural design of a building on its capacity to capture and utilize solar energy, considerations for passive solar design should take place early in the design process in order to produce a meaningful impact.

The main principles of solar design are depicted in Fig. 1.3. Below is a detailed discussion of each of those five principles.

Solar Gain

Passive solar design is based essentially on capturing solar heat gain for utilization in various applications. Passive solar heat gain is primarily employed for space heating. There are three main techniques to capture passive solar energy in a building: direct gain, indirect gain, or isolated gain. These techniques, defined in detail below, represent mainly the way heat is collected, stored, and distributed

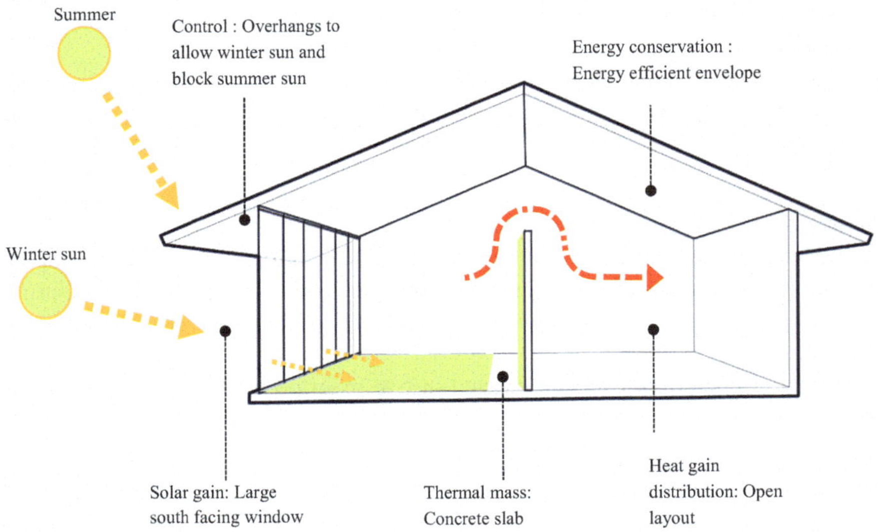

Fig. 1.3 Illustration of the five main principles of passive solar design

within the interior space of the building. The proper implementation of these techniques and their integration within the overall building design affect their efficiency, as detailed under Sect. 1.3—Selected applications.

Storage
Passive storage of solar thermal energy by means of thermal mass is an essential component in the design of passive solar buildings. It allows storing solar heat gained during the day, when solar radiation hits the mass, and releasing it passively during the night. Thermal mass reduces the risk of overheating and allows balancing the diurnal temperature variations, allowing for improved thermal comfort in the building. The storage mass absorbs heat until the air temperature of the space falls below the temperature of the mass. At this point, the heat flow reverses and heat is given up by the mass to the interior space, until the thermal mass and room reach a temperature equilibrium.
Thermal mass can be of advantage in both summer and winter. In summer, the mass acts in reverse mode to winter, contributing to cooling down the space during the day. The mass gets cooled down during the night, particularly when night natural ventilation is permitted, and therefore is able to absorb the heat from the indoor air during the day.
The most effective thermal mass in a house is a solar-exposed concrete slab. Reflective materials or carpets should not cover the slab, to allow an effective absorption and release of passive heat.
There are a number of other envelope components and materials with high thermal capacity that can be used as thermal mass. These are discussed in Sect. 1.4.

Energy Conservation
Building envelope plays a crucial role in the design of passive solar buildings, since it can assist in conserving heat gain and reduce heat loss, during the heating period, while reducing heat gain during the cooling period. To fulfill this role, building envelope should be designed to reduce heat transfer by means of adequate level of insulation, airtightness, elimination of thermal bridging, and high window performance.
Increased insulation level can reduce heating load of a building significantly. Cost–benefit of insulation is, however, characterized by a diminishing return curve, whereby, beyond a certain level of insulation, the cost gained by energy saving is not matched by the increased cost of insulation. Additional measures should be considered, such as appropriate window design. Size and location of windows influence the amount and timing of solar heat gain. Performance of building envelope for improved passive solar design is discussed in depth in Chap. 2.

Distribution
In a passive solar building, the interior layout should be designed to allow distribution of heat gain from the isolated (south) side of the building, or from the location where solar heat is collected, to other interior spaces. Usually, an open floor plan is a good option to facilitate the movement of passive solar heat. Moreover, areas designated for daily activities (such as living areas and kitchens in a residential

unit) should be located along the near south perimeter to take advantage of direct solar heat gain during the day. This is discussed in more detail in Sect. 1.4.

Control of heat gain
Control of heat gain interfaces with maximizing solar heat gain, to avoid overheating. Appropriate shading device should be designed to assist in reducing excess solar heat gain. In northern mid-latitude locations, overhangs are the optimal shading devices for south-facing windows. They allow the low winter solar radiation to penetrate the interior space while blocking direct summer radiation.
Natural ventilation is another method to reduce excess heat gain accumulation. Natural ventilation requires a number of design considerations to be effective. These are discussed in Sect. 1.3.2.
Various other methods of controlling heat gain can be employed in specific applications, such as the plantation of deciduous trees, on south- and west-facing patches. These trees can shade the windows during summer, while in winter, losing their leaves, they allow solar radiation to penetrate into the space.

1.3 Selected Applications

The main passive design applications are for heating, cooling, and daylighting of buildings. Passive design relies primarily on building-integrated components and materials. For instance, for passive heating, windows, walls, floors, and roof are used to collect, store, and distribute heat to the interior space. These same architectural components serve also to passively cool the interior.

The applications surveyed in this section relate primarily to residential buildings, but the main principles apply also to commercial buildings (including office buildings, as well as industrial and institutional buildings). A note on special considerations for such buildings is given in Sect. 1.4.4.

Passive solar design is not a substitute for, but rather a supplement to, mechanical systems, especially for heating in cold climate. It can, however, significantly reduce the size of mechanical heating systems and the quantity of non-renewable fuels required to achieve comfortable indoor temperatures in cold climates.

1.3.1 Passive Heating

Passive solar heating is a cost-effective means of providing heat to buildings, especially for small-scale residential buildings (such as single-family houses). A well-designed passive solar building may provide 45–100% of heating requirements, on a sunny winter day, even in cold northern climate [21].

Provisions for passive solar heating applications, particularly those related to the architectural design concepts such as orientation of the building and window

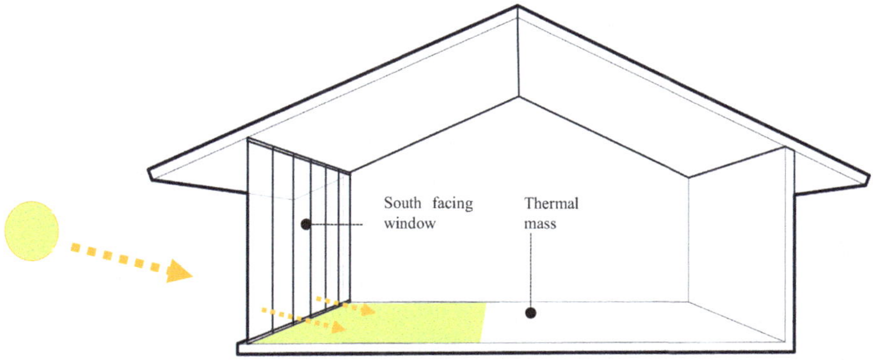

Fig. 1.4 Illustration of direct heat gain

locations, do not affect significantly the initial cost, when included in early design decisions. Moreover, the long-term operational cost savings achieved by passive solar design can outweigh the additional cost in building upgrading, such as added insulation and advanced glazing materials.

There are three main means to collect heat in a building, which are described below.

Direct gain

Direct gain is the simplest method of gaining heat from solar energy, relying mainly on near-equatorial facing glazing (Fig. 1.4). This technique was formulated early in the history of solar architecture and is still considered the most efficient passive method to capture solar radiation. Equatorial facing windows (south facing in the northern hemisphere) admit solar radiation to the interior space, where it is converted into thermal energy. The size and characteristics of glazing play an important role in determining the amount of solar radiation that accesses the space [3, 22].

Direct gain design can be achieved through a wide variety of building materials and combinations of concepts. Specific implementations depend greatly on the site and topography, building location and orientation, building shape (depth, length, and volume), and space use.

South-facing window area governs the amount of thermal mass required to achieve optimal performance, to minimize daily temperature fluctuations, and to increase thermal comfort. Building components serving as thermal mass may include floor, ceiling, and wall elements, and can range from solid materials (concrete, adobe, brick, etc.) to water. This is discussed in more detail in Sect. 1.4 (Building design and its impact).

Direct gain systems have the advantage of being relatively inexpensive to implement since they do not require additional space or equipment.

Some disadvantages of direct gain systems arise from inadequate design of the thermal mass and shading, which may lead to local overheating of the space that is directly exposed to solar radiation, even in cold days. In addition, the glazed area, although it may be of very high performance, will have lower insulation value as compared to the opaque areas. The size of windows needs to be optimized for the specific location, to balance between heat gain and heat loss. Night insulation (incorporated within curtains or blinds) can be added to these glazed areas to reduce heat loss in the absence of solar radiation. Design components applied to enhance the energy performance of the building envelope are detailed in Chap. 2.

Indirect gain

Another strategy of capturing solar energy consists of collecting and storing solar heat in a component of the building and then using natural heat movement (convection and radiation) to warm specific spaces. While, in direct gain, the solar radiation is directly admitted to the interior space, in indirect gain, heat gain is conveyed to the interior through another medium, usually a thermal mass.

A well-known example of such methods is the Trombe wall, also known as a storage wall or solar heating wall. This wall, located on the solar (south) facade, consists mainly of a massive wall of dark color, constituting a thermal mass with high heat capacity, allowing to store thermal energy for a time period consisting of a number of hours (depending on the material). This wall is isolated from the exterior air by a layer of glass and an air gap to prevent heat loss to the exterior [23]. The stored heat is then conducted through the mass to the indoor space (Fig. 1.5). This system is discussed in more detail in Chap. 2. Similar system to the Trombe wall can be applied in roofs.

Indirect gain systems have a number of advantages. For instance, an efficient collection of heat can be achieved since the storage is located in close proximity to the glazing (e.g., within the Trombe wall system). In addition, the existence of thermal

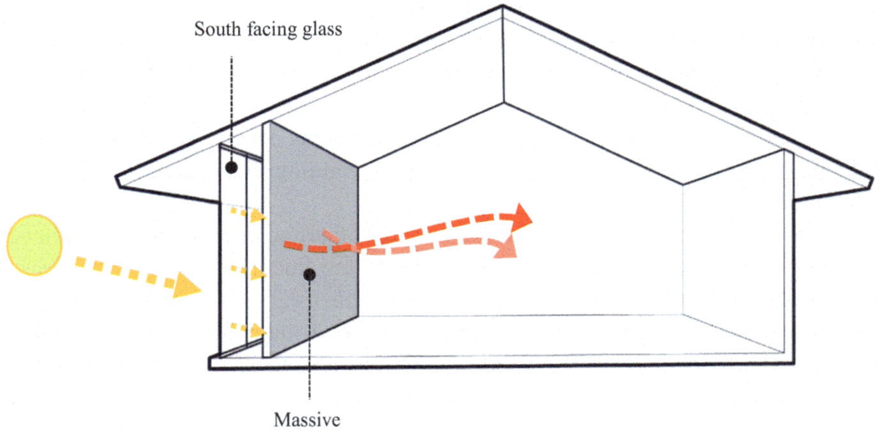

Fig. 1.5 Illustration of indirect heat gain methods—Trombe wall

mass prevents extreme diurnal temperature swings. Adequate space for windows needs, however, to be maintained to allow adequate daylight and outward visibility.

Isolated gain

Isolated gain refers to a design approach by which heat gain is collected and stored in a location distinct from the space to be heated. Ventilation is essential in this method of solar heat gain, in order to distribute heat to desired parts of the building. Consequently, the assistance of some mechanical devices, such as fans, may be required to move air around.

A typical isolated gain system consists of a solarium or greenhouse attached to a building. This can be a separate space designed on the south side of the building with a large glass area and thermal storage mass. This concept can be applied in a single- or multiple-storey building, with specific modifications. A sunspace with appropriate size of thermal mass can be as efficient as a Trombe wall system or direct system.

A forced air convection system can be implemented to circulate air from the sunspace to the living spaces. An effective method to circulate air through the entire mass of the building is to design a hollow core concrete slab, using thus the hollow core as ducts [24]. This system can be used also to cool down the building in hot periods.

A number of advantages are associated with isolated gain systems. For instance, temperature fluctuations resulting from the variation in solar radiation are restricted to the sunspace and do not affect directly the living space within the building. This technique can be used in new buildings as well as in retrofitting buildings that are not designed initially to benefit from passive solar energy (Fig. 1.6).

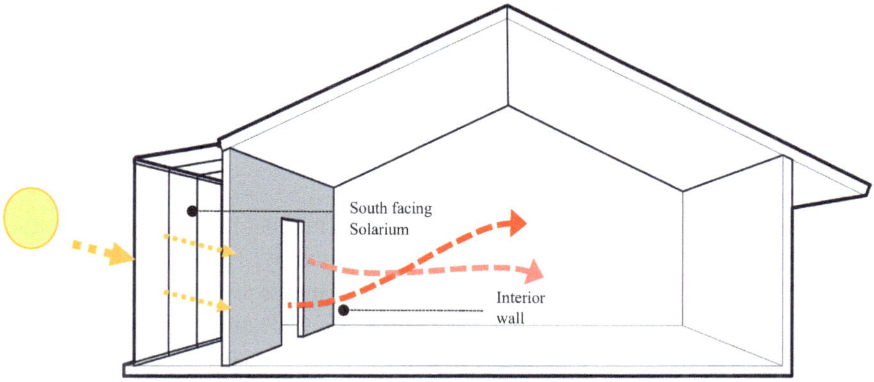

Fig. 1.6 Illustration of isolated heat gain methods

1.3.2 Passive Cooling

Passive cooling employs natural processes to reject heat from inside the building into the atmosphere (by convection, evaporation, and radiation), or into the ground beneath the building (by conduction and convection). Cooling may be required even in cold climate, if only occasionally. Applying passive cooling strategies contributes to reduction of energy and peak demand, allowing thus to reduce, and in some cases dispose of mechanical air conditioning. Certain passive cooling procedures result in immediate benefits, such as reducing heat gain, natural ventilation, and direct evaporative cooling. Other systems rely on absorbing heat from the indoor space and storing this heat in the thermal mass to be discharged to the surrounding [25].

There are three major sources of undesirable summer heat gain: direct solar radiation through building glazing, heat transfer, and infiltration of exterior high temperature air through the building envelope, as well as internal heat gain generated by appliances, equipment, and occupants. In general, the excess solar heat gain from south and near south windows constitutes the major concern of heat gain issues in buildings, but this is usually the easiest to control.

Natural cooling relies on many of the principles and techniques applied in passive solar heating. For example, methods employed to prevent heat loss during the heating period, such as high level of insulation, retard heat gain during the cooling period. As discussed above, thermal mass such as masonry walls and concrete floors absorbs heat when the indoor temperature is higher than the temperature of the mass, and delay rise of the indoor temperature. The thermal mass passively releases the heat absorbed, during the night as the outdoor temperature drops below the temperature of the mass. To increase the efficiency of the mass and allow it to cool down, it should be exposed to the outdoor air (e.g., by nighttime ventilation).

Building spaces, embedded into the ground, benefit from the lower temperature of the ground and thus the difference in temperature between the indoor and outdoor surfaces of the mass. This is particularly relevant for one-storey buildings, the ground floor of multistory buildings, and buildings that are partially termed.

Natural ventilation

Natural ventilation is an effective passive cooling technique. In general, the ventilation is also necessary to maintain a good indoor air quality. This technique exploits prevailing winds on the windward side and natural convection to ventilate a building as needed. Natural convection can be employed to ventilate and cool the building as long as the outdoor air is cooler than the indoor air.

There are three main types of natural ventilation: single-sided ventilation, cross/wind ventilation, and passive stack ventilation. These are presented in Fig. 1.7.

The design of natural ventilation within a building requires understanding of the airflow patterns around this building as well as the effect of neighboring buildings. The objective of natural ventilation is to ventilate the largest possible part of the indoor space, through appropriate interior design and windows configuration. For instance, operable windows can be an integral part of current constructions dominated by large glazed facades. Current practice often does not allow opening of

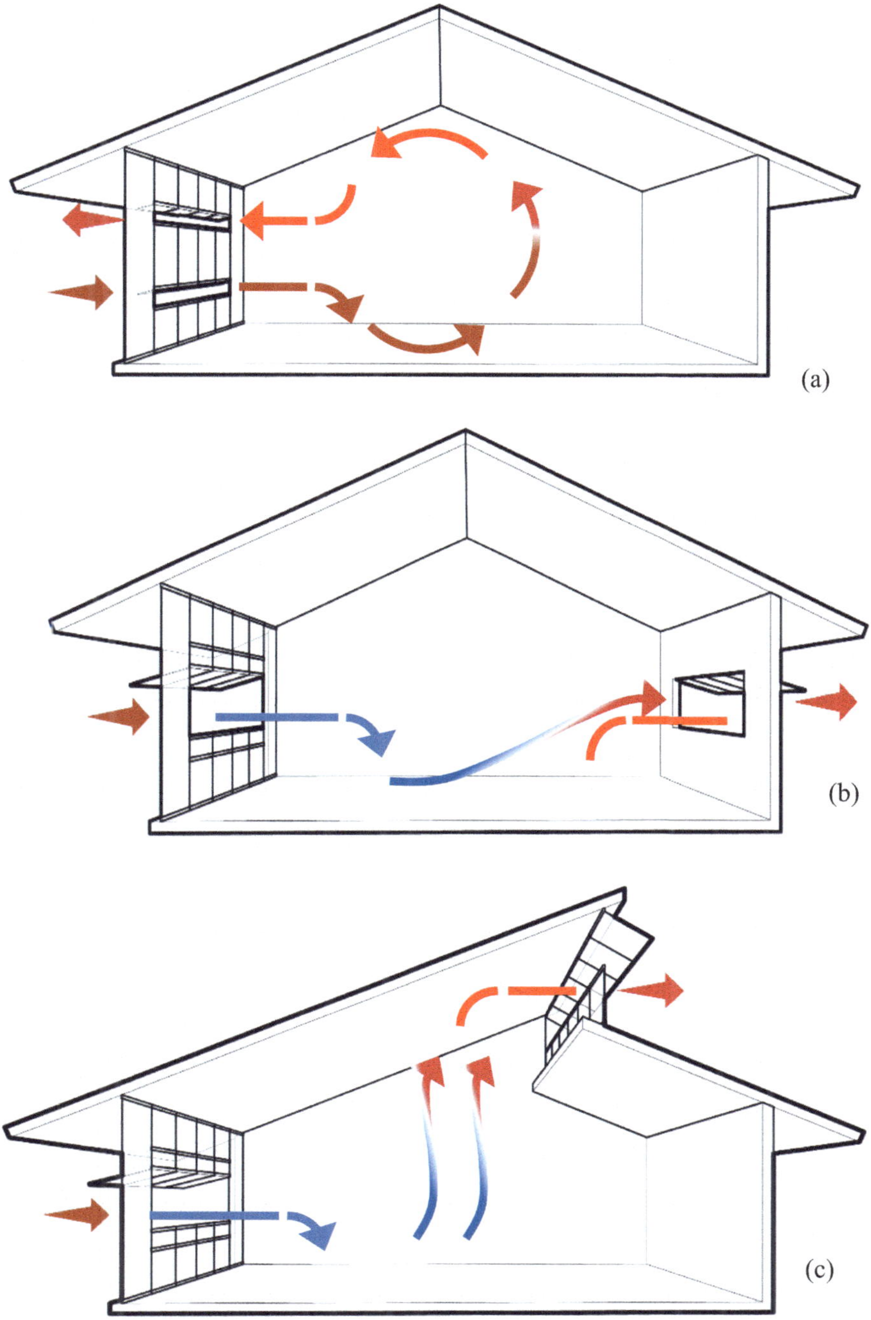

Fig. 1.7 Illustration of the methods of natural ventilation; **a** single-sided ventilation, **b** cross-ventilation, **c** stack ventilation

windows, particularly in office buildings, eliminating thus potential utilization of natural ventilation to supply fresh air to indoor spaces.

A number of techniques can be employed to enhance natural ventilation. These are summarized below.

Solar-induced ventilation

Solar radiation can be used to enhance airflow, allowing efficient natural ventilation. One approach employs the Trombe wall, discussed above, but vented to the outside (Fig. 1.8). Solar radiation striking the mass of the Trombe wall will heat the air in the space between glass and wall, causing this air to rise quickly and escape, drawing thus cooler outdoor air into the building.

Another technique consists of *solar chimneys*. These chimneys are specifically constructed with seasonal dampers, so that the passively heated air can be dumped into the interior space of a building in winter, and rejected to the outdoor environment in summer, to draw cooler outdoor air through the building and ventilate it (Figs. 1.9 and 1.10) [26]. The design principles shown in Fig. 1.9 can be applied in a multistory building.

The solar chimney principle can be applied as a part of the roof design (of single-storey buildings), in a variety of methods, as shown in Fig. 1.10.

Underground-floor ventilation

Another natural cooling strategy is to employ the ground temperature to cool outdoor fresh air, before supplying it to buildings. This strategy consists of drawing outdoor air through tubes buried in the ground, called earth tubes. These tubes are made of material of high thermal conductivity allowing easy thermal transfer between the ground temperature and the air inside the tube [27]. As a result, warm outdoor air gives up its heat to the cooler surrounding earth and cools substantially before being dumped into the building. Earth tubes are buried at

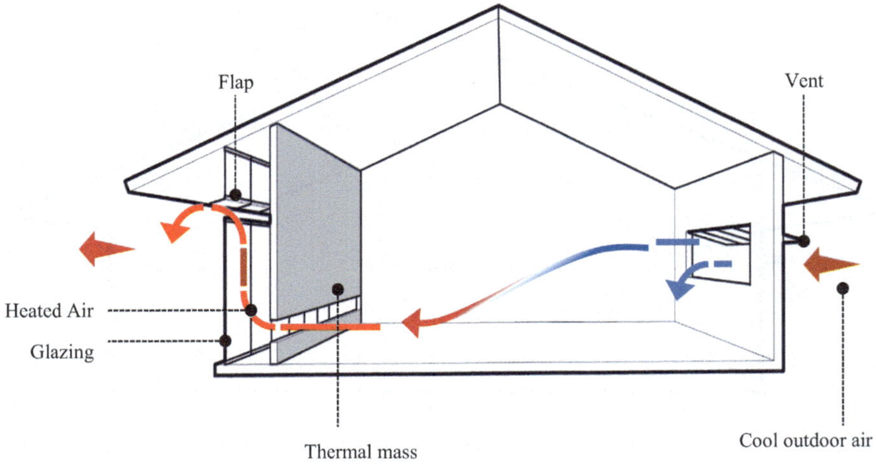

Fig. 1.8 Use of Trombe wall in a cooling mode

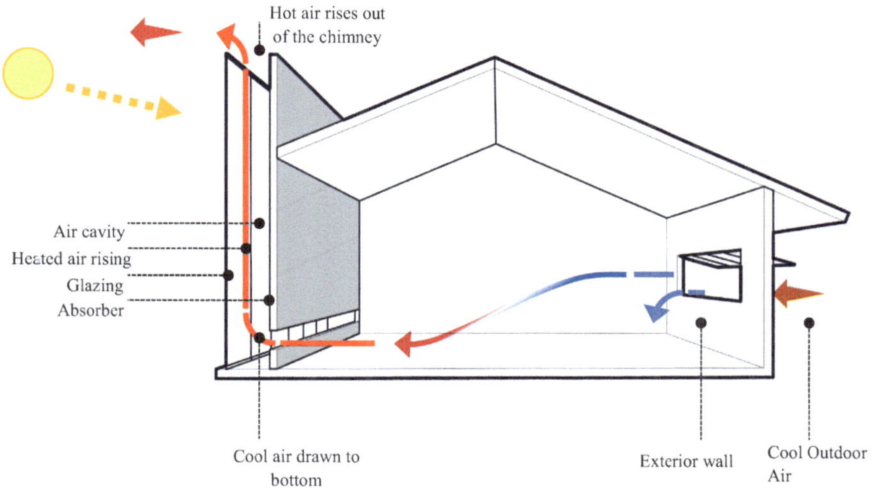

Fig. 1.9 Solar-induced ventilation: Solar Chimney

about one meter depth, to reduce the impact of the warm surface temperature. To prevent the surrounding earth thermally saturating, methods such as landscaping and watering can be employed.

Night ventilation
This method consists of the use of night temperature to cool the indoor air temperature. The efficiency of this technique depends on many factors, including the relative difference in temperature between indoor and outdoor during night hours, the thermal capacity of the building, the airflow rate. Coupling night ventilation with thermal mass can increase the efficiency of natural cooling, by chilling the mass, allowing it to absorb internal heat gained during daytime. Recent studies show a successful application of night ventilation to low-energy buildings [22].

Heat mitigation strategies
An important strategy of passive cooling is to reduce heat gain in a building from solar radiation and from the outdoor environment. Various methods can be implemented to minimize solar heat gain during the cooling season. Increasing the insulation of building envelope plays an important role in restricting heat transfer from outside to inside environment during hot periods. Other strategies consist of blocking direct solar radiation penetrating the indoor space through windows and skylights, reducing heat absorption by the building exterior surfaces, and restricting re-radiation and reflection from the surrounding area.

Controlling solar radiation is one of the most effective methods to reduce direct heat gain. Such methods include shading devices as well as natural features such as deciduous trees. These methods are briefly presented in the following. Shading devices are presented in more detail in Chap. 2.

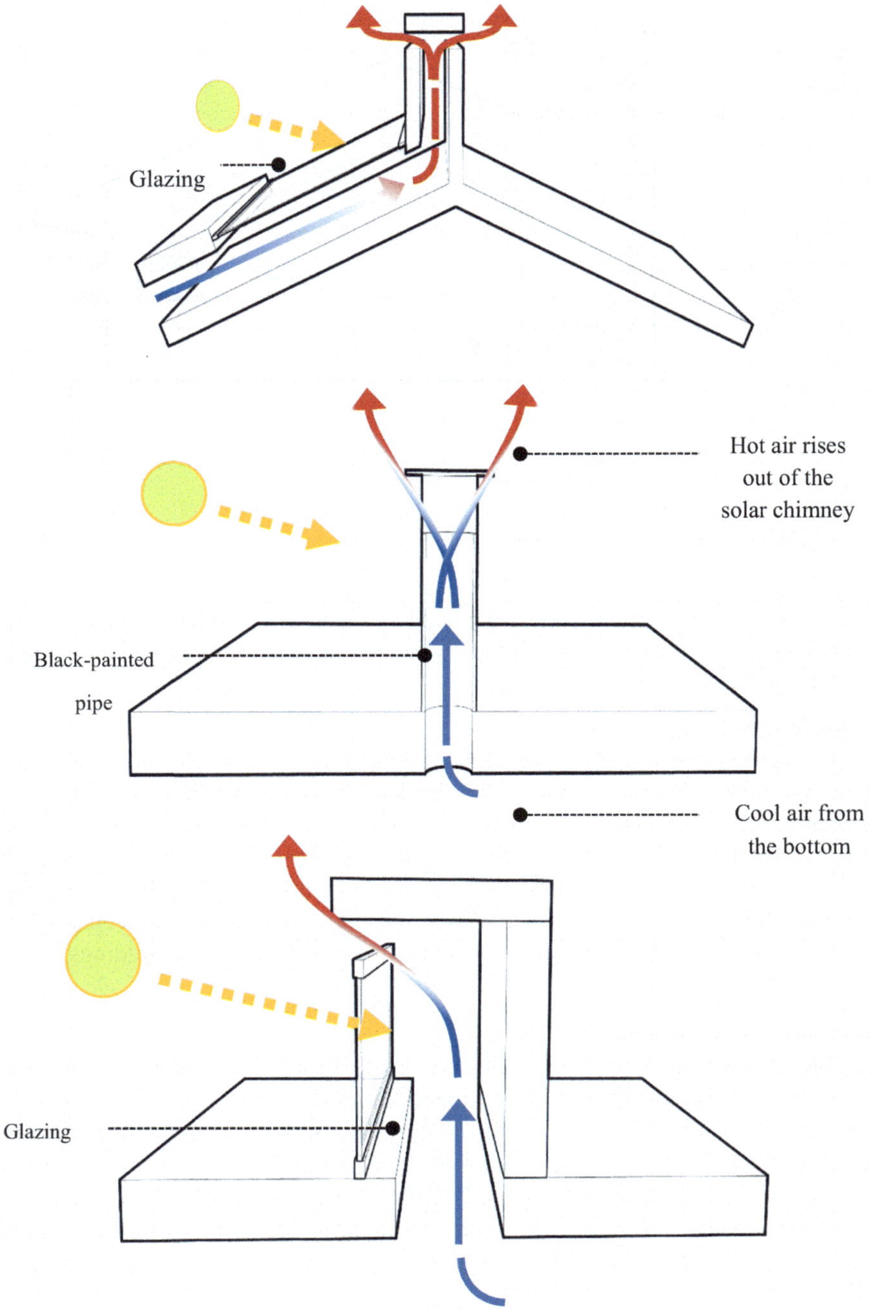

Fig. 1.10 Different methods of application of solar chimney to the roof

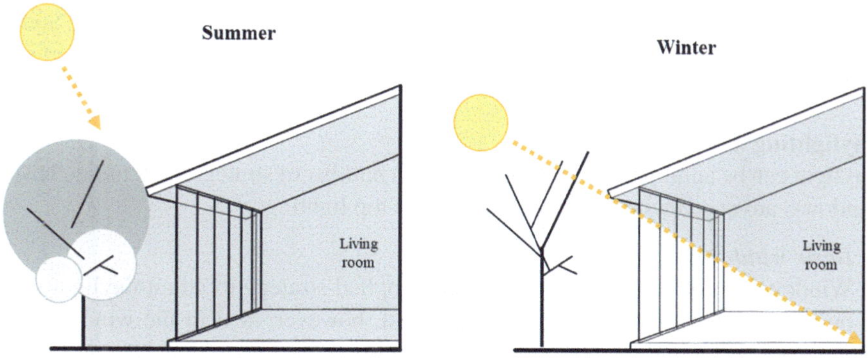

Fig. 1.11 Illustration of the use of deciduous trees for solar radiation control

Window shading
South windows are relatively easy to shade using a horizontal overhang or an awning. Control of solar radiation on east and west orientations is usually more difficult than that of the south orientation and requires a different approach. Vertical shading, in conjunction with shutters and blinds, can be more suitable for such applications.

Use of vegetation
Deciduous trees can be employed to mitigate summer direct solar radiation, on the lower floors of a building (Fig. 1.11). Vegetation can be employed in special designs of facades, as, for example, employing trellis with deciduous plants, or in a double-skin facade (see Chap. 3). Planting vegetation to cover the ground reduces ground reflection and can assist in keeping the earth's surface cooler preventing re-radiation.

1.3.3 Daylighting

Daylighting management is another energy-efficient strategy that depends on the availability of solar radiation and incorporates several technologies and architectural design approaches. Daylighting can improve the quality of light in a space and diminish artificial lighting.

Daylight is more efficient than artificial lighting in providing the required illumination level [28] and, since it encompasses the full-spectrum it matches better the human visual response [29]. Research demonstrates that daylight has positive impact on human well-being, satisfaction, performance, and productivity [30]. However, the level of daylight in a space should be controlled to maintain visual comfort [31]. Daylighting design considerations should be accounted for early in the design stage of a building to achieve optimal quality and quantity of light. A number of design factors affect daylighting performance, including the geometrical shape of the facades and the interior space and windows' dimensions, location, and orientation [32]. Some design considerations are similar to those required for passive heating and cooling of the space.

Daylighting design
Daylight can be admitted to buildings through a number of strategies, including basic windows, advanced windows and glazing, and top lighting.

Basic windows
Windows constitute the most commonly applied strategy to admit daylight into buildings. The illumination level is greatest, however, next to the window and it decreases rapidly as the depth of the building increases, to levels that may be inadequate for common visual tasks. Designing windows on two walls, especially two opposing sides of a single space assists in increasing illumination level in areas that are further away from the windows.
The type of glazing has significant impact on the amount of daylighting that penetrates the space. For instance, windows with clear glazing can admit a large proportion of daylighting into the indoor. A negative aspect of these types of window is their susceptibility to be a source of glare and excessive brightness, due to direct solar radiation. Mounting widows at increased height can reduce such risks, while increasing the penetration depth of sunlight.

Advanced windows
Advanced windows can be employed to modify the quality and quantity of daylighting. Advanced windows strategies include the use of devices to reflect light into the building for deeper penetration of daylight. For instance, light shelves offer an efficient method that allows increasing the daylight zone, by reflecting light onto the ceiling and then into the space (Fig. 1.12).
The use of light shelves is effective primarily on the south orientation. Other orientations need devising other methods to maximize daylight diffusion. Another effective strategy to reflect light onto the ceiling is the use of reflecting Venetian blinds. Dynamic systems respond to the changing levels of daylight and sunlight and thus are more effective than static systems such as light shelves.
Other strategies used to maximize daylighting consist of adopting advanced glazing types. These, however, should be designed to balance between lighting design and controlling heat loss/gain. Advanced glazing is discussed in Chap. 2.

Top lighting
In addition to conventional windows, light can be admitted into buildings through top lighting devices, such as clerestories and skylights. Clerestories are vertical windows located at roof level, while Skylights consist of openings in a roof. Skylights can be either horizontal or pitched; they can also be integrated within sawtooth structure. An effective shading device should be installed along large skylights to reduce heat loss at night and heat gain during the cooling period.
The advantages of top lighting include enhanced illumination over a large area, and more even distribution of light. This type of daylighting is, however, restricted to the top floor of a building.

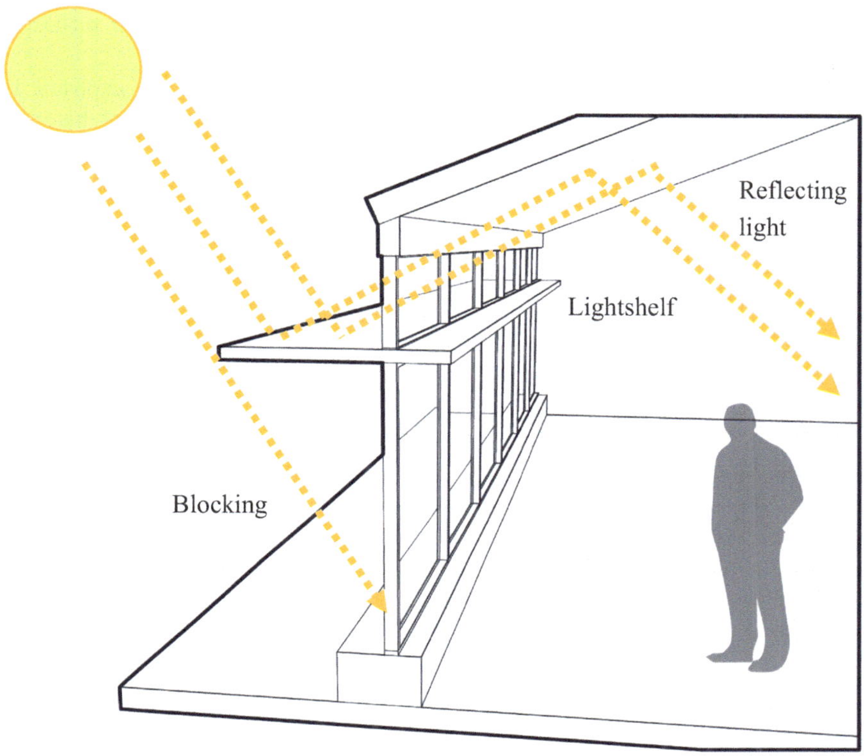

Fig. 1.12 Illustration of light shelf principle

1.4 Building Design and Its Impact

There is a strong interaction between passive solar design and architectural design. A holistic design approach should be adopted, which takes into account at early design stages considerations such as building site setting, building shape, and envelope parameters, including opaque and glazed areas.

Cost considerations form an integral part of the design process. While solar passive features can add up to 15% to design and construction costs [32], this initial cost is offset by significant reduction in building operational cost.

1.4.1 Building Site Setting

Siting

The site selected and the siting of the building in relation to its surroundings play a key role in determining the amount of solar radiation available for utilization. A site should have minimum obstruction of solar radiation during the cold season, including obstruction from surrounding trees and landscape.

The location of the site in proximity to water surface, hills, or its exposure to wind affect the passive heating and cooling performance of buildings, and therefore should be considered in the design of the building.

Orientation

The building should be oriented with the long axis running east–west, so as to have the largest facade equatorial facing (south in the northern hemisphere). This is due to the fact that east- and west-facing buildings can be potentially subjected to overheating during the cooling season, and to reduced heat gain during the heating season, in addition to high solar radiation fluctuations—high radiation and daylighting at early and late hours and low at most of the day.

Deviation from south orientation toward the east or west leads to increase in heating and cooling load. For example, studies show that heating load can be increased by up to 30% if the building is oriented about 60° east or west from south, in a northern, mid-latitude cold climate [33]. With such orientation, not only the solar heat gain during the winter is reduced, but summer cooling loads are significantly increased, especially when the building is oriented toward the west.

1.4.2 Building Shape

General geometry

Several design factors should be considered in the optimization of building geometry for solar capture. Rectangular layout, with long south-oriented facade, is generally considered the optimal shape for energy efficiency. Non-rectangular and particularly self-shading shapes (like L shapes) offer a wider flexibility in architectural as well as solar design, but their efficient design is influenced by several parameters such as the location and dimension of the shading part of the shape [34]. Such issues should be considered at early design stages. Details of building shape and its impact are presented in Chap. 5, for residential single-family buildings as well as for multistory buildings.

Building layout and interior space

The interior layout, form, and design of various zones in a building play a key role in enhancing passive solar design and utilization of passive solar energy. The layout of the interior space should be such that daily activities match the sun's path across the sky. This allows to benefit from solar heat gain during human active hours, reducing thus the need for heating and daylighting.

The internal layout of the building should promote natural movement of the heat from one space to the other. This can be achieved through various architectural

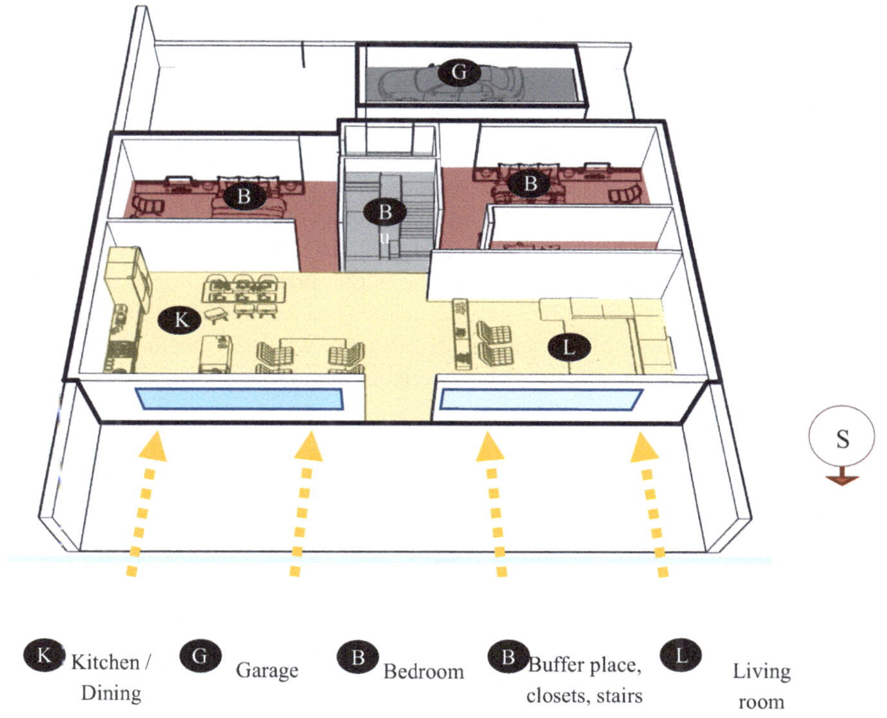

Fig. 1.13 Diagram illustrating the interior layout to maximize solar utilization

designs and concepts, such as open interior layouts, and devising openings to induce airflow between floors and between north and south zones.

For example, in single-family houses, zones allocated for daily activities, such as living areas and kitchens, can significantly benefit from passive solar heat gain, during the day, and thus should be located on the south side of the building. The north side can accommodate areas with less frequent usage such as storage rooms and garage. Such arrangement on the north side has the added benefit of creating buffer zones to reduce heat loss through north walls.

Figure 1.13 presents an example of zoning in a single-family house to increase the efficiency of passive solar design.

1.4.3 Building Envelope

This section presents an overview of the main components of the building envelope design and their impact on capture of solar radiation. In-depth discussion of the design of building envelope to maximize solar capture and energy performance is presented in Chap. 2.

Overall Envelope

Building envelope, including walls and roof, plays an important role in passive solar design. Enhancement of the building envelope can reduce energy demand for space heating in the range of 30–85% [35]. This improvement should target the reduction of heat loss (during the heating period) and gain (in cooling period) through the envelope. Heat transfer through the envelope is due primarily to poor insulation, thermal bridges, and air infiltration. Significant improvement to the building envelope can be achieved through highly insulated walls and windows (including frames) and improved airtightness. Window characteristics and their effects on heat loss are summarized below.

Heat loss by air infiltration is highly significant, constituting a large portion of the total heat loss of a building. High airtightness can be achieved through appropriate construction methods that implement air barriers, sealants, and weather stripping [36], and should be coupled with proper ventilation of the building. Additional details of infiltration and its impact are given in Chap. 2.

Glazing

Windows constitute the most significant building envelope components governing passive heat control, representing a significant source of heat loss, on one hand, but for solar heat gain, on the other hand. Heat loss occurs through both glazing and framing of windows [37, 38]. High insulated windows (low U-value), including glazing and frame, should be selected as a fundamental step to achieve energy-efficient design.

Window characteristics should allow reducing heat loss, while maximizing solar heat gain during the cold season. The design of south-facing windows should balance between low U-value of glazing, high solar heat gain coefficient (SHGC), and high visible transmittance, in order to optimize net energy gains, daylighting, and visibility. SHGC represents the relative portions of solar radiation transmitted by the glazing into the indoor space, including direct radiation and absorbed radiation and subsequently released to the indoor space. SHGC is usually used to measure glazing's ability to transmit solar gains.

The glazing area on the south facade, for optimal solar performance, depends on building characteristics and local climate [39]. In mid-latitude, south-facing windows should cover 30–50% of wall area. Glazing on other facades should be minimized to reduce heat losses in winter and overheating in summer [33].

Shading Devices

Shading devices are essential elements of the solar performance of the envelope. Appropriate solar shading devices can control indoor illumination, glare, and solar heat gains, while saving energy demand for heating and lighting [40]. Shading devices are divided into two main categories, static and dynamic. Static devices are simple and can be efficient in blocking excessive solar radiation, but they have

limited capability of controlling solar gains. Retractable awnings, on the other hand, enable the reduction by up to 80% of summer solar gains [40].
Dynamic shading has more potential in controlling heat gains and adapting this gain to the building energy needs. Extensive studies have examined the potential in energy savings of manually or mechanically operated dynamic shading devices, including internal blinds [40–43].
Exterior insulated roll shutters are effective under northern climate conditions. Roll shutters can reduce heating load, by allowing solar heat gain when it is needed, while blocking radiation in the cooling period, as well as improve thermal conditions in the perimeter zones, near windows [40].

1.4.4 Passive Design in Commercial and Larger Buildings

Commercial buildings have some different characteristics from residential buildings, such as greater internal heat gains from equipment and lighting, higher ventilation requirements, and different occupancy trends. Commercial buildings can benefit from passive cooling, but also from daylighting in the peripheral zone and from preheating air for space heating.

Despite the difference in applications of passive solar design between residential and commercial buildings, many of the design principles mentioned above can be applied. These include the high insulated envelope with reduced thermal bridging, solar control, thermal mass, high-performance glazing, and others.

Heating in commercial buildings can usually be less demanding than in residential buildings, and potential passive heating is compromised by many factors including the size of the building and its layout. Passive heating techniques, such as solar chimneys, solar walls, and others, are available for such buildings. Some of these methods are discussed in Chap. 2. On the other hand, there is a higher risk of overheating during summer, and thus increasing the cooling load of the building, making the design of shading devices and proper ventilation a key design factor in such buildings.

References

1. Paul JK (1979) Passive solar energy design and materials, vol 41. Noyes Data Corporation, Park Ridge, New Jersey
2. Barber S (2012) History of passive solar energy. East Carolina University, pp 1–11
3. Lechner N (2001) Heating, cooling, lighting : design methods for architects, 2nd edn. Wiley, New York
4. Kenworthy JR (2006) The eco-city: ten key transport and planning dimensions for sustainable city development. Environ Urban 18(1):67–85
5. Cahill N (2000) Olynthus and Greek town planning. Class World 93(5):497–515
6. Hawkes D (2013) The environmental tradition: studies in the architecture of environment. Taylor & Francis

7. Calderaro V, Agnoli S (2007) Passive heating and cooling strategies in an approaches of retrofit in Rome. Energy Build 39(8):875–885
8. Ionescu C, Baracu T, Vlad GE, Necula H, Badea A (2015) The historical evolution of the energy efficient buildings. Renew Sustain Energy Rev 49:243–253
9. Perlin J (2013) Let it shine: the 6,000-year story of solar energy. New World Library, Novato, California
10. MOFIDI SSM (2011) Environmently responsive education in urban history
11. Erdman JA, Douglas CL, Marr JW (1969) Environment of Mesa Verde, Colorado, no 7. US National Park Service
12. Olgyay V (2015) Design with climate: bioclimatic approach to architectural regionalism-new and expanded edition. Princeton University Press
13. Saadatian O, Sopian K, Lim CH, Asim N, Sulaiman MY (2012) Trombe walls: a review of opportunities and challenges in research and development. Renew Sustain Energy Rev 16(8):6340–6351
14. Jones HG, Archer N, Rotenberg E, Casa R (2003) Radiation measurement for plant ecophysiology. J Exp Bot 54(384):879–889
15. Sen Z (2008) Solar energy fundamentals and modeling techniques: atmosphere, environment, climate change and renewable energy. Springer Science & Business Media
16. Geiger R (1965) The climate near the ground. Harvard University Press, Cambridge: MA
17. Holland PG, Steyn DG (1975) Vegetational responses to latitudinal variations in slope angle and aspect. J Biogeogr, 179–183
18. Gates DM (1980) Biophysical ecology (Springer Advanced Texts in Life Sciences), 1st edn. Springer, New York
19. Kirkpatrick JB, Nunez M (1980) Vegetation-radiation relationships in mountainous terrain: eucalypt-dominated vegetation in the Risdon Hills, Tasmania. J Biogeogr 7(2):197–208
20. Aguilar C, Herrero J, Polo MJ (2010) Topographic effects on solar radiation distribution in mountainous watersheds and their influence on reference evapotranspiration estimates at watershed scale. Hydrol Earth Syst Sci 14(12):2479–2494
21. American Society of Heating Refrigerating and Air-conditioning Engineers Inc (ASHRAE) (2007) Solar energy use. In: ASHRAE Handbook–HVAC applications. American Society of Heating, Refrigerating and Air-Conditioning Engineers, Inc, Atlanta
22. Chiras D (2002) The solar house: passive heating and cooling. Chelsea Green Publishing, White River Junction, VT
23. Dimassi N, Dehmani L (2017) Performance comparison between an improved and a classical Trombe wall: an experimental study. J Build Phys 40(4):372–395
24. Corgnati SP, Kindinis A (2007) Thermal mass activation by hollow core slab coupled with night ventilation to reduce summer cooling loads. Build Environ 42(9):3285–3297
25. Geetha NB, Velraj R (2012) Passive cooling methods for energy efficient buildings with and without thermal energy storage–a review. Energy Educ Sci Technol Part A Energy Sci Res 29(2):913–946
26. Bansal NK, Mathur R, Bhandari MS (1993) Solar chimney for enhanced stack ventilation. Build Environ 28(3):373–377
27. Florides G, Kalogirou S (2007) Ground heat exchangers—a review of systems, models and applications. Renew Energy 32(15):2461–2478
28. Edwards L, Torcellini P (2002) Literature review of the effects of natural light on building occupants
29. Franta G, Anstead K (1994) Daylighting offers great opportunities. Wind Door Specif Lab, 40–43
30. Nazzal AA (2005) A new evaluation method for daylight discomfort glare. Int J Ind Ergon 35(4):295–306
31. Tabet KA, Sharples S (1990) The interaction of preferred window size with thermal and visual comfort. Energy Environ 4:2648–2652
32. Chan HY, Riffat SB, Zhu J (2010) Review of passive solar heating and cooling technologies. Renew Sustain Energy Rev 14(2):781–789

33. Hachem C, Fazio P, Athienitis A (2013) Solar optimized residential neighborhoods: evaluation and design methodology. Sol Energy 95:42–64
34. Hachem C, Athienitis A, Fazio P (2011) Parametric investigation of geometric form effects on solar potential of housing units. Sol Energy 85(9):1864–1877
35. Charron R, Athienitis A (2006) Design and optimization of net zero energy solar homes. ASHRAE Trans 112(2):286–295
36. U.S. Department of Energy (DOE) (2011) building technologies program: air leakage guide
37. Arasteh D, Hartmann J, Rubin M (1986) Experimental verification of a model of heat transfer through windows. ASHRAE Trans 93(Part 1):1425–1431
38. Winkelmann FC (2001) Modeling windows in EnergyPlus. Proc Build Simul, 1–11
39. Charron R, Athienitis AK (2006) Optimization of the performance of double-facades with integrated photovoltaic panels and motorized blinds. Sol Energy 80(5):482–491
40. Laouadi A (2009) Thermal performance modelling of complex fenestration systems 2(3)
41. Athienitis AK, Santamouri M (2002) Thermal analysis and design of passive solar buildings. James & James, London
42. Foster M, Oreszczyn T (2001) Occupant control of passive systems: the use of Venetian blinds. Build Environ 36(2):149–155
43. Tzempelikos A, Athienitis AK (2007) The impact of shading design and control on building cooling and lighting demand. Sol Energy 81(3):369–382

Chapter 2
Introduction to Building Envelope

This chapter presents an overview of the building envelope's function, and main characteristics and components. It gives an insight into the energy performance of building envelope and methods to enhance this performance. The chapter presents a summary of heat transfer mechanisms through the building and outlines main design considerations to reduce energy loss through the building envelope, while improving the admission of solar radiation into the indoor space. Building-integrated passive systems are discussed, including advanced building envelope systems and components that aim at improving solar energy capture and utilization, while enhancing human comfort.

2.1 Building Envelope Characteristics

The building envelope comprises the part of a building, above or below grade that separates the interior space from the exterior environment, offering shelter from external and climatic elements. Building enclosure components are subdivided into roof systems, above-grade walls, including windows (fenestration) and doors, below-grade walls, and underground space (basement) walls. Figure 2.1 presents an illustration of the building envelope associated with low-rise and higher rise buildings.

Building envelope has a large impact on the built environment, through GHG emissions associated with building envelope materials and methods of construction on one hand, and through its impact on building operations and energy consumption on the other hand. Design of building envelope has also a significant impact on building cost and occupants' health and comfort. Defects in building envelope can lead to significant problems such as deterioration of a building and its structural components, health issues, and increased energy consumption. Building envelope accounts for the majority of building defect claims in North America.

Building envelope can be categorized according to various criteria. This manuscript divides the building envelope into two major categories, according to the size of the building: low- to mid-rise and high-rise buildings. In low- to mid-rise

C. Hachem-Vermette, *Solar Buildings and Neighborhoods*, Green Energy and Technology, https://doi.org/10.1007/978-3-030-47016-6_2

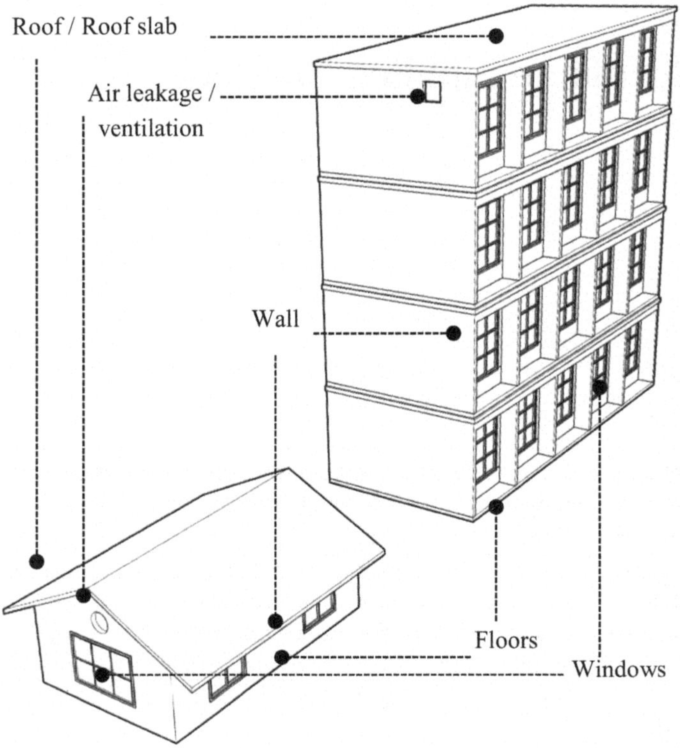

Fig. 2.1 Illustration of building envelope in low-rise and multistory buildings

buildings, an envelope is usually integrated with the structural framework as, for instance, in wooden framed housing units. In high-rise buildings, curtain walls (or variations of these assemblies, see Chap. 3) are often adopted. This type of wall system is externally attached to the structural framework and has very limited load-bearing function. Selected building envelopes for low-rise and multistory buildings are discussed in detail in Chap. 3.

2.1.1 Function and Performance

The building envelope has three main categories of functions: Control, finish, and distribution [1]. Each of those has a number of characteristics as discussed below.

- Control: The building envelope should manage mass and energy flows from the indoor to the outdoor and vice versa, including heat, air, moisture, rain, sound, fire, insects, and access. Figure 2.2 illustrates the main heat and mass transfer flows (e.g., moisture and water vapor diffusion) through the building envelope.

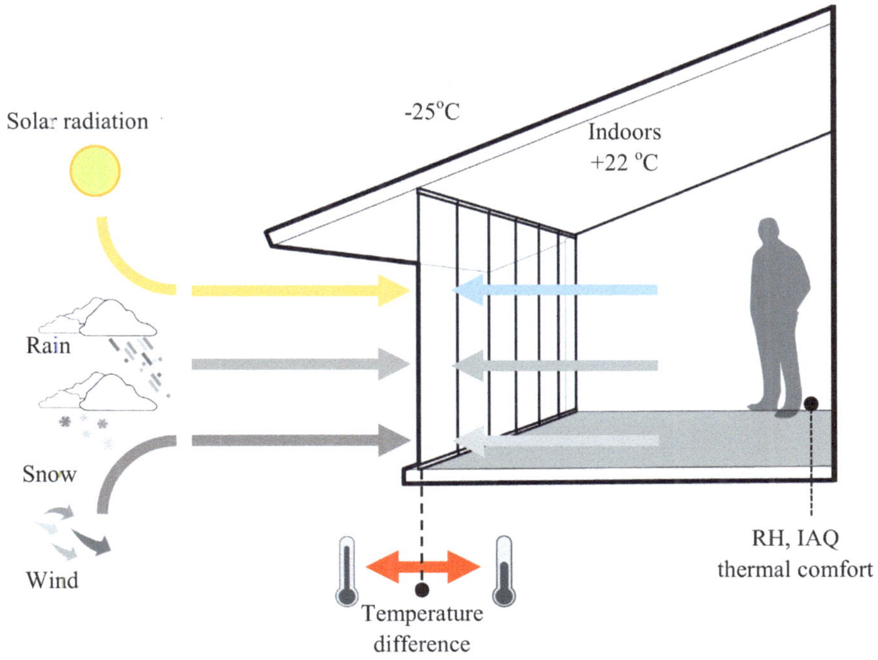

Fig. 2.2 Heat and mass flow through the building envelope

- Finish: The building envelope presents the interface between the indoor and outdoor environments and therefore is usually designed to reflect a specific interior and exterior finish, for architectural and functional purposes. Examples of finish applied to the building envelope include color, texture, reflectance, and patterns.
- Distribution: The building envelope is designed in a majority of cases to protect and house distributed services, such as electricity, communications, plumbing, and others.

Design considerations

The design of high-performance building envelope is affected by a number of factors, related to the construction components of the envelope itself, as well as to the overall design of the building and its surrounding (Fig. 2.3). The construction components of the building envelope include insulation, fenestration, control devices, and components for reducing air leakage and thermal bridging. These factors are discussed in detail in Sect. 2.4.

Building envelope design can be influenced by the building form. For instance, self-shading forms, such as L-shape and U-shape buildings, can affect the design of windows, their relative location in the envelope, as well as the design of window shading system (such as overhangs, fins, etc.). The interior partition and the design of interior zones and their location with respect to various facades play also an important role in designing window size and their orientation. As discussed in Chap. 1, it would

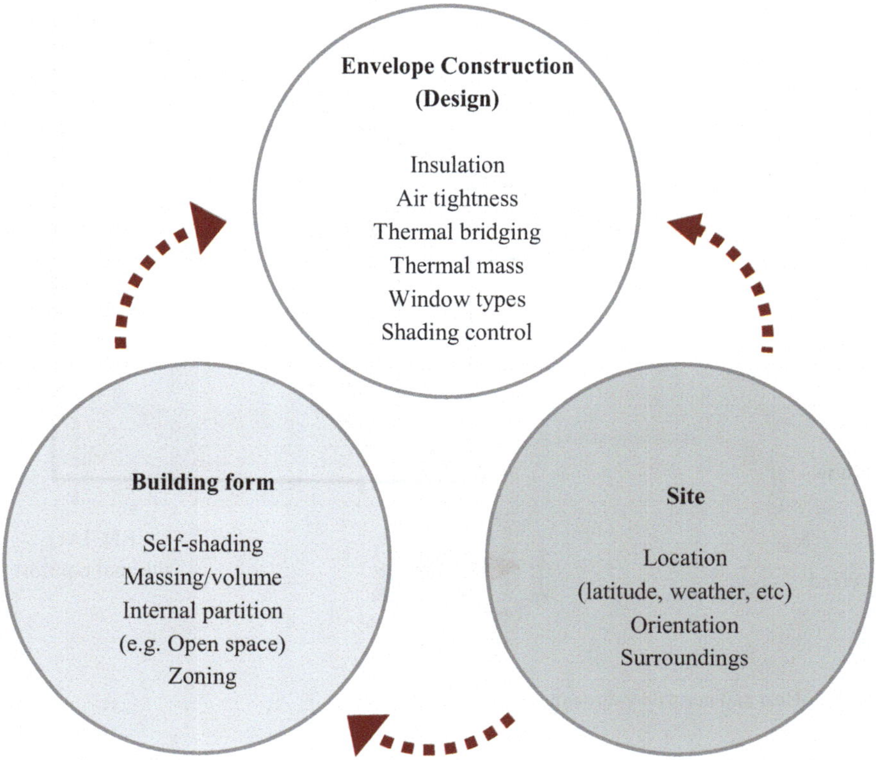

Fig. 2.3 Factors affecting the design of the building envelope

be beneficial to locate zones that can benefit from solar radiation during the day next to equatorial (south[1]) and near-equatorial facades.

The site location and its geography affect design decisions of the envelope. Site factors that need to be considered include the site geographic location, the orientation of the site as well as the surroundings (e.g., type of vegetation and their location with respect to the buildings).

The interaction between the various factors that a building envelope should incorporate is depicted in Fig. 2.3.

2.1.2 Trends in Building Envelope

Technological innovations are targeting the building envelope and particularly facades, in order to maximize the capture and utilization of solar energy and to

[1]The reference is to northern climate. "South" should be replaced by "north" (and vice versa) when referring to southern climate.

regulate and control the intensity of solar radiation admitted into buildings. These innovations include enhancements of building envelope materials, advancement in control, and operations of specific components, as well as automation of some parts of the envelope to respond to specific requirements.

Solar building envelope systems can be classified into one of the two categories: Opaque/semi-opaque, or transparent/translucent [2]. Each of these can in turn be divided into active or passive envelope. This chapter concentrates on passive building envelope and facade characteristics, while active envelopes, especially building-integrated solar thermal collectors (BIST), building-integrated PV systems (BIPV), and hybrid (BIPV/T) systems are presented in Chaps. 4 and 5. Figure 2.4 presents the classification of solar building envelopes, and their characterization as passive or active.

Current trends in building envelope design, while differing in mode of operation, materials, and technology incorporated, tend to focus on solar energy capture and regulation. Two trends of building envelope illustrating such focus on enhancing solar energy potential—*high energy performance* building envelope and *adaptive facades*—are outlined below.

High energy performance

A major component in the design of passive solar buildings is the design of a high-performance energy-efficient building envelope, which allows capture of solar radiation, controls its admission, and reduces heat loss through its assembly.

Building envelope can be designed to increase the energy efficiency of the building through various measures such as high-level insulation, high-performance windows with solar heat gain coefficients (SHGC) appropriate for the given climate, properly sealed structures, and minimization of thermal bridges. These measures are summarized in Sect. 2.3.

A high energy performance building envelope may integrate technologies to generate renewable electric and thermal energy, such as photovoltaic modules, solar thermal collectors, or even small wind turbines. Extensive research is in progress on various methods to increase the energy potential of building envelope for higher efficiency as well as for energy generation. Some of these methods are presented in Chaps. 4 and 5. Codes and standards are being updated concurrently with advancing research and implementation.

Adaptive facades

Adaptive or responsive building envelope relies primarily on the properties of materials and technologies to manage energy and mass transfer between the indoor environment of a building and the external environment. Adaptive building envelopes can adapt their shape and characteristics to simultaneously fulfill multiple functions, such as enhanced thermal performance, adjusted indoor lighting level, glare control, natural ventilation, and others.

A responsive building envelop has several advantages including (i) significant reduction of the energy consumption of buildings associated with heating, cooling, and lighting; and (ii) improvement of the indoor environment quality. Research on

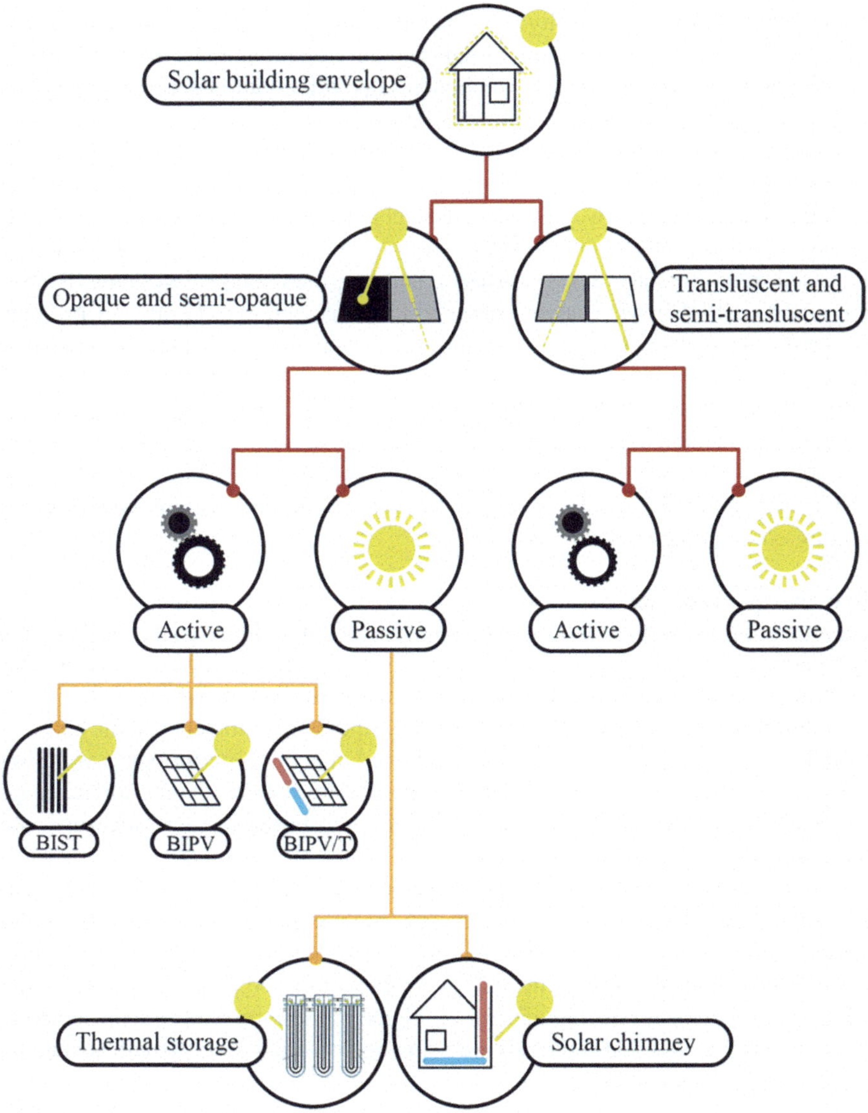

Fig. 2.4 Classification of solar building envelope

some of these facade systems is demonstrating that these responsive properties can significantly improve the performance of the facade compared to a static facade, while enhancing the indoor thermal comfort.

A wide range of technology options are currently available as integrated building envelope components, allowing them to reach desired performance criteria. Some of the most promising technologies include switchable glazing, movable solar shading, phase-change materials (PCMs), and dynamic insulation. Some of these are discussed in Sects. 2.3 and 2.4.

Despite the advantages of the application of responsive facades, deployment of such facade systems and the overall market uptake of these systems are still limited. This is due in part to the complexity and cost of design and implementation, combined with uncertainty regarding potential benefits and risks. The majority of existing applications consist of demonstration projects or conceptual studies.

Generally, there are two major categories of active responsive facades, referred to in pertinent literature as *smart building envelope* and *intelligent building envelope* [3]. A brief definition of each of these facade types is presented below. Additional details are given in Chap. 3.

- *Smart building envelope* relies on the characteristics of materials to respond to various impacts [4]. An example of smart material is phase-change material that changes its state passively allowing storing and releasing of heat to the indoor space (see below for more details).
- *Intelligent building envelope* employs computers to respond to various environmental and indoor conditions [3]. Example of intelligent envelope includes employing computer programs to control the opening and closing of shading devices, or operable windows to allow natural ventilation.

2.2 Overview of Building Envelope Heat Transfer

An understanding of the building envelope design and its impact on passive performance of the building requires the recognition of various mechanisms that govern the heat transfer through the envelope assemblies. An overview of these mechanisms is presented below.

2.2.1 Heat Loss and Gain Through Building Envelope

The indoor space is subject to heat loss or gain through walls, windows and doors, floors, and roofs. In addition to heat transfer through the envelope assembly, heat can be gained or lost due to defects in the construction that may result in infiltration (air leakage) and thermal bridges.

Three main mechanisms—conduction, convection, and radiation—allow heat to be transferred through the building envelope to the interior or exterior of the building. Figure 2.5 illustrates heat transfer through the building envelope, which relies on one or a combination of these mechanisms.

Conduction: This is the flow of heat flow through solid material, as, for example, various wall assemblies. Heat conduction is driven by temperature difference,

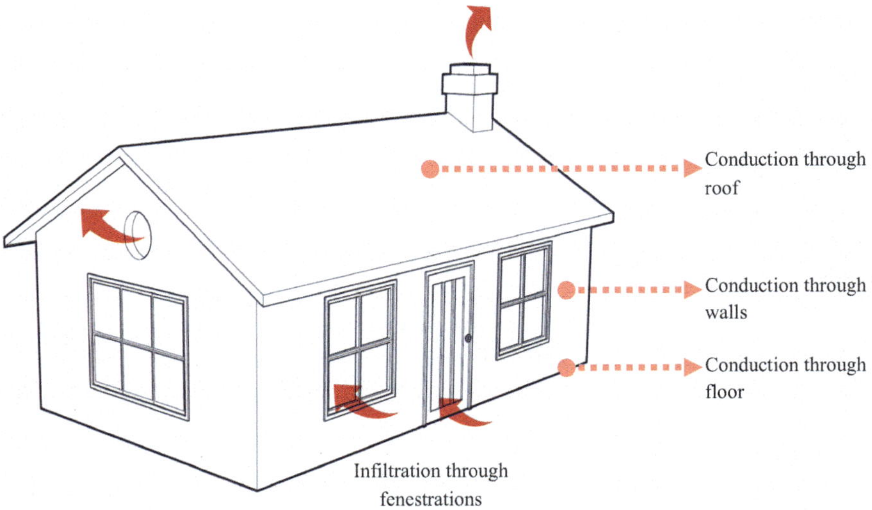

Fig. 2.5 Heat transfer through the building envelope

causing heat to flow from the warmer to the cooler side of the envelope. Conduction constitutes a major source of heating and cooling loss, within a building, resulting in increased demand for thermal energy to maintain thermal comfort.

Building codes and standards address the issue of heat conduction through building materials and components by specifying minimum thermal resistance to heat flow (R-Values) or maximum rate of steady-state heat flow (U-Factors) for building envelope construction assemblies. The ability of different materials to conduct heat is known as *thermal conductivity*.

To reduce conductive heat transfer through the envelope, its overall thermal resistance should be increased. This is achieved through a number of strategies including the use of low-conductivity materials, proper use, and installation of insulation, and avoiding thermal bridges. The impact of thermal bridges on heat conduction and building performance is discussed below.

Convection: This is the transfer of heat from one place to another by the movement of fluids. This is the process by which heat is transported from the envelope exposed surfaces to the indoor or outdoor air.

There are two types of convection—natural convection and forced convection. Natural convection occurs as a gravitational effect when warm air rises and cool air settles, or by heat flow (through air) from warmer to cooler regions in a process similar to conduction in solids. Forced air convection occurs when air is moved by mechanical means, such as fans.

A number of methods are used to reduce heat loss/gain by convective heat transfer through the envelope, such as reducing airflow, reducing air gap size between

various components and elements of the building envelope, and installing air barrier materials. Some of these techniques are explained in Sect. 2.3.
Radiation: This is the method by which heat is transferred from hot objects to cooler objects through outer space, by electromagnetic waves without heating the air. Unlike convection or conduction, radiation does not require any medium (solid or fluid). A significant amount of heat loss (and gain) can occur by means of radiation through windows.

2.2.2 *Solar Radiation and Heat Transfer*

Through walls and roofs
Solar radiation striking walls and roofs is absorbed by the outer layer of these surfaces, and then conducted through layers of the envelope assembly into the indoor space. The amount of heat absorbed by the outer surface of the building envelope depends on the properties of these surfaces, including surface texture and color, and on the intensity of the solar radiation incident on the outside surface. Solar radiation intensity is governed by the orientation of the surface and solar altitude and azimuth angles.

The rate at which energy is conducted through a material depends on the thermal conductivity of the material. For instance, gases are poor heat conductors while metals are good conductors. Materials containing air pockets have higher thermal insulation value since they conduct heat at lower rates than dense materials. Solar radiation heating the exterior layer of a wall increases its temperature. Assuming this temperature is higher than the indoor temperature, the heat is then transferred by conduction through the envelope layers into the indoor surface and then transferred to the indoor air by convection. In case of lower exterior temperature, the solar radiation reduces the rate of heat flow from the interior to the exterior. Figure 2.6 illustrates the processes of transfer of solar heat gain from the exterior side of a wall to the indoor air.

Through windows
Solar radiation interacts differently with glass than with opaque surfaces such as walls and roofs. The characteristics of glass play an important role in capturing and admitting solar radiation (details are provided in Sect. 2.3). In contrast to opaque surfaces, glass can transmit a large portion of solar radiation. Depending on the transmittance of the glass, this portion can reach up to 90% of the total solar radiation striking its surface.

Solar radiation which is not transmitted to the indoor is partially absorbed by the glass and partially reflected to the outdoor environment. Absorbed solar radiation is conducted through the glass and released to the outside or the indoor environment depending on the temperature difference between inside and outside. The absorbed portion of the solar radiation is relatively small compared to the transmitted and

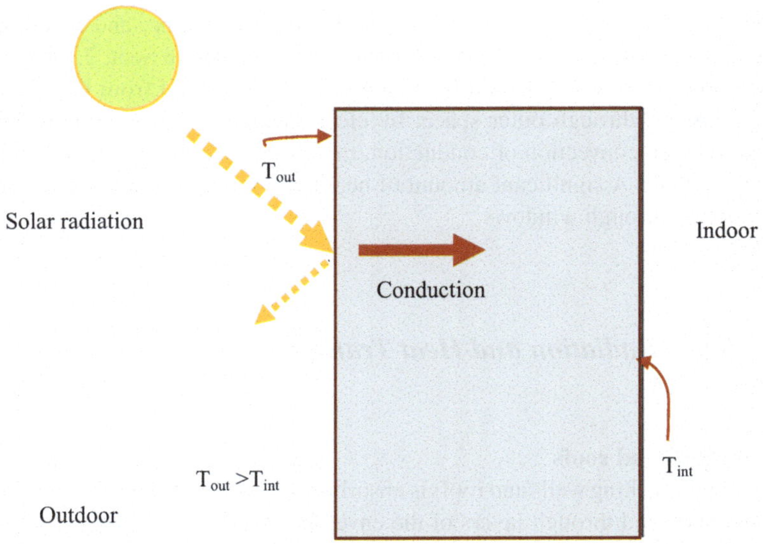

Fig. 2.6 Illustration of solar radiation striking an opaque wall

reflected components. Figure 2.7 illustrates the interaction of solar radiation with a double-glazed window.

Radiant heat admitted to the indoor space through the glass does not directly affect the indoor air. Interior surfaces including walls, floor, and ceiling as well as furniture absorb this radiant heat and then release it to the indoor air, through conduction and convection.

2.3 High Energy Performance: Main Considerations

2.3.1 Overview

There are a number of criteria that a high energy performance building envelope should meet, such as reducing heat loss, reducing air infiltration, and controlling and managing solar radiation for utilization in passive heating and daylighting while blocking the excess or unwanted heat.

The main strategies employed to increase the energy performance of building envelope include the following:

- Increasing the level of insulation;
- Designing high-performance windows, with climate-appropriate solar heat gain coefficients (SHGC);
- Properly sealing the envelope;
- Minimizing thermal bridges.

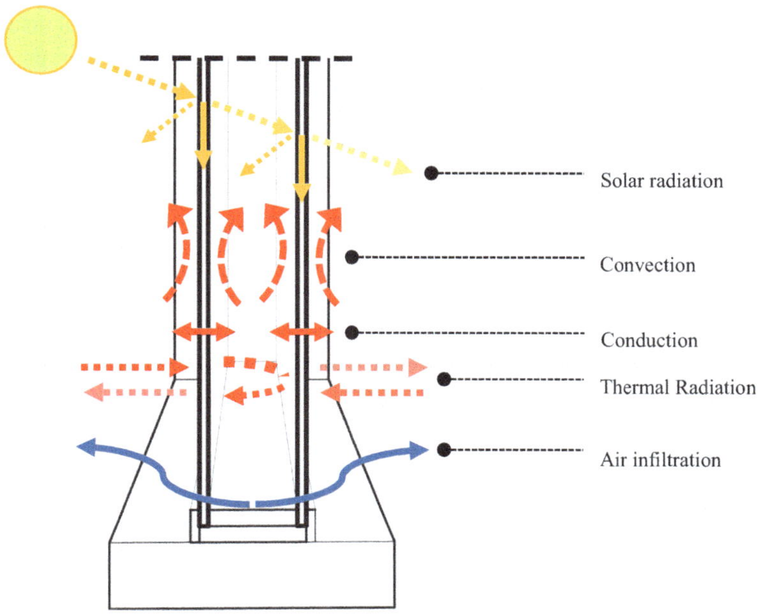

Fig. 2.7 Solar radiation interaction with window

2.3.2 *Thermal Insulation*

Thermal insulation consists of a material or combination of materials of high thermal resistance. Thermal insulation has the ability to restrain heat flow through a material by conduction, convection, and radiation. Applied into the building envelope, it retards heat flow into or out of a building.

Adequate insulation is essential for a high-performance building envelope. The required insulation level depends on the climatic conditions, the type of insulated materials, as well as on the type and function of the building. For instance, in residential buildings, particularly houses, high insulation is of higher priority than in commercial buildings, characterized by a high level of internal heat gain emitted from various sources including occupants, lighting, and equipment (see below).

A proper application of thermal insulation in buildings leads to reduced energy consumption and, consequently, reduced size of mechanical systems, and associated annual energy cost. Additionally, adequate insulation level can assist in prolonging the periods where building can provide thermal comfort, without mechanical systems.

In the design of high-performance building envelope, the amount of insulation needs to be optimized for the specific application, as excessive insulation may not significantly improve performance. Cost–benefit function of insulation is characterized by a diminishing returns curve, whereby, beyond a certain level of insulation, the benefits of energy savings are not matched by the cost of increased insulation.

Several factors should be considered in the process of selecting thermal insulation materials, including the specific application and the properties of the materials. Issues associated with specific types of insulation such as fire resistance, moisture penetration, and emissions of greenhouse gases should as well be considered in the selection process. A large number of insulation materials currently exist and in development. Following is a summary of commonly available thermal insulation materials [5].

- Mineral fiber blankets: batts and rolls (fiberglass and rock wool).
- Loose fill that can be blown-in (fiberglass, rock wool), poured-in, or mixed with concrete (cellulose, perlite, vermiculite).
- Rigid boards (polystyrene, polyurethane, polyisocyanurate, and fiberglass).
- Foamed or sprayed in-place (polyurethane and polyisocyanurate).
- Boards or blocks (perlite and vermiculite).
- Insulated concrete blocks.
- Insulated concrete form (see Chap. 3).
- Reflective materials (aluminum foil, ceramic coatings).

In addition, some novel, highly insulating building components include aerogel and vacuum insulated panels (VIP). These are briefly discussed in Sect. 2.4.

2.3.3 Windows

Windows are considered as one of the most complex building components of the building envelope as it has to balance between some conflicting aspects of design and performance, including building's aesthetics, providing light and fresh air to the indoor environment, and offering views that connect the interior space with the outdoors. On the other hand, windows constitute a source of significant heat loss and gain of the building. A high-performance building envelope should incorporate windows designed to mitigate this effect. High-performance window systems should aim at an optimum balance of the objectives of increasing energy efficiency by reducing the overall cooling and heating loads, while providing daylighting, view, and overall comfort.

Energy-Related Properties of Windows

In the design of high-energy-performing windows, a number of parameters should be considered, including heat transfer factor (U), solar heat gain coefficient (SHGC), visible transmittance (VT), glazing type, number of panes and cavity design (e.g., gas fills, spacers), and framing.

U-factor

The rate of heat loss or gain through the window frame and glazing depends on the temperature difference between the two sides of the window. Heat transfer occurs through the combined effects of conduction, convection, and longwave (thermal) radiation (Fig. 2.7). The U-factor of a window system represents its overall heat transfer rate or insulating value, incorporating the insulating level of glazing, frame, and spacers. Center-of-glass U-factor describes the insulating value of the glazing without the effects of the window's edges.

A number of characteristics affect the U-factor of a window system, including the total number of glazing layers and their dimensions, width of the cavity between the glazing layers, type of gas within the cavity, type of spacer in between glazing layers, and the characteristics of coatings on the various glazing surfaces. Figure 2.8 presents the different heat transfer modes expected through various parts of the window and the wall around it.

Solar Heat Gain Coefficient (SHGC)

Heat can be gained through windows by direct or indirect solar radiation, regardless of outside temperature. The *solar heat gain coefficient* (SHGC) is a dimensionless parameter (value ranging between 0 and 1) employed to measure the ability of a glazing system to control heat gain.

As defined in Chap. 1, the solar heat gain coefficient (SHGC) represents the fraction of solar radiation admitted by a window into the indoor space. This fraction accounts for directly transmitted portion and for the portion absorbed by the glass and subsequently released to the indoor [6]. The SHGC generally refers to the performance of the overall window system, representing an accurate indication of solar gain under a wide range of climatic conditions. The higher the SHGC, the larger the amount of solar heat gained by the window.

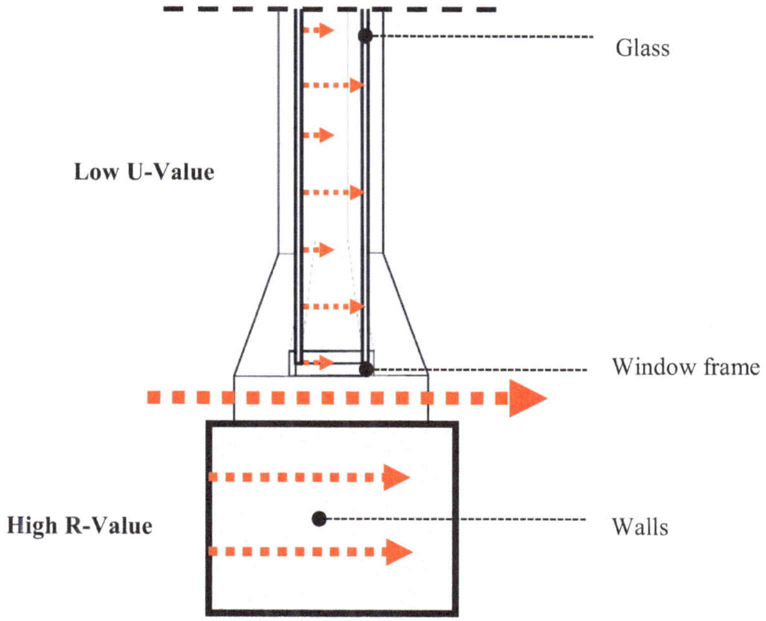

Fig. 2.8 Illustration of heat transfer through window parts. The arrows indicate the heat flow through different parts of the window

Visible Transmittance (VT)

The amount of visible light admitted through the window is represented by the *visible transmittance* [7]. Visible transmittance is expressed as a number between 0 and 1. It can also be termed as visible light transmittance (VLT), expressed as percentage.

Visible transmittance is the main indicator of the amount of daylight admitted to the indoor space. This factor determines views to the outside, as well as the privacy of the occupants. VT plays an important role in controlling issues such as glare and deterioration of interior furnishings.

The ability of a window to transmit visible light has significant impact on energy performance as it directly affects energy consumed by the lighting system. This is particularly significant in office buildings, schools, and other types of buildings where visual tasks prevail. Visible transmittance is influenced by the glazing type, the number of window panes, and the coatings applied. A larger value of VT indicates that greater proportion of visible solar radiation is transmitted to the interior.

Window System Components

Recent advances in window components and technologies allow achieving a high-performance window with relatively large flexibility. A wide range of high-performance window materials are currently available. These include an assortment of glass types, infill gas, spacer materials, and frame systems. In addition to glazing and frame materials, components that provide sun and daylight control constitute an important part of the design of high-performance windows. Below is a summary of window components and their impact on performance.

Glazing

Glass has relatively poor thermal insulating properties, as compared to other building envelope components. Designing window systems with multiple glazing layers, with air cavities in between, is an efficient method of improving the insulating value of the glazing system. The resistance of the window system can be further enhanced by filling the cavity by specific gases (such as argon or krypton, or a mixture of them). This is explained in more detail below.

Commonly applied glass types in a glazing system include: clear, tinted, low emissivity (Low-E) coated glass, and reflective coated glass.

Clear Glass

Relative to all other glazing types, clear glass is of the highest heat conductivity while permitting the highest daylight transmission.

Tinted Glass

Tinted glass, also termed "heat-absorbing" glass, is specially developed to maximize absorption of solar radiation across specific portion or all the solar spectrum.

This type of glazing is mostly applied to reduce glare from outdoor sources and reduce the amount of solar energy admitted to the indoor through the glass. Tinted glass can be efficient in blocking and controlling incoming solar radiation when employed as the external layer of a window system. In addition, tinting changes the color of the window and can increase visual privacy.

During daytime, tinted glass retains its transparent properties from the inside, while reducing the brightness of the outdoor view. Many tint colors are available to conform to various architectural and aesthetic objectives. Commonly applied colors include neutral gray, bronze, and blue-green. These tint colors do not substantially change the color of the actual view, as seen through them.

There are two categories of tinted glazing: traditional tints and selective tints. Traditional tints absorb solar radiation across the whole spectrum and therefore reduce the admission of light and heat, while selective tints absorb portions of the solar spectrum, reducing thus heat without significantly affecting light admission into the indoor space.

Tinted glazing is more commonly employed in commercial buildings. Solar radiation is typically controlled in residential indoor space employing blinds, drapes, or other interior shading device (see below).

Reflective Glass

Reflective coating can be used when significant reductions in solar gain are desired. These coatings increase the surface reflectivity of the material and consist usually of thin metallic layers. Various metallic colors such as silver, gold, and bronze are employed for reflective coatings. The visible transmittance of reflective glass is, in general, substantially reduced. Similar to the tinted glass, reflective glass is mostly applied in commercial buildings, especially for large extent of windows where solar heat gains and/or glare are excessive.

Low-E Glass

Low-E glass products are employed to reduce heat loss while allowing solar heat gain. There are two types of Low-E coating—High solar gain and low solar gain [8–10]. High solar gain Low-E glass, referred to as pyrolytic or hard coat Low-E glass, can decrease heat conduction through the glazing system. This glass is particularly suitable for buildings located in cold climates.

Low-solar-gain Low-E coatings (spectrally selective), suitable in particular to cooling dominated climates, reduce solar gain by blocking admission of the infrared portion of the spectrum. This type of coating reduces heat loss in winter but also reduces heat gain in summer, while providing a higher level of visible light transmission, as compared to tinted and reflective glazing.

Smart Glazing

Smart glass is a special type of glass, which can be activated to change from one status to another. For instance, this glass type has the ability to transform from a clear transparent state to a dark, semi-transparent or opaque state, and vice versa. Smart glazing can be used to control the amount of solar radiation entering the building, preventing thus overheating and issues related to excess daylighting such as glare. Some types of smart glass existing in the building industry are summarized in the following.

Photochromics: Photochromic glass can change its optical property in response to light intensity, reverting to its original state in the dark. The light transmittance of a photochromic window system drops significantly with increased solar radiation [8, 11].

Thermochromics: Thermochromic glass responds to change in the temperature of the glass, due to solar intensity and outdoor and indoor temperatures. This type of glass, which change state as a function of solar radiation and indoor and outdoor temperatures, can regulate the amount of solar radiation admitted to the building interior, when space cooling loads become excessively high [8].

Liquid Crystals: Liquid crystal glazing can switch from transparent state to a diffuse white state. In the diffuse state, these glazing types are capable of glare control, allowing thus to replace shading devices.

Electrochromics: Currently, the most promising smart glass technologies consist of the electrochromic (EC) glass. Electrochromic coating is composed in general of multiple layers (typically five layers), deposited on a glass or plastic substrate. Electrochromic glass typically switches between clear state and blue-tinted transparent state. Switching from one state to the other is operated electrically.

In building application, electrochromic windows can be operated in various methods, including manual switch, remote controller, or as a simple automatic system. The operation of electrochromic windows can be as well integrated within a central management system that controls simultaneously other building systems operation such as lighting and mechanical systems [8]. Figure 2.9 presents an illustration of an electrochromic window (switched-on and switched-off).

Gas fill

The thermal performance of glazing units can be significantly improved by reducing the conductance of the air space between the layers. Gas fill can be applied to the cavity between glazing layers to minimize heat transfer between the interior and exterior of the window. Gases typically used are argon or krypton. These gases are inert, non-toxic, clear, and odorless. The use of argon and krypton gas fill demonstrates significant improvement in thermal performance [7, 9, 12].

Argon is currently more commonly used than krypton due to lower cost. Similar to an air cavity, the optimal width of an argon-filled cavity is about 12 mm. Krypton has superior thermal performance to argon and can be especially advantageous for narrow cavities. An optimal krypton cavity width is about 9 mm.

Spacers

Another component of window system that can contribute to heat loss is spacers. The spacer is primarily used to separate the panes of glass and to provide sufficient surface for the application of sealants. In addition, the spacer system provides a number of functions including accommodating thermal expansion (due to daily and seasonal change in temperature) and pressure differences, averting leakage of water or water vapor, sealing the cavity, preventing thus gas loss, and reducing the potential of water condensation at the edges [7, 13].

In high-performance windows, *warm-edge* spacers play significant role in restricting heat transfer. Warm-edge spacers are characterized by low conductivity, as compared to conventional aluminum and stainless-steel spacers. Such low-conductivity spacers reduce thermal bridging at the edge of the window, contributing thus to improving the overall performance of the window. Examples of

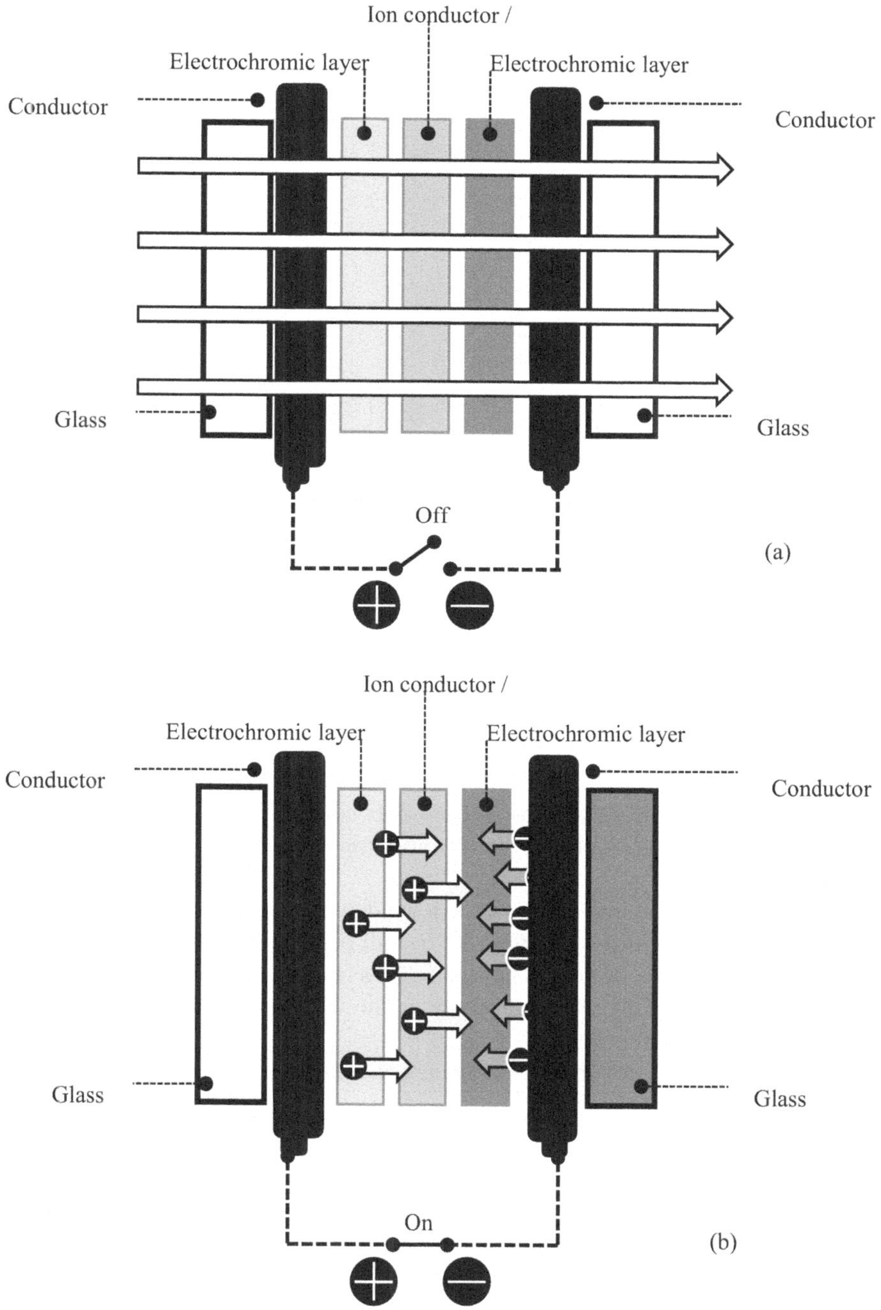

Fig. 2.9 Illustration of an electrochromic window, **a** switched-off (or clear status) **b** Switched-on

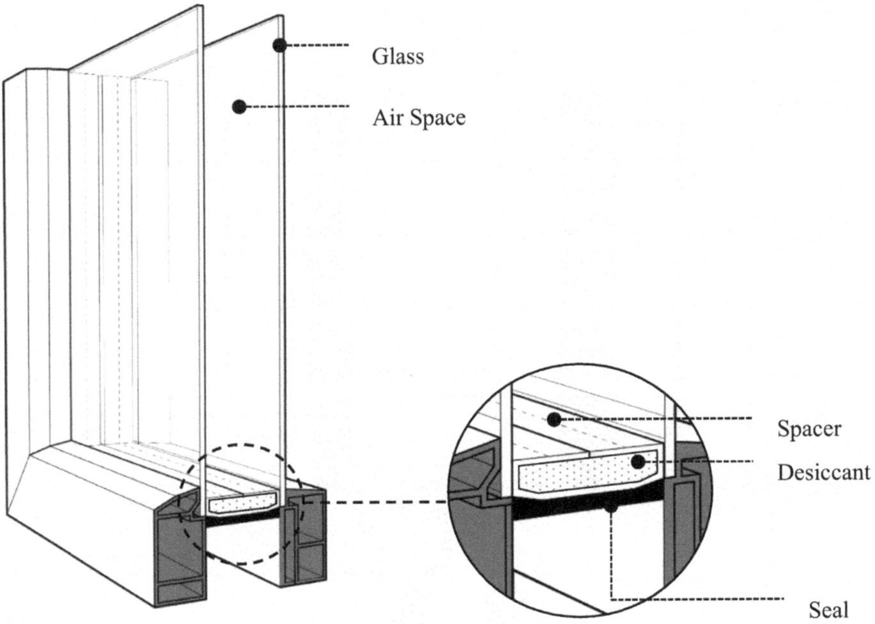

Fig. 2.10 Illustration of a warm-edge spacer

warm-edge spacer materials include flexible foams, thermoplastics, plastic/metal hybrids, and others [7]. It can consist of a rectangular tube filled with a desiccant (such as zeolite spheroids) that prevents condensation between the panes. Figure 2.10 presents an illustration of warm frame spacer.

Framing System

Window frame is an important part of a window system and can contribute significantly to heat transfer and thus the overall performance of the window. The area occupied by the frame can be significant as compared to the total area of the glazing, increasing the impact not only on the window performance but on the building envelope in general [9, 14].

Some of the most commonly implemented framing systems are summarized below.

Aluminum Frames

Aluminum window frames present a number of advantages such as durability, lightweight, and ease of shaping into complex forms required in some window framing. The high thermal conductivity of aluminum is one of the major disadvantages of this material, forming thermal bridge at the window's perimeter. In cold climates, in addition to the heat loss through thermal bridging, aluminum frames can cause condensation on the inside surfaces of the frames.

An effective, most commonly applied solution to restrict heat conduction through aluminum frames is to implement a "thermal break" to the frame. A thermal break refers to breaking the frame components into two pieces, interior and exterior,

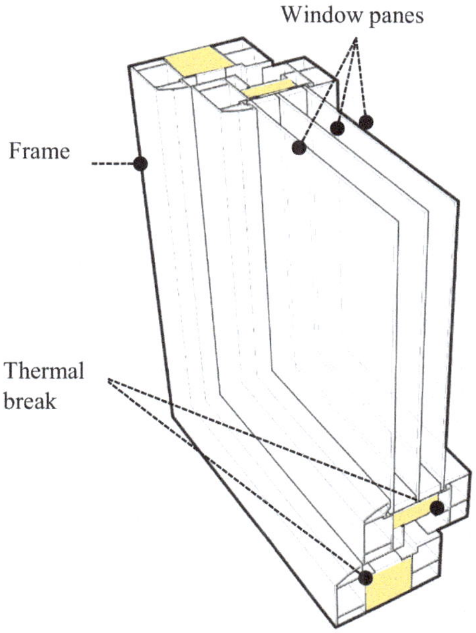

Fig. 2.11 Thermal break within the window frame

joined by material of low conductivity. Employing this method allows to reduce the U-Value of the frame by more than half. Figure 2.11 presents an example of application of thermal break into an aluminum window frame.

Wood Frames

Wood is a traditional window framing material, favored in residential applications, particularly in wood-framed houses. Similar to aluminum, wood is easily wrought into various shapes required for window framing. Wood frames allow higher thermal resistance than aluminum frames. One major disadvantage of this type of frame is its susceptibility to material degradation and need of regular maintenance (such as painting), and thus is not as durable as other materials. Durability can be improved by employing chemically treated timber with water-repelling materials

Vinyl Frames

Polyvinyl chloride (PVC) frames commonly known as *vinyl frames* are characterized by thermal resistance similar to wood framing and thus improve the thermal performance of window systems (as compared to aluminum frames). The performance of these frames can be enhanced by devising small hollow chambers within the frame structure, which reduce heat loss by convection.

Additionally, PVC material has the advantage of being moisture resistant and low maintenance. For instance, vinyl frames do not require painting and do not need a finishing coat, which reduce potential deterioration. Various coatings and surface treatments are increasingly available, allowing a wide selection of frames.

Hybrid Frames

Hybrid frames, combining different construction materials, are gaining popularity in the construction industry. Examples of hybrid framing system are vinyl-

or aluminum-clad wood frames. Cladding wood frames with vinyl or aluminum allows maintaining the benefits of wood materials, while reducing exterior maintenance needs. Existing wood veneer framing features a variety of interior finishes, to accommodate different tastes and possibilities.

Window location and its impact

In the northern hemisphere, under any climatic zone, south windows are the most optimal to allow solar heat gain during the winter period. North windows are the source of significant heat transfer, while presenting reduced potential for useful solar gain in winter. East and west windows can admit solar radiation primarily in summer, when it is not needed, increasing thus cooling loads, while in winter they are conductive to heat loss. However, non-south-facing windows still serve non-energy-related functions, such as daylighting and view [15]. The summary below discusses window location and their impact (for the Northern Hemisphere).

North windows: Diffuse northern light is desirable in general, and in particular, for office and institutional buildings. Large north-facing windows can, however, lead to significant heat loss, and thus they should have high insulation value. Since north windows receive relatively little direct sun in summer, they do not require shading.

East and west windows: The area of east and west windows should be reduced to the minimum required for the interior functions. These windows, exposed to the morning and afternoon sun do not admit sufficient solar radiation in winter to benefit from passive useful heat gain, while on the other hand can have significant negative impact on increasing cooling loads during summer. These windows, particularly west windows, require efficient shading systems.

Reducing heat gain of west windows can be implemented by employing glazing with a low shading coefficient such as tinted glass or some types of Low-E glass, providing thus some shading while allowing almost clear views.

South windows are a key component of any passive solar system. Glazing area should be properly designed to allow enough solar radiation in winter while not causing overheating and therefore cooling burden in summer.

In houses in cold climate zones, an optimal range for window size is between 30% and 40% of the south facade. This amount of glazing should be accompanied by an adequate shading system and thermal mass to reduce diurnal temperature swing and potential overheating.

Glazing that combines high solar heat gain coefficient (SHGC ≥ 0.5) and insulation should be adopted to exploit passive solar gains in the heating season. Employing an appropriate shading device, such as overhangs or controlled reflective blinds (see shading devices below), can prevent overheating during the cooling season.

Shading and miscellaneous devices

Shading devices are major components in the design of solar building envelopes for adjusting the amount of solar radiation access over the year. A large range of designs can be adopted to accommodate architectural and functional objectives [16, 17]. Common designs of shading devices include horizontal overhangs and vertical fins.

Window materials with optical characteristics are often employed to regulate solar radiation admitted to the indoor space, such as tinted glass, reflective glass, as well as more advanced solar technologies such as semi-transparent PV modules. Shading devices in general can be divided into two major categories, fixed and dynamic. Shading systems should be designed to match the orientation of the window, to accommodate the corresponding insolation patterns.

A summary of various types of shading devices and their applications is presented below.

Fixed Shading devices

Fixed shadings include devices such as overhangs, vertical fins, canopies, balconies, protruded window frames, egg-crate louver, and others. This category of shading systems is static, not allowing thus for adjustment to respond to variations of climatic conditions and building requirements.

The geometric design of fixed shading design, including tilt angle and dimensions, is usually determined according to the sun incidence angle of the summer solstice. Shape and materials, including colors of fixed shading devices, can be explored to achieve diverse architectural effects.

Horizontal overhangs

Fixed horizontal overhangs block high-angle solar radiation but allow low-angle solar radiation to enter the indoor space. This type of shading device is mostly beneficial for south-facing windows in northern climate, allowing useful heat gain during the heating period, while reducing potential overheating in summer.

The ideal depth of the overhang depends on the height of the window, and the desired shaded portion of the window. Horizontal overhangs can be designed in a multitude of methods, to produce various effects such as reducing the depth of the overhangs, or, for instance, allowing only diffuse radiation.

Figure 2.12 presents some horizontal overhang design options.

Fixed vertical fins

Low-angle direct sunlight, such as occurring on the east and west orientations of a building, is more difficult to block. Vertical elements perform better than horizontal shading elements, on east and west facades (Fig. 2.13). However, effective shading by vertical fins may be at the expense of daylight and view.

Hybrid horizontal and vertical fins

A hybrid combination of vertical and horizontal shading device can be also designed (Fig. 2.14). Such elements combine the properties of horizontal overhangs—suitable for south windows—and of vertical fins which are more appropriate for east/west windows, so it can be used for various window orientations, with similar advantages.

This type of shading device may create a challenge in terms of integration with the architectural design as well as optimal view to the exterior.

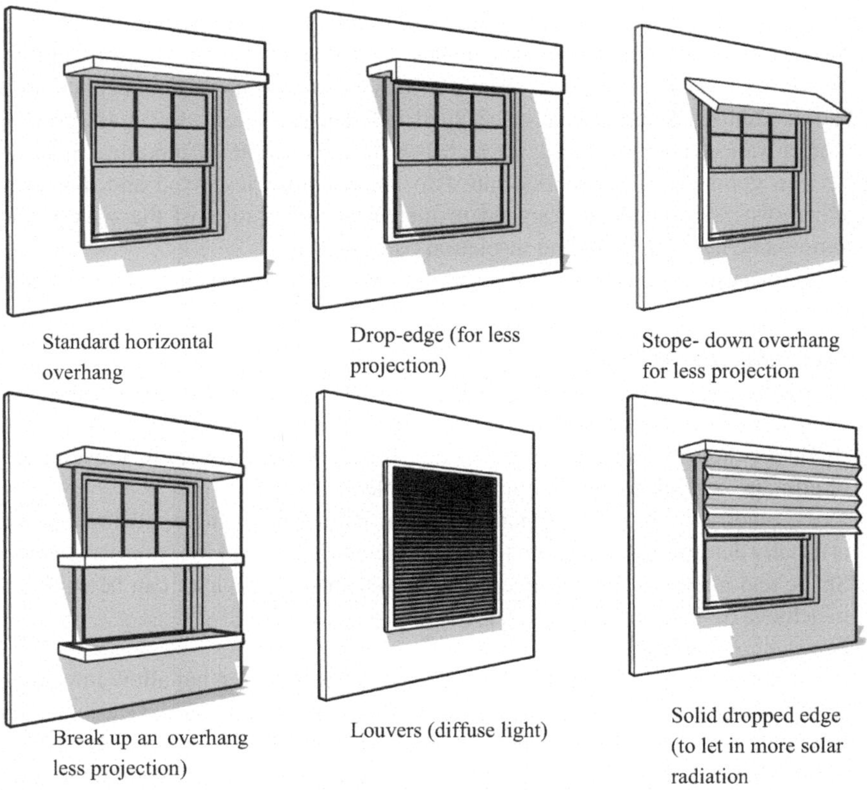

Fig. 2.12 Illustration of different overhang designs

Fig. 2.13 Vertical fins

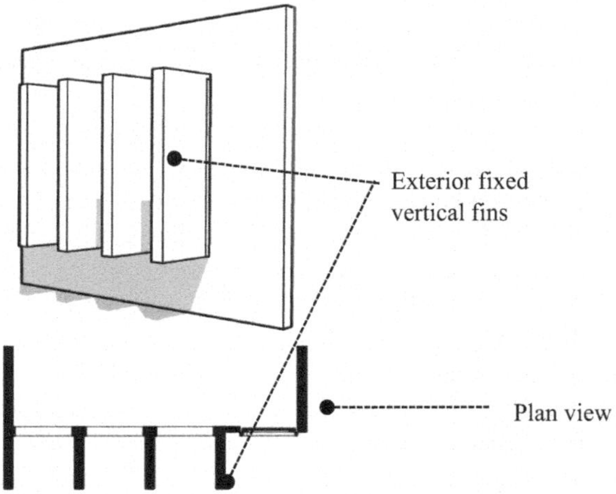

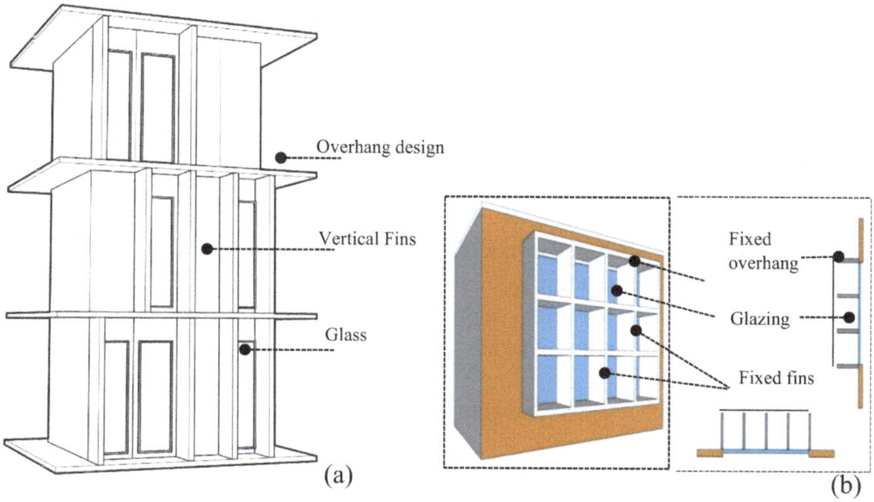

Fig. 2.14 An example of hybrid horizontal and vertical shading **a** on a full floor level (within a multistory building), **b** on a window level

Mobile shading

Mobile shading devices include a wide range of devices such as shutters, Venetian blinds, roller blinds, and curtains. Mobile shading can be easily adjusted to meet the requirement of the indoor space, such as indoor temperature and illuminance levels. This type of shading system can result in significant improvement in the energy consumption of buildings, for annual heating and cooling as compared to fixed systems. Mobile shading device can be controlled manually or mechanically.

Interior Shading Devices

Interior shading devices are, in general, less expensive, and more controllable than external devices. Although interior shadings are efficient in regulating the direct sunlight and in reducing glare, they cannot control passive heat gain, since the solar radiation admitted through the window heats the interior shade and the surrounding indoor air. Internal shadings are commonly employed to provide privacy and thermal comfort while allowing the occupant varying degrees of control of the solar radiation accessing the space.

There is a large variety of interior shading options, including roller shades, Venetian blinds, and drapery. More advanced options consist of blinds situated between the panes of an insulated glazing unit (in multi-pane window). The amount of daylight and passive solar heat gain admitted to the indoor environment are significantly affected by the shading material and their location with respect to the glazing (between the window panes or on the interior side of the window).

Roller shades allow to regulate daylight, transmit solar heat gain, and view to the exterior and privacy by various design options, including texture and materials of the shade. In Venetian blinds, the fins' angle can be adjusted to control daylighting and solar radiation at different times of the day, while allowing partial view to the exterior. The performance of between-pane shade depends mainly on the window cavity width and whether it is a ventilated cavity.

Motorized Shading Systems

Motorized shading systems are attracting a lot of interest, particularly in commercial buildings and buildings designed with large extent of glazing. The growing interest in motorized shading systems is due to their potential to increase energy efficiency, while improving the thermal comfort of the indoor environment. These types of shades are operated employing motors, mounted typically in the roller tube or headrail of the shading system. Automation of motorized shadings needs to consider balancing between energy management needs, human thermal and visual comfort, and functional tasks.

Mobile insulation

As discussed above, heat loss and gain through windows can be significant, affecting the energy demand of the building. Heat transfer occurs through window frame, glass, and infiltration around window frames. Design and application of movable insulated panels that cover the window occasionally, during critical periods, can reduce heat loss. Such applications can be beneficial for buildings with large extents of glass.

There are various creative methods to design moveable insulation. For instance, they can consist of panels sliding on a track across the glazed area. Moveable insulation can be controlled manually or mechanically. They can also be equipped by sensors to allow them to be automatically activated by the indoor temperature setpoint, illumination level as well as building occupancy.

2.3.4 Thermal Bridging, and Air Infiltration

Thermal bridging

Thermal bridges in building enclosures can be defined as localized areas with higher thermal conductivity than the adjacent areas. Figure 2.15 presents examples of thermal bridging due to discontinuity of insulation layer.

A typical thermal bridge in a building enclosure would be a location where a material of high conductivity, such as a structural component or metal flashing, penetrates the insulation layer [18]. The presence of a thermal bridge in a building assembly can cause higher heat transfer, resulting in higher energy consumption. Another issue associated with thermal bridges is condensation or frosting on colder interior surfaces, which could lead to mold growth and associated health concerns.

The rate of heat flow through a thermal bridge depends on a number of factors including the temperature difference, the thermal conductivity of the bridge material, and the cross-sectional area of the thermal bridge.

Elimination of thermal bridges is a major component of the design of energy-efficient building enclosures, particularly in cold climate. A key approach toward

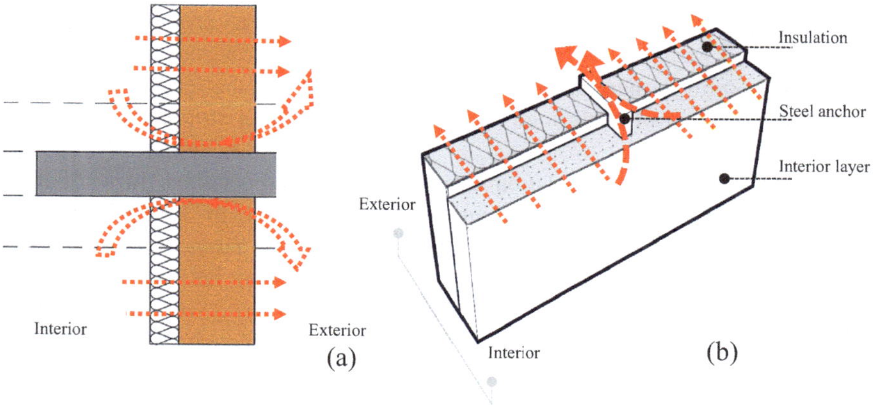

Fig. 2.15 Examples of thermal bridges; **a** associated with a slab, **b** associated with structural elements

this goal is to ensure the continuity and alignment of the insulation layer, within the building envelope (see Fig. 2.16). In addition, thermal bridges can be mitigated by overlapping the layers of insulation where direct continuity is not possible.

Air infiltration and ventilation

Infiltration consists of unwanted and uncontrolled ventilation. It is generally defined as the flow of outdoor air into a building through cracks, leaks, and other unintentional openings in the envelope. Air leakage can be responsible for a significant amount

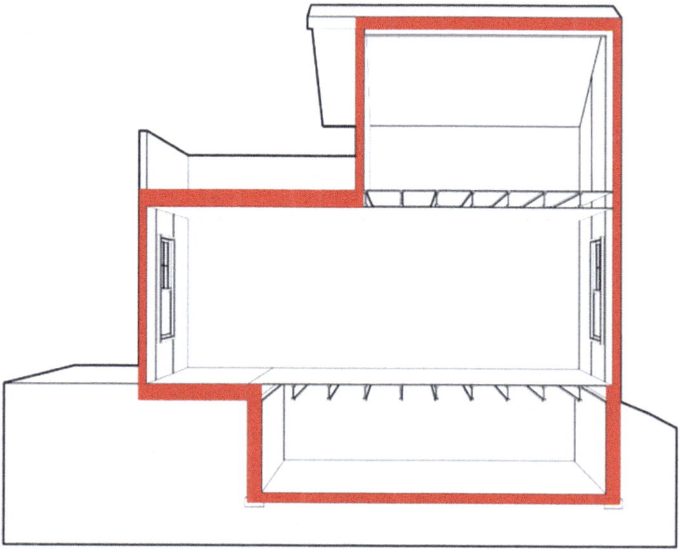

Fig. 2.16 Schematic illustration of the continuity of insulation layer

of heat loss and gain in a building. Air leaks through walls and roofs affect energy demand in two ways: by allowing airflow between the indoor and outdoor spaces, and by transporting water vapor to places where it may condense. Air leakage can be significantly controlled through tight sealing and weather stripping of windows including sash and frames.

The sole application of insulation layer over cracks and openings does not eliminate air infiltration through them. A number of factors affect the rate of infiltration such as the overall tightness of the building envelope, the temperature difference between the indoor and outdoor, wind velocity, and building height governing the stack effect.

Methods to reduce air leakage are numerous including effective weather stripping of windows and doors and sealing and caulking all cracks and penetrations such as electrical outlets and light switches that could be a source of uncontrolled air leakage into or out of the conditioned space. Air barriers can be installed to reduce air infiltration, through the building envelope, preventing thus outdoor air from infiltration to the space or exfiltration of conditioned indoor air from the building. Air barriers should, however, block air while allowing potential trapped-in moisture to escape the building envelope.

Well-insulated airtight buildings can be prone to issues such as poor indoor air quality and moisture accumulation, and thus need to be sufficiently ventilated employing natural or mechanical means (or both in a hybrid system). Ventilation rate should be adequately designed, according to building type and occupation level, to provide a good indoor air quality. Adequate ventilation can assist in preventing high indoor moisture levels, resulting in moisture condensation on window surfaces as well as concealed condensation within the building envelope during heating season.

2.4 Building Envelope Integrated Passive Systems

This section discusses advanced building envelope passive solar systems. Such building envelope systems are mostly based on employing thermal mass coupled with various technologies, solar heated air, and advanced high insulated material.

2.4.1 Thermal Mass

Currently, thermal mass is receiving increasing attention in the design of low-energy, high-performance buildings. Incorporating thermal mass within the building envelope retards and attenuates the impact of external climatic conditions. Thermal mass plays a significant role in balancing diurnal variations of the indoor temperatures. The benefits of thermal mass are derived from its capacity of storing excess heat from incident solar radiation and from internal loads of the building, and releasing this stored heat to the indoor air during cooler periods.

Thermal mass is characterized by its time lag, defined as the time difference between the outdoor temperature peak and the indoor temperature peak. The thermal mass allows to increase this time lag. Each wall and roof orientation has a different time lag, since for each orientation the peak heat gain from solar radiation occurs at a different time of day, according to the sun's position [19].

Thermal mass concept constitutes the basis of the advanced passive solar design envelope systems presented below.

Trombe wall

Since its conceptualization in 1881 by Edward S. Morse, the Trombe wall has been further popularized by Felix Trombe and Jacque Michel [20]. Since then a plethora of modifications and alterations of this system were developed to adapt its application to different climate conditions and indoor requirements. All subsequent alterations have been a direct offshoot of the classic Trombe wall. Discussion of the Trombe wall, its function, drawbacks, and the upgrades and altered configurations are presented below.

Function and application

As discussed in Chap. 1 (Sect. 1.3.1), the main components of a Trombe wall consist of (1) a massive dark color wall, with high heat storage capacity, (2) exterior glazing, and (3) an air gap between the glazing and the thermal mass (Fig. 2.17) [21].

The glazing admits direct and diffuse solar radiation into the air cavity, allowing to raise the temperature of the air within this cavity and assisting in increasing the temperature of the thermal mass [20]. The external dark-colored surface of the thermal mass wall absorbs simultaneously the incident direct solar radiation, stores it, to be released later on (according to the time lag) to the interior space, by convection and radiation.

There exist slight variations of the basic Trombe wall system, employing various techniques of natural or forced ventilation, as well as techniques of increasing the thermal capacity of the thermal mass by employing special materials such phase-change materials (discussed below).

Advantages and disadvantages

The Trombe wall system presents a number of benefits, including reducing building's energy consumption, decreasing moisture and humidity levels of adjacent occupied spaces in locations of high humidity, and moderating diurnal variations in temperature. The implementation of thermal mass can assist in shifting electricity consumption from peak to off-peak electricity demand, which can result in cost saving and reduced stress on the local grid. Most scenarios of the implementation of Trombe wall system are relatively low cost and low tech, as the components of the system are standard construction elements (glazing and masonry). In addition, it is rather simple to retrofit an existing uninsulated masonry wall into a passively heating Trombe wall with the addition of glazing [22]. The effect of Trombe walls on thermal comfort is not limited to the directly attached space but extends

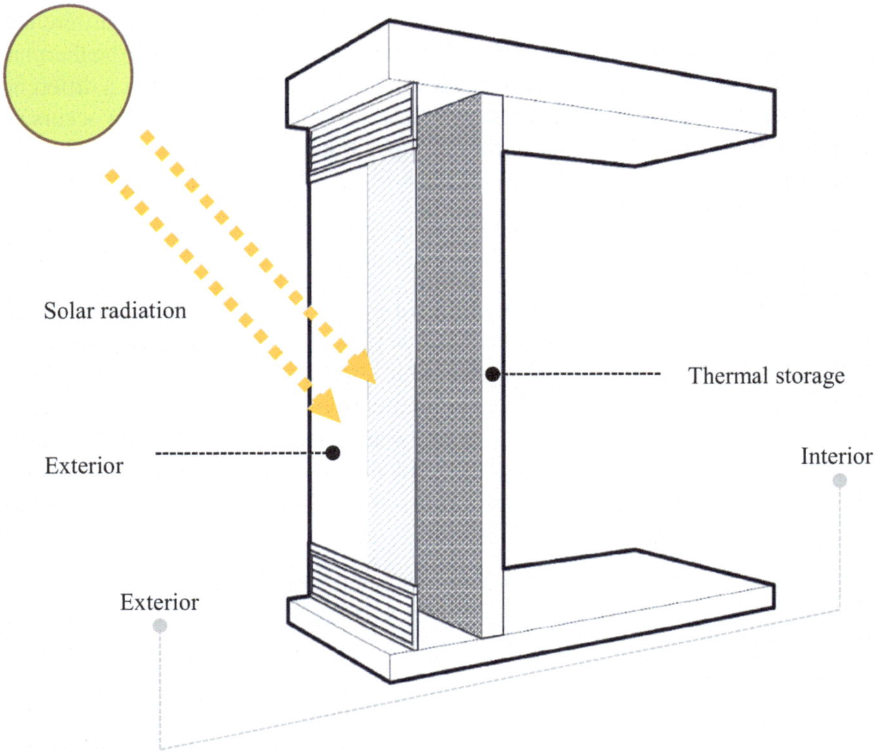

Fig. 2.17 Basic composition of a Trombe wall

to adjacent spaces by convection. Adjustments can be made to mitigate disadvantages, and the invention is increasingly becoming the subject of studies that are exploring potential efficiency upgrades and compatibility with new technologies. The benefits of this system are climate dependent. In regions of hot summers and relatively mild winters, summer overheating can override the benefits during winter. Moreover, in climates that typically have extended periods of cloud cover, the massive wall can become a heat sink working in a reverse manner by transferring heat from the interior to the exterior [23]. Additional disadvantages of the Trombe wall system relate to the restriction of light and view. Access to light can be addressed by designing a clerestory above the wall. The access to view can be addressed by creative solutions that allow openings in the Trombe wall, alternating thus between direct and indirect solar gains [24]. Figure 2.18 presents an example of Trombe wall with punched window.

Design considerations

A number of design factors affect the performance of the Trombe wall, mostly related to the properties of the materials employed and methods of construction. These include the thickness of the thermal mass, type of mass material, the color of the exposed exterior surface of the wall, as well as type of glass used within

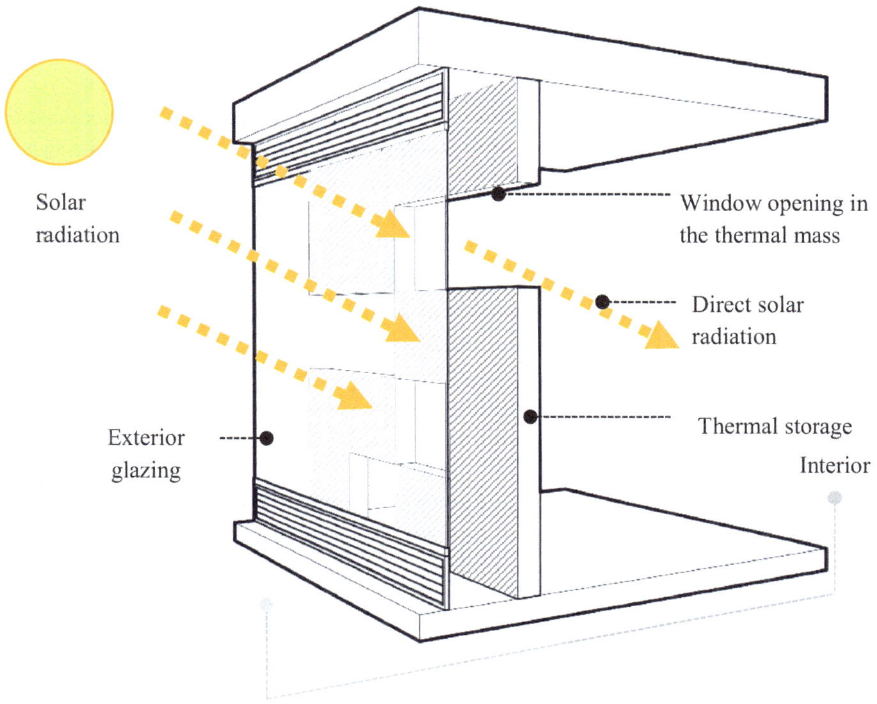

Fig. 2.18 Drawing of an integrated design of Trombe wall and direct gain through window opening (in the thermal mass)

this system. Methods of construction are presented in the discussion of available configurations of Trombe wall below.

Materials: The thermal mass wall can be built from concrete or masonry. Other materials with high storage capacity can be employed as well. Some applications explore the utilization of water or phase-change materials for their thermal capacity.

Thickness: Wall thickness affects the timing of maximum and minimum temperatures, as well as the overall heat capacity. While the wall thickness may involve some cost consideration, the main effect of this design factor is related to thermal storage capacity, which governs thermal comfort [20]. An excessively thick wall can have negative impact, since it can significantly delay the heat transfer from the thermal storage, affecting thus thermal comfort of the indoor space.

Color: Dark colors for the exterior surface of the thermal storage wall enhance the efficiency of the wall system, due to their higher solar radiation absorption as compared to lighter shades of colors.

Glazing: Double glazing enhances the performance since it restricts loss of solar gain to the exterior. Applying low emissivity (Low-E) film on the glazing reduces reradiating solar gain back to the environment, increasing thus the efficiency of the system. Combining double glazing with Low-E film can be highly beneficial for overall collected energy.

Upgrades and Configurations

A number of upgrades can be applied to the Trombe wall to enhance its performance, ranging from simple to more complex methods. These modifications are driven by the overall attempt to store and use solar gain more efficiently while also reducing the risk of heat loss.

Vents

In addition to direct conduction through a thermal wall, solar heat can be moved to adjacent spaces by providing a natural convection loop. The addition of vents that can be opened and shut at upper and lower locations in the glazing and in the thermal mass enables heat convection between the exterior and interior spaces. Vented Trombe wall systems enable more efficient circulation of the preheated air in winter and reject overheated air from the interior in summer (Fig. 2.19).

Composite Trombe wall

The composite Trombe wall, also known as insulated Trombe wall, consists of adding a layer of insulation at the interior of the wall. This insulation layer is separated from the mass by a second air gap. Vents in the insulation wall allow control of heat flow along the thermal mass into the interior space via convection (Fig. 2.20) [20]. The composite system is suitable for cold climates or cloudy climates due to the enhanced thermal resistance of the assembly [23].

Miscellaneous technologies

Additional innovations in the application of Trombe walls consist of further enhancing this wall system performance through the integration of various solar technologies. One promising technology is the incorporation of phase-change materials (PCM), as thermal mass. PCM is defined as material that changes phase (e.g., from liquid to solid) at specific temperature. The physical phase of this material changes when it is heated by solar radiation. During this phase change, it stores a significant amount of heat that is released to the indoor space when the material cools and returns to its original phase [20]. PCM can be particularly beneficial in applications where thick masonry walls cannot be accommodated. For instance, a 150-mm-thick massive wall can be substituted with a 35 mm of phase-change material wall. Phase-change materials are discussed in more detail below.

Other materials are being examined for their capabilities of replacing the traditional thermal mass wall. For instance, water can be superior to masonry in terms of thermal capacity and storage of heat. However, water applications require additional careful considerations to ensure proper, durable sealing.

Phase-change materials

A phase-change material (PCM) can be defined in the building context as a material that melts (changing phase) at a temperature close to the operational temperature of a building. This characteristic allows the material to store and release thermal energy of the latent heat associated with the process of changing phase. PCM materials can

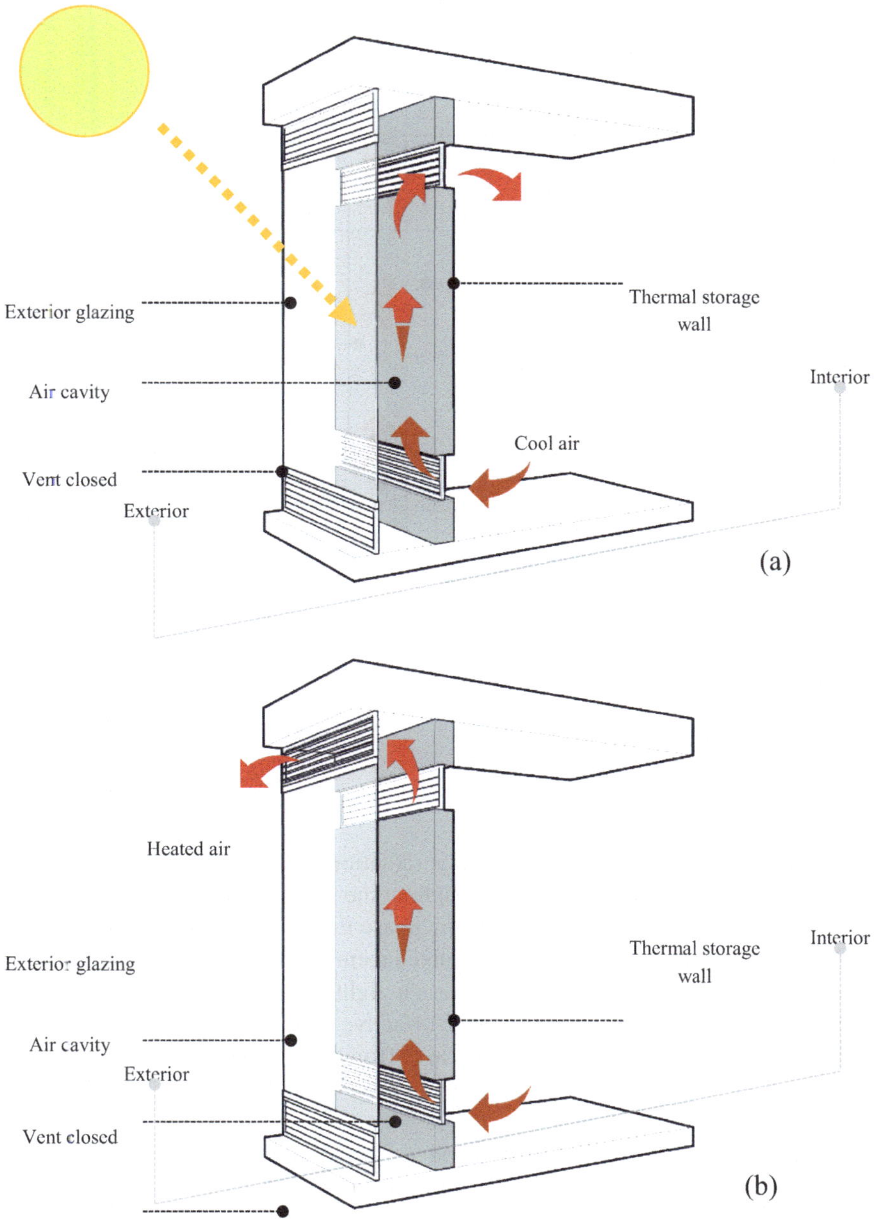

Fig. 2.19 **a** Vented in winter mode; **b** Vented Trombe wall in summer

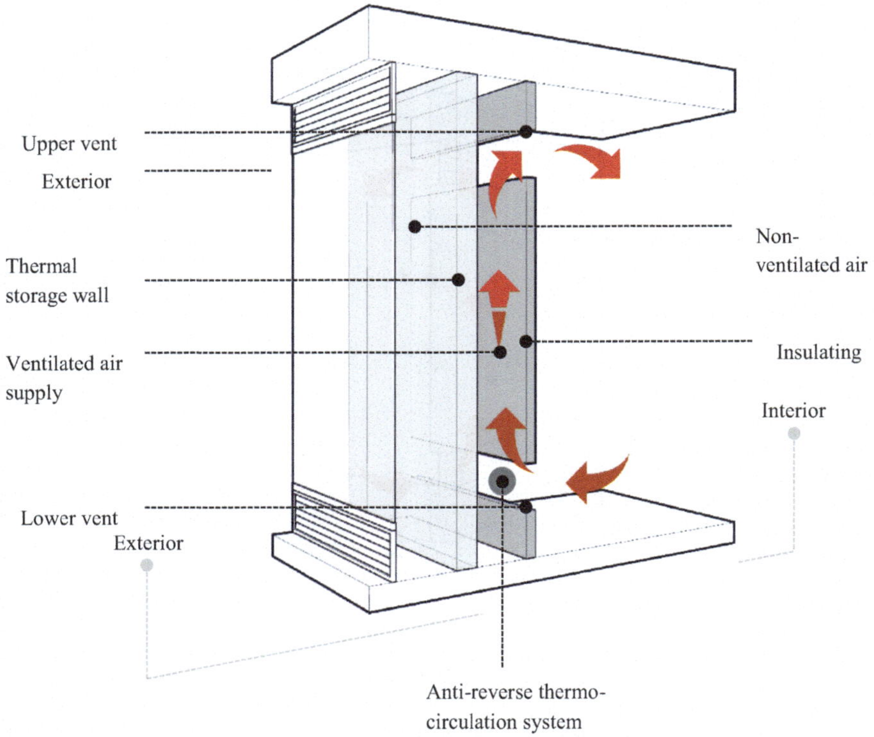

Fig. 2.20 Composite solar wall (based on [23])

absorb and store excess heat gain from solar radiation and other internal sources, and subsequently release the stored heat throughout the solidification process, when the indoor temperature drops below that of the phase transition [25]. PCM can thus be employed for the regulation of diurnal change in temperature and improving thermal comfort, serving as alternative thermal mass in lightweight building construction.

PCM can be integrated into a number of active and passive solar systems. For instance, it can replace the massive wall in a Trombe wall. PCM can be integrated into underfloor air distribution systems, window systems, as well as in the building envelope.

Experimental assessment of PCM application in building construction [26–29] shows that PCM integrated into walls can stabilize the indoor temperature fluctuations within the comfortable range. This can lead to a significant reduction in the thermal load of buildings and hence decreasing energy consumption by the mechanical system, for heating and cooling.

PCM in Facades

PCM can be used in facades to replace some of the window panels. Relying on the translucency of the melted phase of the material during the warm hours of the day, it allows to fulfill the vision and daylight transmission requirements of windows. On the other hand, the heat storage capacity of the PCM improves the heat balance of the window, through moderate heat gains with very low heat losses throughout the day, as compared to conventional window panes.

PCMs are manufactured from one of the three main groups of raw materials: eutectic, organic, and inorganic materials [26]. Paraffin-based organic materials are mostly employed applications. This is due to their optical properties, since they are transparent during their liquid state and translucent in their solid state.

The main limitation of PCM as window elements is the visual effect associated with the solid phase, presenting inhomogeneous appearance. More uniform visual appearance can be realized employing some masking methods, such as screen-print glazing. Consequently, facade panels with PCM are best employed in situations where the visual contact with the outdoor surroundings is not paramount.

PCM windows can be advantageous in providing uniform illumination and diffuse daylighting to the interior space, simultaneously increasing the thermal performance of these windows [30].

Another innovative application of PCM panels is moveable shutters. Research on incorporating PCM within an internal slat-blind shading device reports that such application can reduce cooling load significantly, while providing some advantages during the heating period as compared to other types of blinds [30].

2.4.2 Ventilated Concrete Slab

An advanced technology for utilizing thermal mass consists of employing hollow core concrete slab for active or passive ventilation [31]. This application consists of applying forced air through channels within precast concrete slab units (floors or roofs) (Fig. 2.21). Forcing air through ventilated slab system improves the capabilities

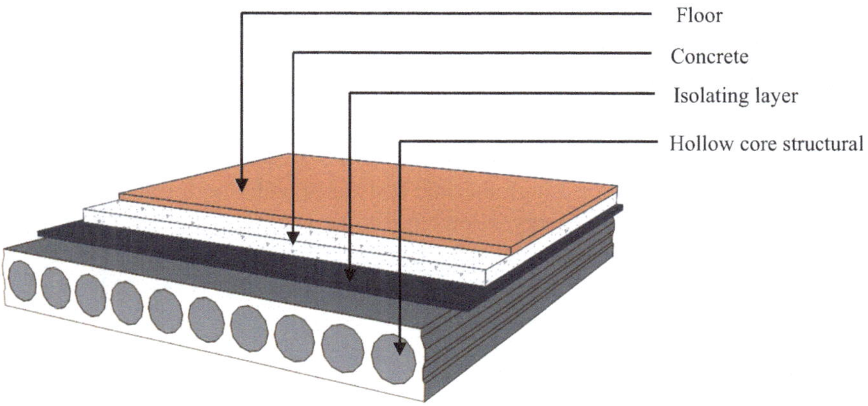

Fig. 2.21 Illustration of the main components of a ventilated concrete slab

of the building's mass to store or remove thermal energy, to fulfill the building requirement. Since this system employs some active measures to circulate air through the slab, it is considered as hybrid technology—not wholly passive strategy.

The cavities in the concrete slab are used as conduits to circulate air, leading to heating or cooling of the concrete mass. These cavities can be designed to circulate air through the whole floor slab.

This technology is drawing increasing attention especially in the design of low-energy, high-performance buildings. Studies on the application of hollow-core ventilated concrete slab report not only significant energy savings but also higher comfort levels [e.g., 32, 33].

2.4.3 Advanced Insulation Materials

A number of advanced insulating materials are in continuous development. The application of such materials, such as aerogel and vacuum insulated panels (VIP), demonstrates a significant potential to reduce energy consumption in buildings. Aerogel and VIP characteristics and applications are briefly summarized below.

Aerogel envelope

Silica aerogels are an innovative alternative to traditional insulation due to their high thermal performance, although the costs of the material remain high for cost-sensitive industries, such as the building industry. Research is continuing to improve the insulation performance and lowering the production costs of aerogels [27].

Three types of aerogel insulation are currently in use: (i) materials which exploit solely the high thermal characteristics of silica aerogels, (ii) granular aerogel-based translucent insulation materials, and (iii) transparent monolithic aerogel.

Opaque aerogel insulation materials

Opaque aerogel insulation materials consist of flexible aerogel textile-like blankets characterized by a thermal conductivity of around 2–2.5 times lower than traditional thermal insulation materials. The product may be used to reduce thermal bridges due to studs in wood-framed or steel-framed building envelopes [34].

Translucent aerogel insulation materials

Aerogel is of special interest as a translucent or transparent insulation material due to its combination of a high thermal resistance and a high transmittance of solar radiation and solar energy, allowing it to be incorporated in windows [35, 36]. Research has been conducted in the last decade on the development of highly insulated windows, based on granular and monolithic aerogels [37, 38].

Vacuum insulated panels

Vacuum insulated panels (VIPs) constitute another high-performance thermal insulation technology [39–41]. Such technology basically consists of a micro-porous core structure, such as fumed silica, which is evacuated in order to drastically reduce the gaseous thermal conductivity. The evacuated core is sealed in a thin, gastight envelope bag, which aims at protecting the panel from environmental factors and handling procedures.

Several layers usually constitute the VIP envelopes. These are an external protective layer, a barrier layer employed to shield the core material from gas transport (including water vapor), and the internal layer which seals the core material. The overall envelope joins these different layers by means of an efficient adhesive.

The layers of the VIP envelope are made of films, commonly consisting of plastics or metal foils. Metal foils are relatively strong and impermeable for water and gas transfer; however, they are highly conductive to heat. Such issues are addressed by employing a combination of metal and plastics for these films, in metallized plastics and laminated foils.

In general, VIPs possess more thermal resistance than most known building materials, including advanced technologies such as aerogels. This has the advantage of enabling significant reduction of the thickness of the thermal insulation material within the building envelope. The thermal resistance of VIPs is expected to decrease over time due to unavoidable interaction with air and moisture. VIPs cannot be adapted or cut at the building site, limiting thus its application [42].

References

1. Loonen RC, Trčka M, Cóstola D, Hensen JL (2013) Climate adaptive building shells: state-of-the-art and future challenges. Renew Sustain Energy Rev 25:483–493
2. Quesada G, Rousse D, Dutil Y, Badache M, Hallé S (2012) A comprehensive review of solar facades. Opaque solar facades. Renew Sustain Energy Rev 16(5):2820–2832
3. Thun G, Velikov K (2012) Responsive envelopes: characteristics and evolving paradigms. https://www.researchgate.net/publication/258440713_Responsive_Envelopes_Characteristics_and_Evolving_Paradigms. Accessed 5 Feb 2019
4. Addington M, Schodek D (2004) Smart materials and new technologies for architecture and design professions. Architectural Press, Jordan Hill, Oxford
5. Al-Homoud MS (2005) Performance characteristics and practical applications of common building thermal insulation materials. Build Environ 40(3):353–366
6. Marinoski DL, Güths S, Pereira FO, Lamberts R (2007) Improvement of a measurement system for solar heat gain through fenestrations. Energy Build 39(4):478–487
7. Carmody J, Haglund K (2012) Measure guideline: energy-efficient window performance and selection (No. DOE/GO-102012-3656). National Renewable Energy Lab. (NREL), Golden, CO (United States)
8. Carmody J, Selkowitz S, Lee E, Arasteh D, Willmert T (2004) Window system for high-performance buildings
9. Carmody J, Arasteh D, Selkowitz S, Heschong L (2007) Residential windows: a guide to new technologies and energy performance. WW Norton & Company
10. Sadineni SB, Madala S, Boehm RF (2011) Passive building energy savings: a review of building envelope components. Renew Sustain Energy Rev 15(8):3617–3631
11. Granqvist CG (1990) Chromogenic materials for transmittance control of large-area windows. Crit Rev Solid State Mater Sci 16(5):291–308
12. Wilson HR (2004) High-performance windows. Freidburg Solar Academy

13. Van Den Bergh S, Hart R, Jelle BP, Gustavsen A (2013) Window spacers and edge seals in insulating glass units: a state-of-the-art review and future perspectives. Energy Build 58:263–280
14. Gustavsen A, Grynning S, Arasteh D, Jelle BP, Goudey H (2011) Key elements of and material performance targets for highly insulating window frames. Energy Build 43(10):2583–2594
15. Chiras DD (2002) The solar house: passive heating and cooling. Chelsea Green Publishing
16. Kirimtat A, Koyunbaba BK, Chatzikonstantinou I, Sariyildiz S (2016) Review of simulation modeling for shading devices in buildings. Renew Sustain Energy Rev 53:23–49
17. Konstantoglou M, Tsangrassoulis A (2016) Dynamic operation of daylighting and shading systems: a literature review. Renew Sustain Energy Rev 60:268–283
18. Lawton MRP (2014) Design guide: solutions to prevent thermal bridging. Schöck Isokorb, p 35
19. Russell MB (2001) Influence of active heat sinks on fabric thermal storage in building mass. Appl Energy 70(1):17–33
20. Sadineni S, Srikanth M, Boehm R (2011) Passive building energy savings: a review of building envelope components. Renew Sustain Energy Rev 15:3617–3631
21. Dimassi N, Dehmani L (2016) Performance comparison between an improved and a classical Trombe wall: an experimental study. J Build Phys 40:372–395
22. Omrany H, Ghaffarianhoseini A, Ghaffarianhoseini A, Raahemifar K, Tookey J (2016) Application of passive wall systems for improving the energy efficiency. Renew Sustain Energy Rev 62:1252–1269
23. Bilgen E, Zrikem Z (1987) Theoretical study of a composite Trombe-Michel wall solar collector system. Sol Energy, 409–419
24. Lechner N (2001) Heating, cooling, lighting: design methods for architects, 2nd edn. Wiley, New York
25. Grynning S, Goia F, Rognvik E, Time B (2013) Possibilities for characterization of a PCM window system using large scale measurements. Int J Sustain Built Environ 2(1):56–64. ISSN 2212-6090
26. Baetens R, Jelle BP, Gustavsen A (2010) Properties, requirements and possibilities of smart windows for dynamic daylight and solar energy control in buildings: a state-of-the-art review. Sol Energy Mater Sol Cells 94:87–105
27. Baetens R, Jelle BP, Gustavsen A (2011) Aerogel insulation for building applications: a state-of-the-art review. Energy Build 43:761–769
28. Barreneche C, Solé A, Miró L, Martorell I, Fernández AI, Cabeza LF (2013) Study on differential scanning calorimetry with two operation modes and organic and inorganic phase change material (PCM). Thermochim Acta 553:23–26
29. Becker R (2014) Improving thermal and energy performance of buildings in summer with internal phase change materials. J Build Phys 37:296–324
30. Weinläder H, Beck A, Fricke J (2005) PCM-façade-panel for daylighting and room heating. Sol Energy 78:177–186
31. Navarro L, De Gracia A, Colclough S, Browne M, McCormack SJ, Griffiths P, Cabeza LF (2016) Thermal energy storage in building integrated thermal systems: a review. Part 1. Active storage systems. Renew Energy 88:526–547
32. Dincer I, Rosen MA (2002) Thermal energy storage (TES) methods. In: Dincer I, Rosen MA (eds) Thermal energy storage: systems and applications. Wiley, New York, NY, pp 93–212
33. Kato Y (2007) Chemical energy conversion technologies for efficient energy use. In: Paksoy HO (ed) NATO Sciences Series, II. Mathematics, physics and chemistry, thermal energy storage for sustainable energy consumption: fundamentals, case studies and design, vol 234. Springer, Dordrecht, pp 377–391
34. Kosny J, Petrie T, Yarbrough D, Childs P, Syed AM, Blair C (2007) Nano-scale insulation at work: thermal performance of thermally bridged wood and steel structures insulated with local aerogel insulation. In: Proceedings of the ASHRAE/DOE/BTECC conference on thermal performance of the exterior envelopes of whole buildings X, Clear Water Beach, Florida, December 2–7, 2007, pp 1–6

35. Kaushika ND, Sumathy K (2003) Solar transparent insulation materials: a review. Renew Sustain Energy Rev 7:317–351
36. Wong IL, Eames PC, Perera RS (2007) A review of transparent systems and the evaluation of payback period for building applications. Sol Energy 81:1058–1071
37. Schultz JM, Jensen KI, Kristiansen FH ((2005)) Super insulating aerogel glazing. Sol Energy Mater Sol Cells 89:275–285
38. Schultz JM, Jensen KI (2008) Evacuated aerogel glazings. Vacuum 82:723–729
39. Simmler H, Brunner S, Heinemann U, Schwab H, Kumaran K, Mukhopadhyaya P, Quénard D, Sallée H, Noller K, Kücükpinar-Niarchos E, Stramm C, Tenpierik MJ, Cauberg JJM, Erb M (2005) Vacuum insulation panels, study on VIP-components and panels for service life prediction in building applications (Subtask A), final report for the IEA/ECBCS annex 39 HiPTI-project (High Performance Thermal Insulation for Buildings and Building Systems)
40. Baetens R, Jelle BP, Thue JV, Tenpierik MJ, Grynning S, Uvsløkk S, Gustavsen A (2009) Vacuum insulation panels for building applications: a review and beyond. Energy Build 42(2):147–172
41. Binz A, Moosmann A, Steinke G, Schonhardt U, Fregnan F, Simmler H, Brunner S, Ghazi K, Bundi R, Heinemann U, Schwab H, Cauberg JJM, Tenpierik MJ, Johannesson G, Thorsell T, Erb M, Nussbaumer B (2005) Vacuum insulation in the building sector. Systems and applications (Subtask B), final report for the IEA/ECBCS annex 39 HiPTI-Project (High Performance Thermal Insulation for Buildings and Building Systems)
42. Simmler H, Brunner S (2005) Vacuum insulation panels for building application: Basic properties, aging mechanisms and service life. Energy Build 37(11):1122–1131

Chapter 3
Selected High-Performance Building Envelopes

This chapter presents selected high-performance building envelope systems, as key components in the design of solar buildings and communities. Building envelopes are divided into two main categories, low-rise (up to three floors) and multistory buildings. For the low-rise buildings, advanced, high energy performance, building envelope systems such as double-stud walls, structural insulated panels, and insulated concrete forms are presented. Multistory building envelope systems discussed in this chapter include double-skin facades and climate-adaptive facade systems, with special focus on *intelligent* envelope systems. The multifunctionality of facade systems in multistory buildings, as a key component in advanced high-performance envelope, is highlighted.

3.1 Low-Rise Building Envelope Systems

Designing and constructing a high energy performance building envelope constitutes an important stage in the design of solar buildings. Diverse design methods and building components are employed to enhance the energy performance of building envelops (as discussed in Chap. 2). The most common high-performance building envelope systems employed in northern climate low-rise buildings are double-stud walls (DSW), structural insulated panels (SIP), and insulated concrete forms (ICF). The discussion below summarizes the main characteristics of these systems, their method of construction, as well as expected performance.

3.1.1 Double-Stud Wall

The double-stud wall (DSW) system is designed to address energy performance deficiencies associated with the typical, conventional exterior residential wall system.

C. Hachem-Vermette, *Solar Buildings and Neighborhoods*, Green Energy and Technology, https://doi.org/10.1007/978-3-030-47016-6_3

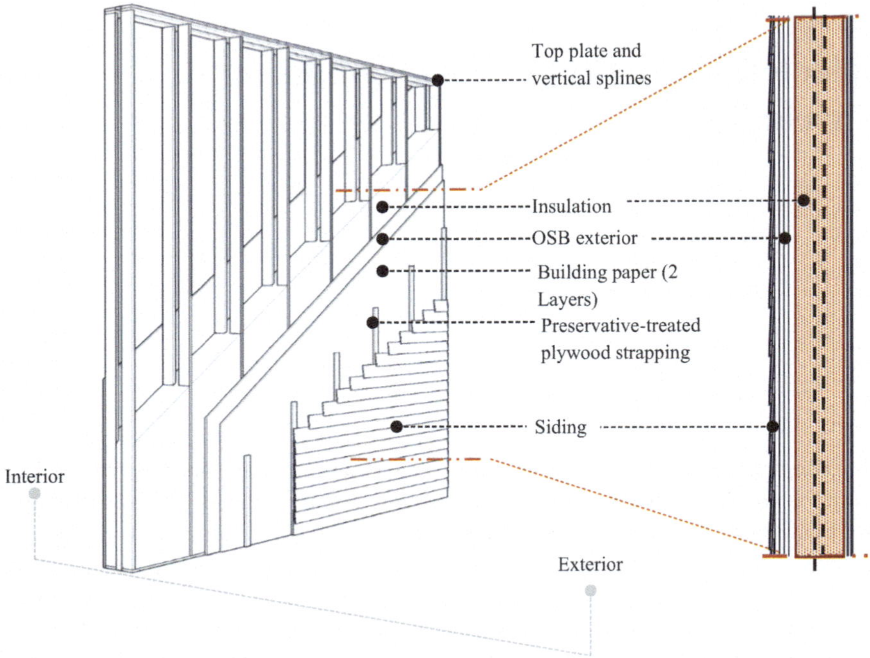

Fig. 3.1 Illustration of a double-stud building envelope

This typical single-stud wall system consists of load-bearing wooden frame comprising studs, and a wall cavity with batt insulation infill between studs. The exterior surface of plywood or OSB sheathing covered by finishing layers is fixed directly to the studs. The finishing layers can vary over a range of options including paper and fiber-cement siding (see Fig. 3.1), timber sidings, brick veneers, and others.

A number of issues can be associated with this typical design, such as the insulation level and potential occurrence of thermal bridges. For instance, this wall system restricts the type and size of insulation layer, thereby limiting the overall thermal resistance of the building. In addition, the placement of insulation in between, rather than over the studs, may lead to greater thermal bridges around the studs. The double-stud wall is designed to address these issues, improving thus the overall energy efficiency in residential wall construction.

Construction and performance

Construction

Double-stud wall (DSW) system relies on traditional methods of building envelope construction. DSW consists of two parallel stud frames with the space in between filled entirely with a continuous insulation layer (Fig. 3.1). The external frame wall is usually constructed as a load-bearing system, which is built and sheathed in a similar method to a typical exterior frame wall [1]. Windows and sidings are mounted employing conventional techniques. The internal wall frame is constructed after the building is enclosed, at some distance from the external load-bearing wall. The size of the space between these two frames can vary depending on the required insulation level [1, 2].

The studs of each of the double wall frames can either be aligned or staggered [3]. The finish of the outer layer of the DSW system is generally treated in a similar method to the single-stud wall. Other construction details of DSW such as structural bracing, building paper, and siding installment are similar to conventional, framed wall systems [1]. To install window and doors in the DSW system, plywood boxes are inserted within the designed openings, allowing to bridge the gap between the two framing systems. This ensures that openings are flush with the exterior surface [4]. Figure 3.1 shows a double-stud wall assembly, consisting of an outer structural wall with standard cladding and an inner framed wall.

A variety of double-stud wall configurations can be implemented, to reduce the impact of increased envelope thickness on the available interior inhabitable space. Diverse insulation materials can be implemented in the cavity. Dense blown cellulose is the most commonly material used in DSW assembly. Often, a layer of spray foam against the outside sheathing provides a reliable air seal and thermal break [1, 5].

The main benefit of a double-stud wall system consists of its high thermal performance, as compared to conventional wall assembly. This is due primarily to the thicker insulation layers inside the wall. In addition, thermal bridges through double-stud walls are significantly reduced, raising the overall thermal resistance value of the wall system.

Research has shown, however, that DSW standard assembly can present issues of losing thermal energy due to thermal bridging around its rims. These shortcomings can be addressed by covering the rim joist by insulation. Variants of DSW are continuously developed to address such issues, while increasing potential thermal resistance of the whole assembly.

Risks and Limitations

Although double-stud wall (and its variants) presents numerous benefits in terms of energy performance, the DSW system suffers two primary shortcomings as compared to conventional single-stud walls. These two limitations consist of issues of moisture penetration, and the overall lifetime cost efficiency of the assembly system.

Moisture penetration

Moisture penetration in a residential wood-framed structure can be controlled utilizing vapor barrier. This prevents moisture from moving through the wall assembly and potentially damaging or corroding electrical systems, destroying surface finishes, or allowing harmful molds and rot to develop. Although this method of mitigation applies as well to DSW systems, there are certain instances where a vapor barrier is not sufficiently effective in mitigating the diffusion of moisture through the assembly [1]. Since DSW assemblies are thicker and highly insulated, a significant difference in temperature between interior and exterior faces can occur, with the exterior sheathing nearly matching outdoor air temperatures

[6]. Consequently, the exterior sheathing becomes susceptible to the diffusion of interior moisture if exfiltration—the leaking or diffusion of interior air through an assembly—occurs [7].

Moisture diffusion into the exterior sheathing can be mitigated by installing vapor retarder on the warm side of the wall assembly (the interior side). Additional measures to prevent moisture migration within the DSW assembly include selection of materials with adequate drying capacity, including insulation materials, and creating a vented cladding system. For instance, plywood or structural fiberboard have higher drying potential than conventional OSB boards and can be less vulnerable to moisture issues [8]. Each of these techniques, however, has its own limitations and needs to be considered in the context of specific applications.

Cost

In comparison to the conventional single-stud wall commonly implemented in the residential sector across North America, the DSW assembly involves higher costs due to a number of factors including increased construction time, framing and insulation costs, and quality control. Additional attention should be given to window and door openings, as well as the connection of the two stud frames to control differential deformations.

Construction time and labor costs play a significant part in the increase in overall construction cost of DSW. Another important cost consideration relates to increased insulation cost. Given that the double-stud wall concept involves the installation of insulation within the two wall frames, the amount and consequently cost of insulation can significantly exceed that of a single-stud assembly.

Although the cost of double-stud walls exceeds the cost of conventional single-stud wall system, the increased construction costs are likely to be more than offset by lifetime savings in energy costs.

3.1.2 Structural Insulated Panels (SIPs)

Structural insulated panels (SIPs) are manufactured "sandwich" panels, most commonly composed of two exterior layers of oriented-strand-board (OSB) and a rigid core of expanded polystyrene foam insulation (EPS) coated with a structural adhesive [9, 10].

One of the main characteristics of SIPs is that it reduces the conventional multi-layered building envelope into the three basic layers mentioned above. Structural adhesive is generally applied to hold together the layers within the SIP assembly. Figure 3.2 presents a typical design of wall system employing SIPs.

SIPs combine the structural performance of framing with the insulating properties of the EPS foam to create a single envelope panel that provides the function of structure, insulation, air barrier, and vapor barrier. SIP panels can be employed for all envelope components, including external walls and roofs. In roof applications, bearing support members are employed to support the SIP roof system (Fig. 3.4).

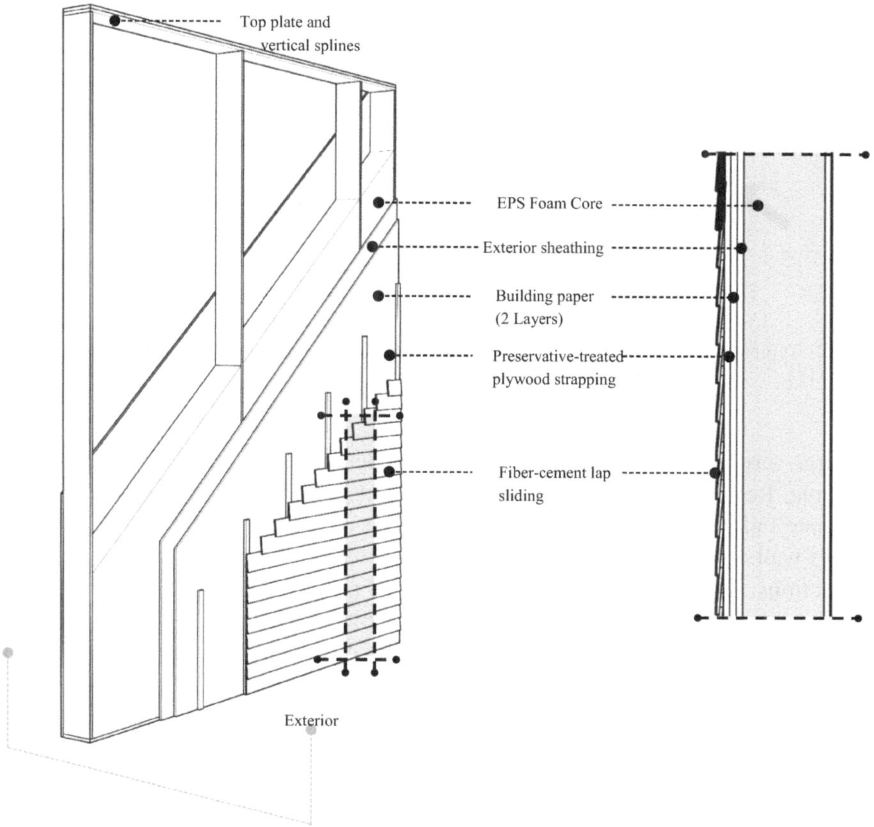

Fig. 3.2 Illustration of the basic design of SIP wall assembly

Such bearing elements may include ridge beams, rafters, trusses, as well as bearing walls. A roof SIP panel must span from one bearing support to another.

The structural insulated panels bonded together provide a high energy performance of the building envelope with reduced thermal bridging and increased airtightness, as compared to conventional building envelope [11]. Diverse types of exterior cladding/roofing materials can be deployed on the exterior face of SIP to complete the envelope assembly, providing thus environmental protection (Fig. 3.2)

Types of SIPs

Design of SIP varies in size, insulation type and resistance value, structural strength, sheathing material, and other parameters. Currently, a multitude of sheathing materials such as plywood, magnesium-oxide board, fiberglass mat gypsum sheathing, and fiber-cement panel are increasingly employed to replace the traditional OSB sheathing. Methods of connection between panels can vary as well. The most common connection strategies include OSB thin spline, foam block spline (or mini-SIP spline), and dimensional lumber spline. Figure 3.3 illustrates the typical panel-to-panel joints.

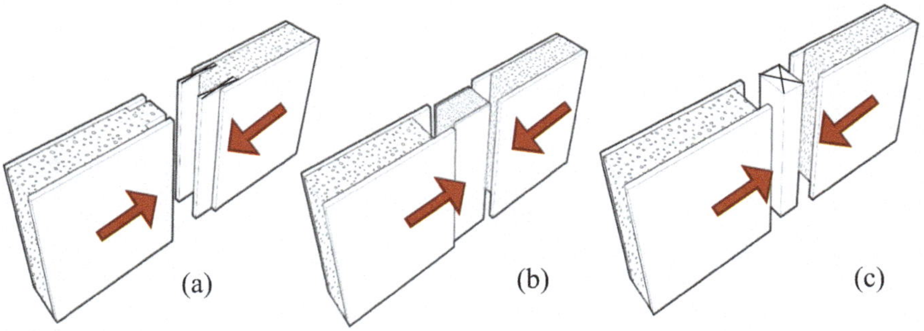

Fig. 3.3 Typical panel-to-panel joints, **a** OSB thin spline, **b** Mini-SIP splines, **c** Dimensional lumber spline [12]

Large variations of panel types, construction details, and techniques are currently available. Regardless of construction detail, joints should be carefully designed to eliminate failure potential. Critical joints are the connections at the roof peak, the roof-to-wall connection, where floors are supported on the walls, at the foundation connections, and around windows and doors (Fig. 3.4). In addition, in SIP construction panel-to-panel connections need to be meticulously considered since they represent the most failure susceptible details of SIP. Overall, ensuring a tight fit between the panels, in conjunction with adequate sealant application, will allow the SIPs to work together as a single unit, increasing thus the performance of the building envelope.

Advantages

SIPs are expected to outperform conventional wood construction due mostly to their uniform insulation and airtight construction [9, 14]. The thermal insulation level of SIP construction can be controlled by changing the insulation foam type and thickness.

Other advantages of SIP systems relate to the construction process and its efficiency. SIPs rely on prefabricated components resulting in fast on-site assembly and reduced labor, in addition to limited waste of construction material [15]. Such considerations result in an overall reduction of temporal costs.

Potential drawbacks

A number of concerns need to be identified and taken into consideration in the implementation of SIP construction. SIP on-site construction requires a high level of expertise and training to properly seal critical joints, to eliminate potential significant issues, such as water penetration. Water penetrating the system can work its way down the joint lines, saturating the wood around. This typically occurs on roofs in a distinctive pattern (see Fig. 3.5). If the water does not dry, material can rot, presenting a high potential of mold growth, as well as structural damage.

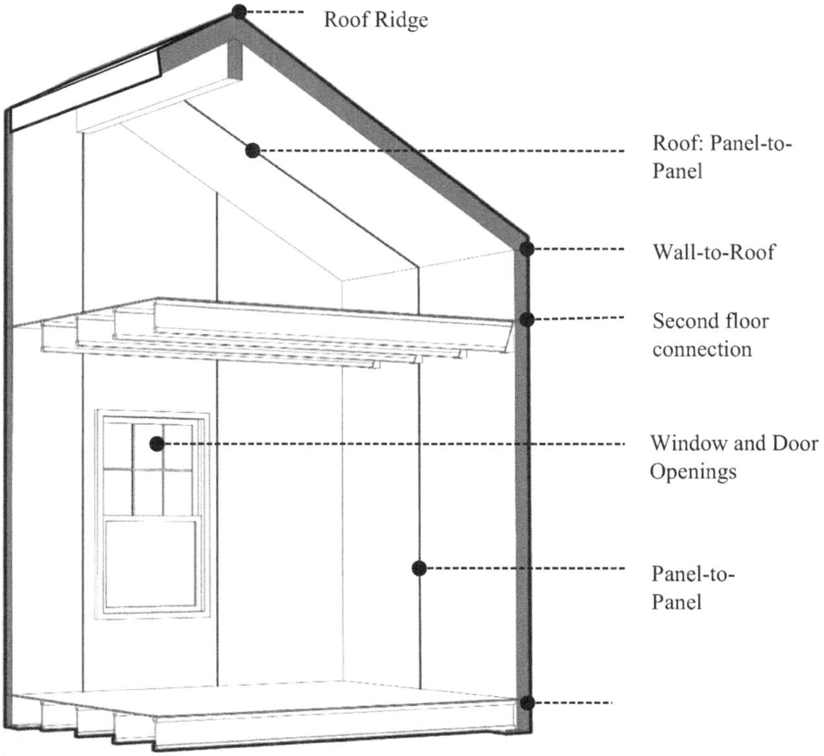

Fig. 3.4 Connection areas of primary concern (Based on [13])

3.1.3 ICF Construction

Insulated concrete forms (ICFs) consist of stackable formwork made from expanded polystyrene foam, which is filled on site with concrete [16]. ICF's installation is widely used in multi-unit residential buildings (MURBs) projects, commercial projects, low-rise building applications (student housing, coop housing, low-income housing), and single-occupant residential construction [17].

Main components

ICF wall assembly is composed of two main components: insulated formwork and concrete infill. The concrete core with adequate steel reinforcing constitutes the structural element of these systems, providing strength as well as thermal mass. The EPS panels provide permanent form for the concrete core and continuous layer of thermal insulation, as well as a substrate for diverse finishes (such as stucco, brick, and sidings) [17]. Siding materials are attached to the ICF employing furring strips (see below) or powder-driven fasteners. The interior finish employed with ICF (in North

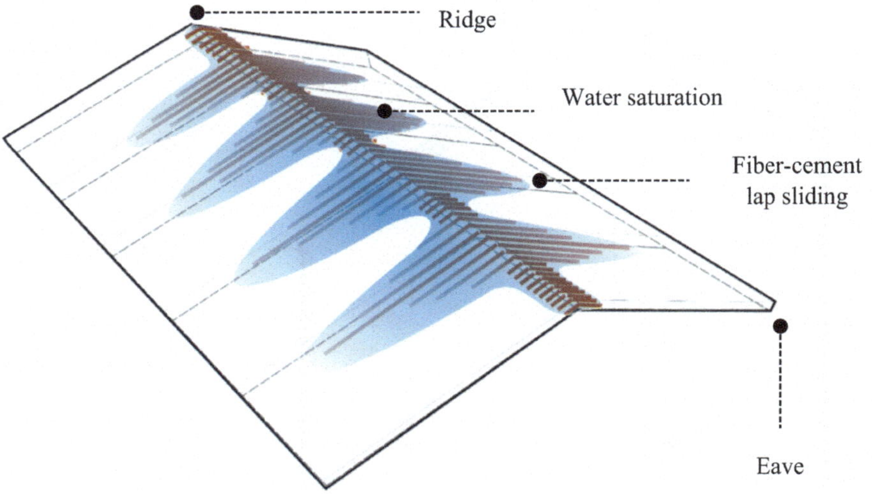

Fig. 3.5 Illustration of potential of water saturation areas around the ridge in a SIP roof construction

American construction) consists typically of attaching a gypsum board, allowing to obtain a uniform appearance with the interior walls.

The openings for doors and windows in the ICF systems are achieved by constructing the perimeter of these openings employing treated woods and omitting the ICF blocks within this perimeter [16]. The preservative-treated wood elements prevent the concrete from obstructing these openings. These wood elements are left in-place for the future attachment of windows and door frames.

Formwork components can be small "knock-down" units or whole wall configurations (Fig. 3.6a and b). Knock-down units consist of relatively small block-like units that can be stacked atop one another on site (see Fig. 3.6b). A wide range of ICF units' sizes and methods of connection exist. The whole wall configuration option represents a prefabricated wall panel, requiring only on-site concrete fill. The insulated formwork, in both configurations, consists mainly of two components: rigid foam boards and cross-ties (commonly referred to as the "web"). These are detailed below.

- **Insulated Foam Board**: The rigid foam board is composed usually of expanded polystyrene (EPS). Various thicknesses and layer combinations of expanded polystyrene can be employed on either side of the concrete core. For instance, to increase the thermal resistance, thicker layers can be utilized. Additional layers can be inserted within the cavity, in the form of plywood, OSB, or extruded polystyrene (XPS), to achieve higher thermal resistance.
- **Cross-Ties:** Cross-ties can be made from polypropylene or a combination of virgin or recycled thermoplastics. New cross-ties are typically made from high-density polyethylene (HDPE). In the ICF "knock-down" units, a horizontal rebar

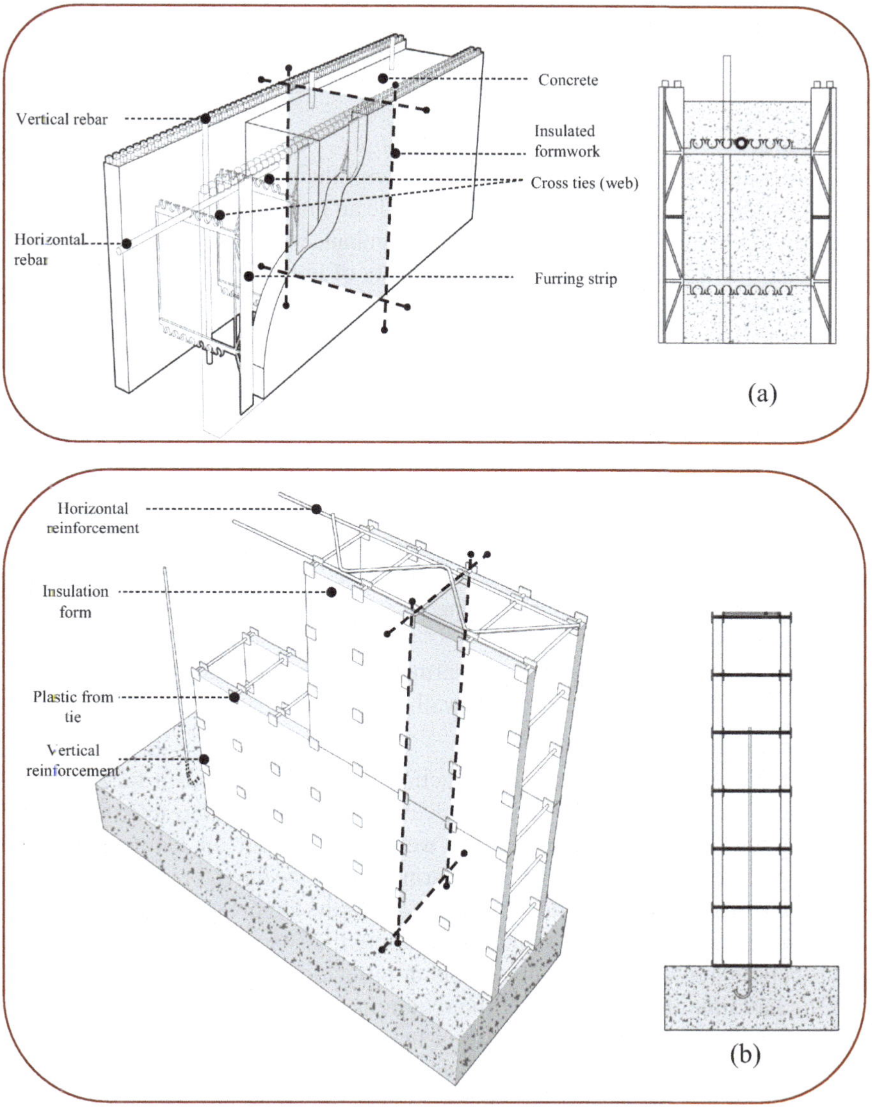

Fig. 3.6 ICF wall configuration, **a** illustration of a knock-down ICF unit, **b** whole wall configuration

is placed on top of seats, located within the molded cross-ties. The seats allow the rebar to remain fixed within the formwork during the concrete casting. Furring strips are attached to the cross-ties to enable fastening additional substrate of finishing material. These furring strips are located within the exterior surface of the EPS ensuring thus thermal resistance continuity (Fig. 3.6).

During the fabrication process, the foam boards and cross-ties are molded as a single piece. Such process ensures the thermal continuity of the ICF assembly, minimizing the potential of thermal bridges.

Advantages

There are a number of advantages associated with the utilization of ICF system in building constructions. These advantages include an improved thermal performance and speed of on-site installation. Three factors enhance the effectiveness of thermal performance of ICF construction—the continuity of insulation, reduced air infiltration, and providing thermal mass. Some of the advantages of ICF construction are summarized below.

Continuous insulation and airtightness

Unlike traditional frame construction assemblies, which require specific treatments of some areas such as the connection of two walls in the corner, to avoid thermal bridges, ICFs by nature offer thermal continuity. The method of construction of the ICF units provides continuity of the concrete and insulation layers, providing thus a consistent thermal resistance without thermal bridges. However, since ICF systems only operate in vertically stacked configurations (i.e., walls), thermal continuity should be maintained in connections where walls meet floors and roofs.

Another key factor in the enhanced thermal performance of ICF is the air barrier provided by the monolithic concrete wall.

Thermal mass

Concrete is one of the densest construction materials, and thus it facilitates the storage of thermal energy and its slow release over time, regulating thus the internal temperature of the wall assembly, as well as the indoor air temperature. The concrete core absorbs heat during the warm hours of the day, stores it, and releases it when the ambient temperature cools off (below the temperature of the concrete), typically at night time. Consequently, applying the benefits of thermal mass, the ICF building envelope passively assists in mitigating heating and cooling loads of the building, thereby reducing the requirements of the building's mechanical system, and thus the overall energy consumption.

Research is continuing to focus on enhancing the performance of the thermal mass of the ICF assemblies. For instance, adding a reflective film on the EPS–concrete interface can re-radiate the thermal energy back into the mass, allowing it to store the heat for a longer period. A radiant heating/cooling system can be integrated within the concrete of the ICF system. This would create an active thermal building envelope which can maintain its temperature at certain level, contributing to the overall thermal efficiency.

Speed of Installation

The ICF envelopes are designed to fulfill the various requirements of traditional envelope layers including resisting heat, air, and moisture transfer. This eliminates the need to install additional components on site, usually required in a traditional wall assembly, such as structure, insulation, and air barrier. Similar to other advanced building envelop, ICF construction requires skilled implementation to achieve manufacture-specified results.

Additional characteristics

In addition to increased thermal performance and relative speed of construction, ICF construction system is characterized by the following properties:

- Potential to significantly reduce sound transmission.
- Reducing the number of layers required in a traditional wall assembly system, since it combines structure, insulation, vapor retarder, air barrier, and sound barrier into one component.
- Restricting on-site construction waste (when planned properly).
- Potential use of recycled materials, including recycled plastic for the internal polypropylene webs (accounting to 40–60% of the ICF units' weight), and recycled steel for the reinforcing rebar (inserted into the ICF walls).
- Contributing to healthier indoor air quality, since ICF assembly does not support mold growth, does not rot, controls dust and allergen contamination, and does not off-gas harmful chemicals.
- It presents numerous performance benefits related to safety and hazard mitigation, fire resistance, and enhanced structural performance, predominantly in extreme wind environments.

Drawbacks

The barriers hindering the upsurge of insulated concrete forms application include a lack of knowledge of the properties and characteristics of this building construction method, and to some degree the relatively higher initial cost of the ICF construction as compared to conventional construction. In addition, the potential moisture penetration, if ICF is not properly implemented, can present an issue. These drawbacks are briefly summarized below.

Cost

The cost and economic impact of ICFs should be considered together with their role and impact on the holistic operation of the building. For instance, ICFs increased capital cost should be considered in a cost–benefit analysis against the long-term HVAC savings. Research into the thermal performance of the ICF system demonstrates that the increased airtightness of the envelope and the continuous insulation layers it provides (discussed above), reduce the thermal energy requirement of the building, consequently reducing the sizing of HVAC equipment as well as equipment runtime significantly [18].

Moisture penetration

The structural concrete core of an ICF wall\as well as the EPS layers are not affected by water penetration. Moisture and excess humidity can, however, penetrate the exterior sidings or interior finish, and, depending on the drying capacity of the materials employed, can potentially create mold and mildew, as well as rot in some building components such as drywall and wood framing. Condensation

on the interior finish can be prevented by installing a vapor retarder between the insulation and interior finish (drywall).
ICF below-grade construction may be exposed to water infiltration. Potential solution for this issue consists of the application of waterproofing membrane, which can be spray-applied or self-adhered, on the exterior side of the ICF walls. This should be coupled with an effective and reliable drainage system, to abolish any likelihood of water infiltration.

3.2 Multistory Buildings

High-performance multistory building envelopes require a different design approach than that of low- and mid-rise buildings. In addition to the structural performance, a number of energy-related functions need to be accommodated by the envelope. The building enclosures discussed below are selected in view of their potential to accommodate some of these considerations, while capitalizing on enhancing the capture, utilization, and control of solar energy.

3.2.1 Double-Skin Facade

A double-skin facade (DSF) is an exterior wall system comprised of two layers separated by an air cavity. The main characteristics of this building envelope system are its potential to dynamically respond to variations in indoor and outdoor environmental conditions. For instance, according to the building requirements, this facade system can be designed to ventilate the cavity to the exterior, or to preheat air for interior space heating.

The typical construction of DSF consists of a pair of glass skins separated by an air gap, which can range from 150 mm to over 900 mm. The size of the air cavity is dictated by the mechanical systems and shading devices that might be installed within it, as well as the maintenance requirements [19]. Although the typical DSF comprises two glass skins, other types of opaque or translucent materials can be integrated within the design of DSF systems. Such design decisions can be made according to the climate requirements, in order to alleviate the impact of outdoor conditions on the indoor space. Employing different types of materials within the skin involves the consideration of specific spatial configuration of the air cavity (see below). Employing different advanced materials in DSF configurations is addressed in more detail in Chap. 5.

Double-skin facade can provide an effective way to control and capture solar gain for utilization in buildings, as compared to other facade systems used in contemporary construction such as a curtain wall system [19, 20]. The air cavity between the two layers of the DSF system can be naturally ventilated through operable glazing apertures, or it can be mechanically ventilated (see Fig. 3.7). This air cavity acts as a buffer zone between the interior and exterior environment, tempering variations in temperature, wind, and sound. Hybrid methods that allow natural and mechanical ventilation are often implemented, presenting more feasible and flexible, energy-efficient solution especially in cold climate. The ventilation strategy employed in the DSF system is dependent on the climate, the width, and the size of the cavity.

The design of air cavity is discussed in more detail below.

Various categories of double-skin facades exist, based on the cavity design (sealed or operable) or the methods of air ventilation (mechanically or naturally ventilated).

The main parameters that affect the performance of DSF systems relate mostly to the spatial configuration of the cavity, the method of ventilation employed, and

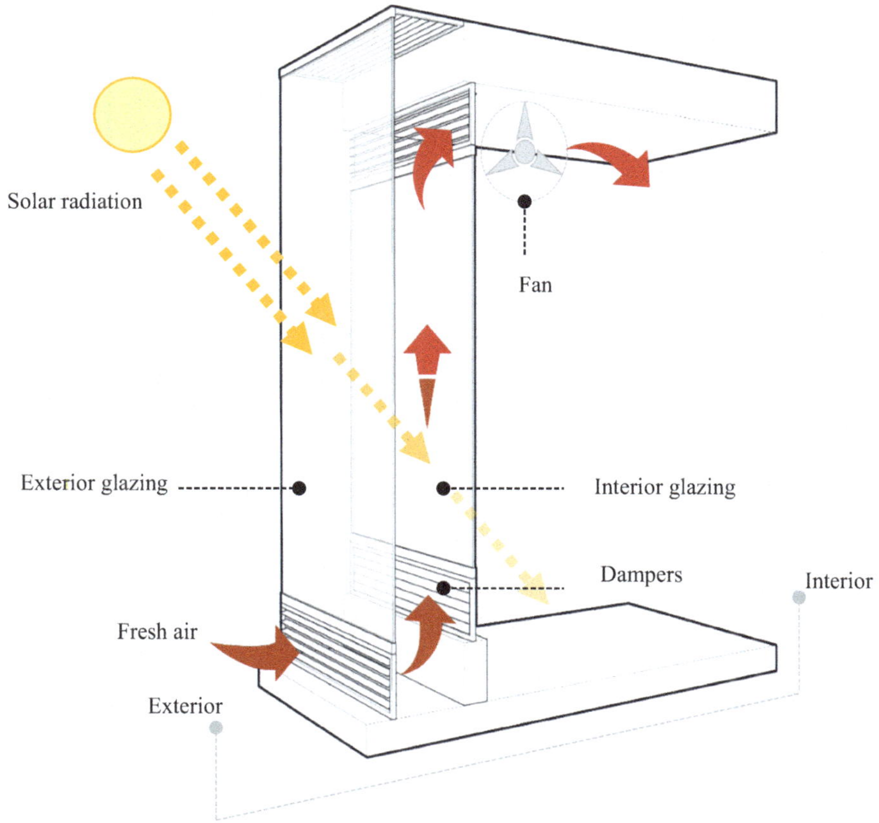

Fig. 3.7 Illustration of advanced mechanically ventilated DSF system (spanning one floor of a multistory building)

the cavity width [21, 22]. Those are presented in the illustrations of Fig. 3.8 and discussed below.

Spatial configurations

There are four spatial configurations of a double-skin facade: Multistory, box window, corridor box, and shaft-box configurations. These configurations are presented in Figs. 3.8a and 3.9 (in more detail) and are briefly summarized below.

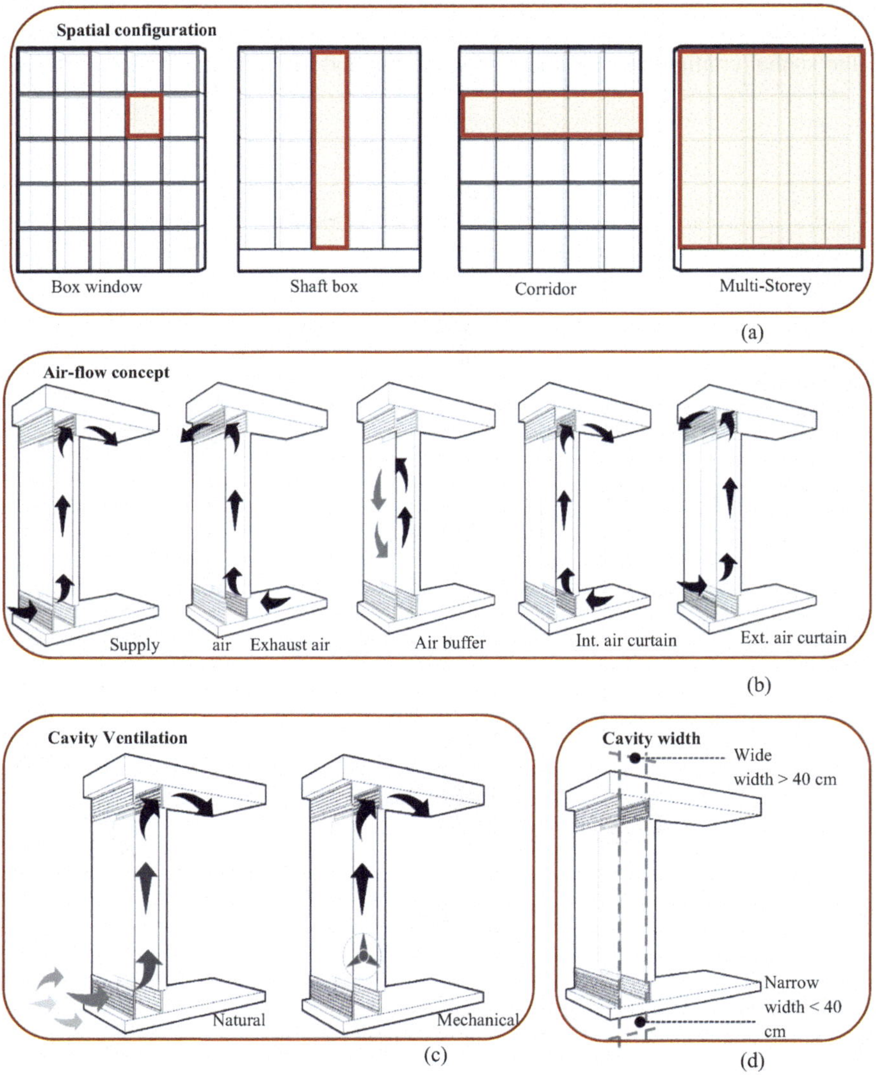

Fig. 3.8 Classification of DSF according to various parameters, **a** Spatial configurations, **b** Cavity airflow, **c** Cavity ventilation, **d** Cavity width

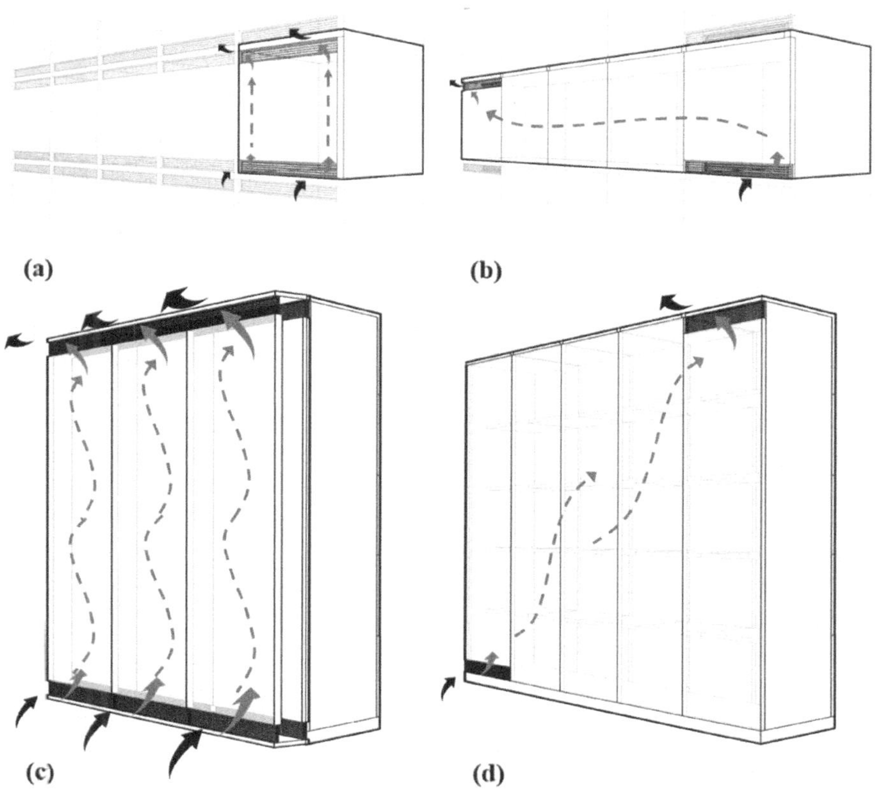

Fig. 3.9 Various spatial configuration of the ventilated air cavity of DSF system; **a** Box window, **b** Corridor box, **c** Shaft box, **d** Multistory

Box window

Box window configuration (Fig. 3.9a) consists of stand-alone units, where each unit consists of two parallel windows with an air cavity between. Vertical and horizontal separations are implemented around each unit. This configuration allows high level of control for heat gain and ventilation of individual air cavities, and adjacent indoor spaces.

The box window concept is derived from one of the original examples of double-skin facade systems consisting of a window set into a solid opaque wall with a wooden shutter on the outer surface, which can be closed, creating an air gap between the shutters and glazing [23]. The outer shutters were later replaced by a second glazed layer in order to keep the visual connection between the exterior and interior and to allow the penetration of solar radiation, while maintaining higher insulation provided by the sealed air gap created between the two glazing skins (the outer and the inner) [24].

More advanced variations of the box window concept are currently available. For example, section of glazing can be boxed in by an additional glass facade that protrudes from the first facade plane. In contrast with the original window box concept consisting of separate units (and non-continuous skins), another variation consists of continuous skins over the building facade, but divided horizontally at each floor plate and vertically between units creating isolated sections of the air cavity (see Fig. 3.9a) [24].

The main advantage of the box window is the control that each unit has over its own air cavity environment, which affects the adjacent building indoor air space. This separation allows for the highest level of acoustic isolation. The shortcoming of such design is that due to the multiple divisions involved, the efficiency of ventilation is reduced, in addition to increased construction and operation costs [24].

Corridor and shaft box

These two types of configurations are similar in that they are divided only into two directions, vertically for the shaft box (Fig. 3.9c) and horizontally for the corridor box (Fig. 3.9b).

The corridor configuration relies mainly on horizontal separation at each floor creating a single air cavity along each floor of the building. This concept is simpler than the window box configuration since it requires fewer separations. The relatively longer horizontal cavity allows for greater staggering of air inlets and outlets, which can benefit the ventilation within the cavity. This design is associated, however, with reduced acoustic separations between neighboring units [24].

The shaft box is divided horizontally at specific locations, with continuous air cavity along the height of the building.

Multistory

This configuration consists of an undivided cavity along the entire facade, with free air movement throughout it (Fig. 3.9d). This configuration is less complex in terms of design than the three other configurations. Since it does not require separations, it requires lower levels of control and reduced acoustic separation.

The temperature in the undivided air space DSF configuration can vary significantly from the bottom to the top of the air cavity, depending on the height and size of the air cavity. To avoid hot air accumulation in the upper areas of the multistory undivided air cavity, the system should be vented at the top [20]. Nevertheless, the design should ensure that the higher floors do not experience overheating in both ventilated and unventilated conditions.

The multistory undivided air cavity can be architecturally designed with larger space, to provide atrium spaces for occupants. A common example is the utilization of this space as an indoor garden. Research highlights that an optimal width of the air space, to reduce the total energy consumption of the building, is 380 mm [25]. As the cavity size increases, energy consumption also increases. Glazing types, or the inclusion of other types of materials, play a major role in the energy performance of the double-skin facade system.

The Air Cavity

The air cavity plays an important role in the performance of double-skin facades. The air cavity design within DSF systems is divided into naturally ventilated, mechanically ventilated, or hybrid (naturally and mechanically) ventilated. Various methods can be used to circulate air in the cavity. The two major DSF systems of air movement in the cavity are buffer system and open system (see Fig. 3.10). The buffer system is divided into three subsystems—static, external air curtain, and internal air curtain, while the open system includes two subsystems—exhaust (or extract) and supply. A DSF air cavity system can combine the double functions of extracting and supplying air, to respond to various indoor requirements [21, 22]. Such systems are termed *Twin-Face*. These are summarized below.

Buffer System
This type of cavity isolates the indoor space from the outdoor space employing three main variants of airflow—static air, external air, and internal air (Fig. 3.10 c-e). In a static air buffer (Fig. 3.10 c), the air cavity is completely closed, forming a sealed buffer zone between the indoor space and outdoor environment. The trapped air is heated by solar radiation to provide an additional shield from the exterior in the cold winter months.

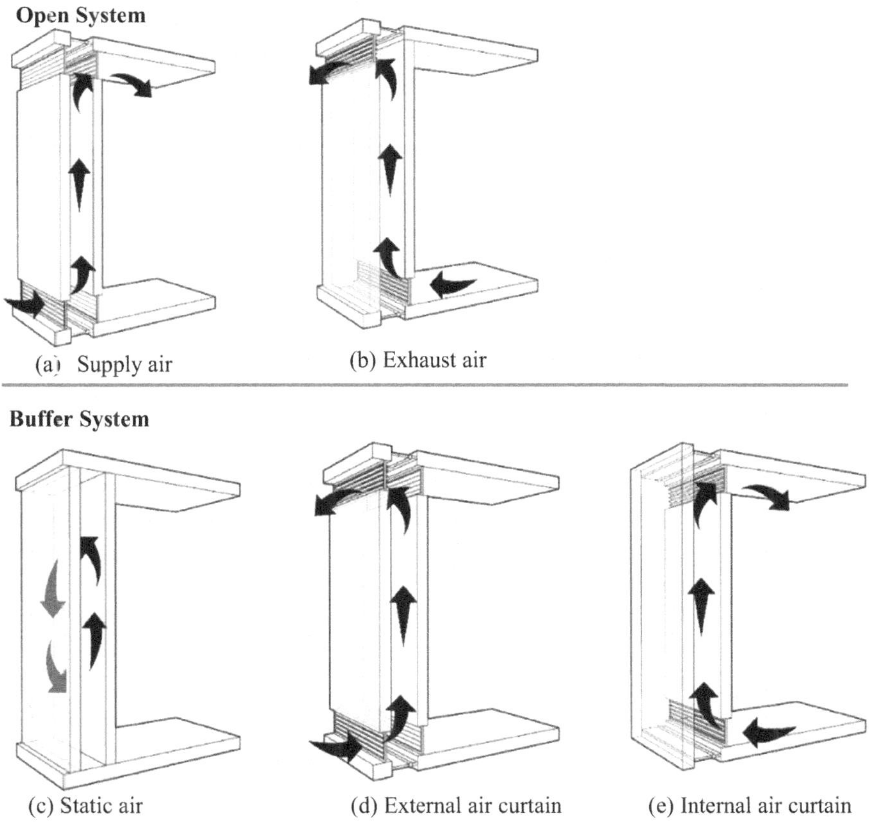

Fig. 3.10 The two major systems of air cavity and their subsystems

In the external air circulation variant, where the cavity is cut off from the interior of the building (Fig. 3.10 d), cooler air is allowed to enter at the bottom of the cavity and warmer air is extracted through the top, providing cooling in the hot summer months.

The internal air circulation variant (Fig. 3.10 e) consists of circulating the indoor air within the cavity. Cool air drawn from the interior space picks up solar heat gain from the cavity and is returned back to the interior space. This variation can be useful during the heating period, allowing passive heat gain. However, it is less common than the other two variants described above [20].

Since the buffer system is designed to isolate the interior of the building from the outdoor environment, the required level of ventilation is provided to the indoor environment by other means; typically, employing a mechanical heating and ventilation system.

The air cavity of the buffer system can be equipped with shading devices to control solar radiation and thus reduce the potential of overheating. A dynamic shading system allows appropriate regulations of the solar radiation admitted into the indoor space, to optimize the HVAC system and reduce energy consumption [20]. The design of the double-skin facade should provide adequate clearance in the air cavity for maintenance and cleaning.

A disadvantage of the buffer system is that by isolating the indoor space it disables the use of the cavity for natural ventilation and for support of the HVAC systems through preheating or precooling of the fresh air supplied to the building. These functions should be handled by alternative means.

Open System

Exhaust Air

The exhaust air system of double-skin facade is not widely employed due to the complexity of its design, and its reliance on a mechanical system for proper functioning. This type of DSF system comprises two skins of different constructions, thermal priorities, and specific tasks. The exterior layer is typically constructed as double-glazed, curtain wall suspended from the building structure. The interior skin is made up of a single-pane aluminum storefront or curtain wall system.

The exhaust air system relies on a mechanical system to exhaust air from the indoor occupied space, through the double-skin facade, to the outdoor environment. The outside skin (double or triple glazed) serves as the thermal control layer of the DSF, which controls the air movement through the facade and prevents condensation from forming on the exterior skin. A schematic of this airflow can be seen in Fig. 3.11a.

One of the shortfalls of the exhaust air system is its inability to provide natural ventilation, and thus the building relies on mechanical systems to achieve the required ventilation. As such, this type of DSF does not actively contribute to reducing the energy consumption of the building. Fresh air and adequate level of ventilation are provided through the mechanical systems as the DSF air cavity is strictly used for exhausting air from inside the occupied spaces of the building.

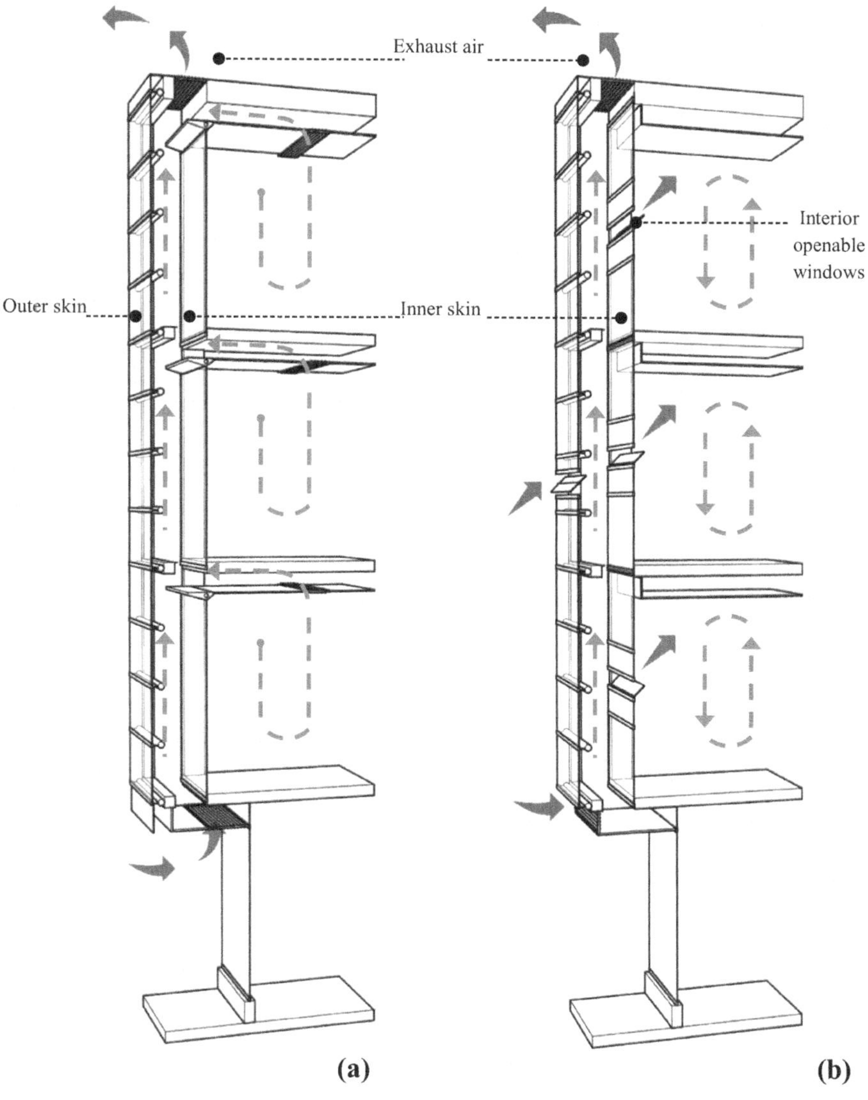

Fig. 3.11 Open air cavity, **a** Exhaust air system, **b** Twin-face system

This system is mostly useful in locations where natural ventilation does not constitute a viable option for the building, due to some environmental and climatic issues. Such issues include high wind locations, high noise level, outdoor air pollution, and others. The system still provides the advantage of an improved thermal performance of the facade in hot and cool climates [20].

Supply Air System
The supply air system is the most common and preferred double-skin facade due to its benefits in supplying preconditioned fresh air, allowing to reduce energy

consumption of a building (Fig. 3.10a). Unlike the exhaust air system, the thermal control layer of the supply air system is on the interior. The interior layer consists of a double-glazed curtain wall system that is thermally broken. The outer layer composition is a single-glazed outer skin. This can be aluminum-framed or a frameless glazing system to reduce the visual obstructions created by an aluminum-framed system.

Additional advantage of this type of DSF system is its capability to provide natural ventilation, to the indoor environment while optimizing daylighting admission. The exterior single-glazed skin is mainly employed to protect the air cavity contents (air, shading devices, etc.). It minimizes heat loss or gain in the cavity and, depending on the season, allows preheating the air and natural ventilation [20, 25]. The exterior skin is also equipped with grills or vents that allow fresh air to pass from the outdoor environment into the cavity. The inner skin is equipped with operable windows. Depending on the indoor space requirements, operable windows allow the preconditioned air to enter the space. The operation of the windows can be controlled by the occupants, as well as through a central building management system.

By allowing the interior layer to open and close, the building can utilize passive cooling and heating to offset the demand on the HVAC system. This assists in reducing the overall size of the mechanical system.

Twin–Face system

The twin-face system combines the double function of supplying and exhausting air, to and from the indoor space, according to the specific requirements of a building (Fig. 3.11b). This allows overnight cooling, when outdoor air temperatures are below the indoor air temperature, by exhausting warm air from the building and allowing the cool night air to enter the occupied space [23]. This feature allows the twin-face system to perform successfully in warm climates or warm periods of the year, while having the capacity to temper the cavity in the cooler times of year.

The twin-facade system is highly efficient in handling acoustics and sound transfer from the exterior to interior environment. Staggering the openings in each layer of the skin prevents sound waves from directly penetrating the indoor space, which dampens exterior noise significantly. Staggering the openings allows as well to neutralize the high wind pressure differences, which increase with building heights, permitting the implementation of natural ventilation in high-rise buildings.

Examples of the application of this DSF design show high adaptability to various design considerations, for instance, employing the insulated double glazing as the exterior skin, and the single glazing as the interior skin [26]. Many examples employ automation to control the opening and closing of the vents, according to the time of the day and the time of the year, as well as the exterior temperature. Such control can be integrated within hybrid——natural and mechanical ventilation strategies of buildings.

3.2.2 *Climate Responsive Building Envelopes*

A climate responsive envelope interacts, naturally and through artificial control, with climatic and building conditions to produce specific desired effects on the indoor environment [27]. It has the capability to alter some of its functions, features, or behavior in response to changing building requirements and outdoor and indoor conditions [28].

A responsive building envelope comprises features such as real-time sensing, kinetic climate-adaptive components (e.g., for shading purposes), and adaptive (*smart*) materials. This type of building envelope provides the option of user override of the automated response. Responsive building envelope may include interactive characteristics which permit building systems to self-adjust, while allowing physical manipulation of some building components to control environmental conditions. One example is computer-controlled exterior shading device capable of repeatedly and continuously adjusting its configuration to respond to solar radiation intensity and incidence angle [28].

Climate responsive building envelope can be achieved at either a macro-scale or micro-scale. The macro-scale would be considered a *kinetic envelope*, meaning the building shell changes via moving parts. On a micro-scale, changes affect the internal structure of the material. Commonly, these materials allow light transmission of the whole solar radiation spectrum, or of a selective portion of this spectrum. For example, adaptive (*smart*) windows have the ability to modulate light levels and solar radiation admission into the space [29].

Effective control is a central aspect of climate responsive building envelope. Control of the climate-adaptive system can be achieved through either extrinsic or intrinsic mechanisms. Extrinsic control relies on sensors, processors, and actuators, such that the current configuration can be compared with and adjusted to the desired state. Intrinsic control is automatically triggered by stimuli such as temperature, relative humidity, and precipitation. Intrinsic control may also be known as *direct control* because it does not have an external decision-making component [30, 31].

This section presents a general review of various responsive building envelope systems, focusing on intelligent envelopes and related innovations. Research related to adaptive building facades is rapidly expanding, and technological and material research is being performed on a variety of scales. Related concepts of the utilization of solar energy in active applications, such as the generation of heat and electricity, are discussed in more detail in Chaps. 4 and 5.

Adaptive Building Envelope

Adaptive (*Smart*[1]) building envelope refers mainly to envelopes that rely on building materials and components to adapt to a range of climatic conditions [32]. Such materials are characterized by their ability to alter passively their physical attributes including their shape, without requiring mechanical means or energy source [33].

---

[1] *Smart* is a term employed in the literature.

Alteration in the properties of smart material is a purposeful change in response to external stimuli [34].

Examples of adaptive building envelopes include the application of aerogel or phase-changing materials, such as micro-encapsulated wax and salt hydrates in window glazing.

Another example of adaptive materials is electroactive polymers, which can alter their shape and size when stimulated by an electric current. An example of electroactive polymer technology is *Shape Shift*, which is a building skin for shading and ventilation. In this dynamic structure, movement of the panels is induced by a high-voltage, low-current circuit. Another example is thermobiometal, which is an alloy of two metals allowing deformation from an initial shape under temperature change.

An additional example of adaptive material is a glazing system that changes its transparency by means of technologies such as electrochromic, photochromic, and thermochromic glass. This allows windows to dim specific (or all) wavelengths of light, or letting light pass through, depending on the incident light intensity and specific needs of the indoor environment [35]. Such materials are discussed in Chap. 2.

Despite having a range of potentially advantageous features, adaptive materials are still limited in their performance, which may not cover the full range of outdoor climate conditions.

Intelligent Building Envelope (IBE)

While adaptive building envelope relies on the intrinsic properties of the employed materials, *intelligent building envelope* (IBE) relies heavily on extrinsic control, computation, and automation, and thus it allows larger range of response.

Intelligent building envelope entails the implementation of a range of technologies, including building automation, adaptive materials assemblies, automatic shading devices, automatic ventilation dampers, and others [33]. The goal of an intelligent building envelope is to optimize the building's energy balance and human comfort under varying climate conditions. Predictive models are often employed to design the responses of the envelope.

Intelligent building envelope can be characterized as possessing the following three components: *building management system*, *learning ability*, and *environmental data collection* [36]. These are described in the following.

- *Building management system*: It consists of a processing unit, receiving all of the information from the various sensor outstations, and determining the appropriate control response to the actuating elements, comprising
  - Input system that receives information by means of information receiver;
  - Processing and information analysis;
  - Output system that reacts to the input by an appropriate response; and
  - Time consideration that makes the response happen within the needed time.
  - Input system that receives information by means of information receiver;
  - Processing and information analysis;

- Output system that reacts to the input by an appropriate response; and
- Time consideration that makes the response happen within the needed time.

- *Learning ability*: It is the ability to learn how to respond to various conditions, by calculating optimum setting for shading, heating, and lighting.
- *Environmental data Collection:* It consists of real-time data collected by sensors.

The main functions that IBE can fulfill include enhancement of daylight level and penetration into the living space, control of solar radiation (e.g., overhangs, sun control devices; insulation (e.g., nighttime shutters), ventilation (e.g., automatic dampers), collection of heat (e.g., solar collectors), heat rejection (e.g., exhaust air), attenuation of sound (e.g., acoustic dampers), optimizing the generation of electricity (e.g., adjustable photovoltaics), and exploitation of pressure differentials (e.g., ventilation chimneys).

Specific response mechanisms can be implemented within the IBE system, according to the required performance. Some of these are reviewed below.

- *Responsive artificial lighting* refers to the ability of the artificial lighting system to deactivate or dim itself in response to indoor daylighting levels;
- *Daylight controllers* aim at controlling various motorized devices that can regulate solar radiation admission into the space, providing thus optimum configuration of these devices, corresponding to solar angles;
- *Sun controllers* aim at controlling sun-blocking building components, such as computer-controlled blinds, to mitigate damaging effects of direct solar radiation, including overheating and glare;
- *Occupant control* allows manual overriding by users, to adjust the performance of the building envelope (when needed);
- *Electricity generation* aims at optimizing renewable electricity generation by controlling the orientation of mobile PV panels (e.g., shades—see examples below) and/or wind turbines
- *Ventilation controllers* aim at maximizing the use of natural ventilation and consequently minimizing energy consumption by mechanical ventilation;
- *Heating and temperature controllers* use passive solar strategies for water and space heating, while tracking the sun for maximum exposure;
- *Cooling devices* use computer-controlled nighttime natural ventilation for pre-cooling of thermal mass (such as opening of vents).
- Method of operation

  The *input system* includes sensors that constantly collect data of both indoor and outdoor environments and send it to the building management system for analysis. Based on the analyzed information, a *building system integrator* orders the system to respond to specific requirements, at the specific time.

  The *management system* has predefined strategies for addressing the situation that needs to be handled, such as adjusting the lighting level, shading, ventilation, energy management, and others. When a situation matches a particular predefined target, the management system sends signals to activate specific motorized components (e.g., Sun shading, daylight redirection, daylight transport, adaptive glazing system, etc.), enabling thus to adjust the impacts of outdoor conditions on the indoor environment. However, these computerized adjustments may fail to reach the precise requirements to achieve a particular comfort level. In such situations, the internal environmental systems (i.e., HVAC) can be activated to bridge this gap. Energy consumption by the HVAC system is, however, minimized, due to the contribution of the automated systems in meeting part of the requirements. An intelligent control mechanism requires a learning capacity, enabling the system to learn from experiences, in trying to reach optimal performance. Modeling intelligence is a crucial component to achieving the objectives of an intelligent building skin [29]. The operation of such autonomous operation, employing preprograming algorithms, can be overridden by occupants' requirements [37].

Examples of Intelligent Building Envelopes

High-performance building envelope design can employ a combination of the intelligent envelope aspects described above. Most of the currently existing examples of IBE couple a double-skin facade system with various automated functions such as natural or hybrid ventilation, employing the air cavity space to ventilate or heat the space (see Sect. 3.2.1). Other examples employ the movement of shading components to optimize the impacts and benefits of solar exposure solar collectors, including photovoltaic and thermal collectors, are also employed within the IBE systems to generate thermal and electrical energy (integration of thermal collectors is detailed in Chap. 4).

Most intelligent building envelope systems are custom designed for specific projects. This feature limits the general applicability of the system to a wider range of building types and functions. Some examples that incorporate various aspects within the IBE systems are discussed below.

Kinetic systems

A number of examples incorporate intelligent kinetic shading devices as part of the building envelope. Such applications generally incorporate a computerized building system that automatically adjusts the sunshades for optimal thermal comfort and solar radiation admission into the indoor space. Some options allow individual occupants to override the system, as desired.

The Terrence Donnelly Centre for Cellular and Biomolecular Research (designed by Architects Alliance, University of Toronto, Canada) incorporates a double-skin south facade, with automated blind layer and automated air dampers in the cavity [38]. The automated blind can be adjusted seasonally or daily to provide the optimal balance of views and shading, blocking excessive solar radiation, and making large expanses of glass more feasible from an energy standpoint. The multilayer approach also offers the benefit of improved sound control. Although occupants can control the degree of ventilation in each office, when a window is opened, a sensor automatically switches off the heating and cooling supply to that space, thereby increasing energy efficiency and avoiding energy waste.

This building illustrates the combination of double-skin facade and automated shading devices, coupled with automated ventilation system.

Incorporation of PV modules in the shading systems is demonstrated in the Technical University Darmstadt's 2007 Solar Decathlon House [39]. The house incorporates an exterior building skin comprising computer-controlled wooden louvers with integrated photovoltaic system. The shutters have then the double function of generating electricity while shielding the interior of the house from excessive solar radiation. An illustration of these shutters is presented in Fig. 3.12. Active solar systems including PV panels are discussed in more detail in the following chapters.

Some more complex intelligent kinetic systems employed to reduce the solar radiation incident on building envelopes present interesting architectural features. Al Bahr Towers in Abu Dhabi (designed by Aedas Architects, completed in 2012) incorporate a responsive facade designed according to a traditional Islamic lattice known as Mashrabiya and to origami [40]. The massive fiberglass-coated Mashrabiya is supported by an individual frame, located in front of the curtain wall. The kinetic elements are programmed to reduce solar heat gain and glare, which constitute a major issue for buildings in the United Arab Emirates (UAE).

At sunrise, the Mashrabiya stays closed on the east elevation due to the intensity of solar radiation incident on this facade, while opening on the west orientation. At sunset, the reverse occurs. At 25-storey tall, the Al Bahr Tower features the world's current largest computerized responsive facade. Figure 3.13 presents a schematic illustration of this Mashrabiya system, in both open and closed states.

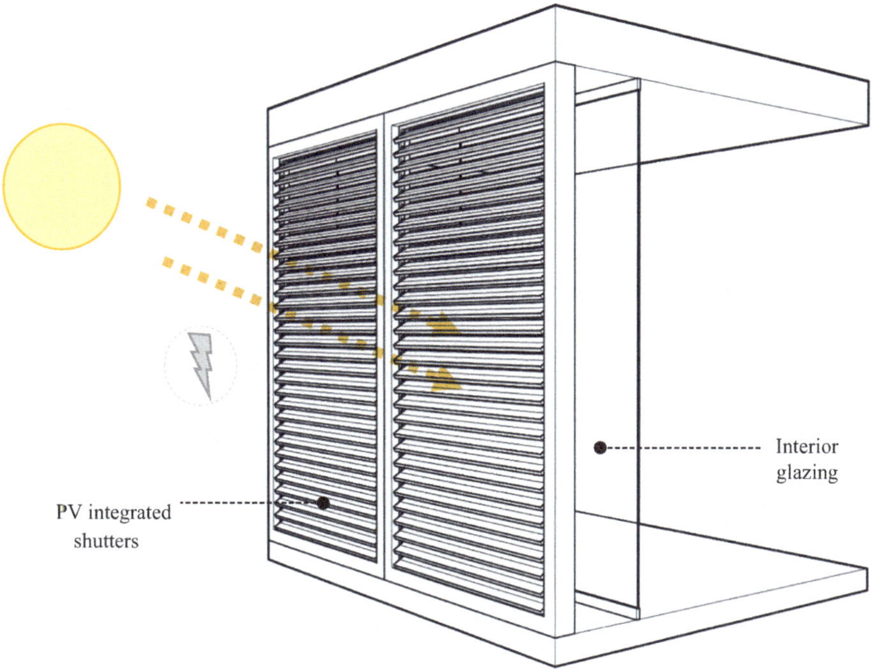

Fig. 3.12 Illustration of the PV shutter system employed by Technical University Darmstadt's 2007 Solar Decathlon House

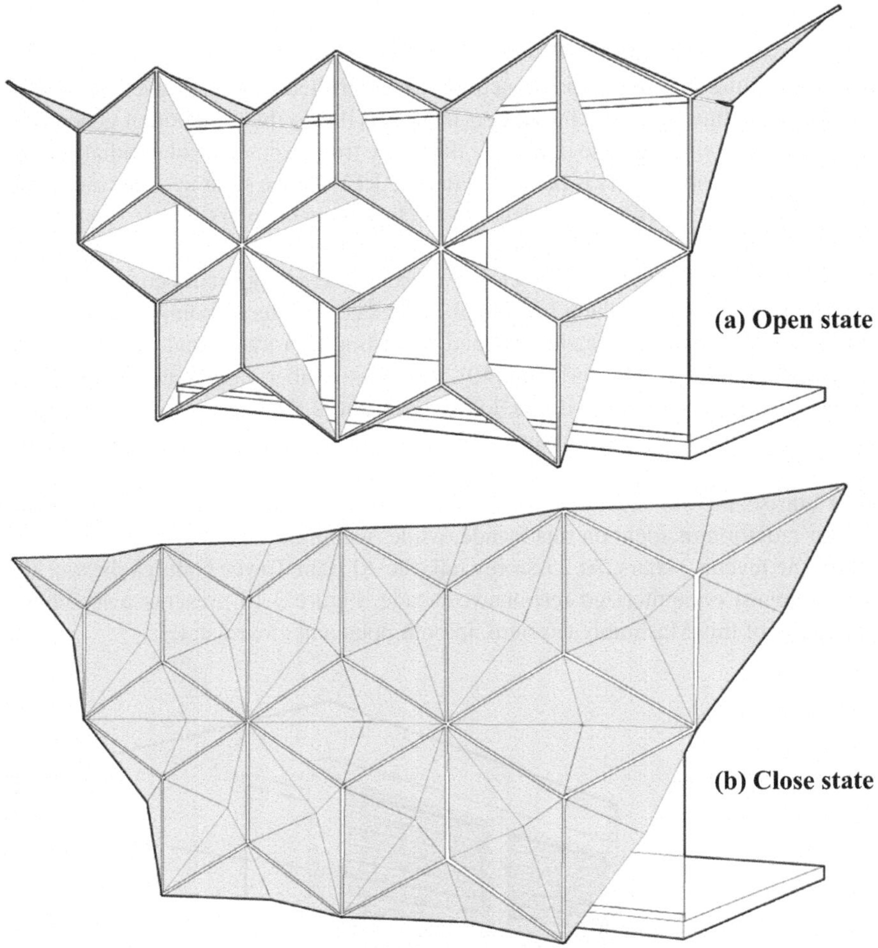

Fig. 3.13 Illustration of the kinetic shading system of the Al Bahr Towers in Abu Dhabi, **a** Open state, **b** closed state

A similar example, applied in colder climate, is the University of Southern Denmark's Communications and Design building [41]. Completed in 2014, the building implements a kinetic facade, designed to be climate responsive. The facade of the project (designed by Henning Larsen Architects) consists of 1,600 triangular motorized movable panels, which are connected to heat and light sensors.

Each panel opens or closes to allow solar control and optimal daylight, reducing the energy requirements for artificial lighting, cooling, and ventilation. Figure 3.14 presents an illustration of the motorized shading system implemented in the building.

Advanced materials

Other examples of automations involve the control of building materials. For instance, materials like ethylene tetrafluoroethylene (ETFE) are coupled with specific gases

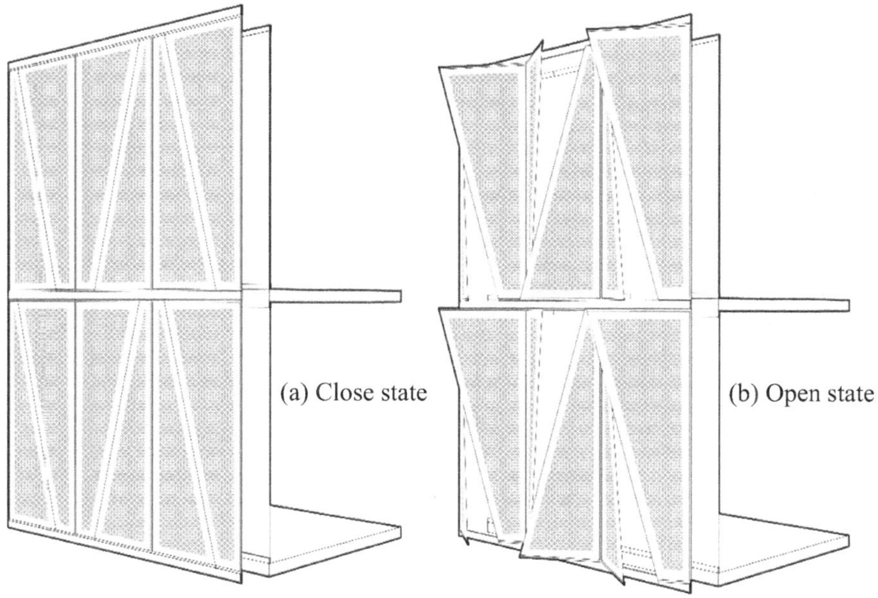

Fig. 3.14 Illustration of the shading system implemented at University of Southern Denmark's Communications and Design building, **a** closed state, **b** open state

(such as nitrogen—see below) to provide solar shading effect. ETFE is a translucent polymer sheeting that can be utilized instead of glass [42, 43]. ETFE is usually installed within a metal framework, where each unit can be lighted and manipulated independently.

The Media-TIC Building (Barcelona) presents an example of the application of ETFE cladding in advanced building envelope implements. The ETFE cladding comprises three layers of plastic, fixed within triangular frames.

Protection of solar heat gain is achieved using a configuration of the ETFE cladding known as diaphragm, whereby the ETFE layers are inflated. The air chambers within the inflated cushions increase the thermal insulation of the building envelope. To protect the building from direct solar radiations on the southwest facade, ETFE cushions are inflated with air coupled with nitrogen, creating a fog-like effect that reduces the solar radiation admitted to the indoor space. When the solar shading is not needed, the bags are deflated and become more transparent, allowing natural diffused light into the building [43]. Figure 3.15 presents an illustration of the ETFE cushions employed in the responsive building envelope of the Media-TIC Building.

Each of the EFTE bubbles, which inflate or deflate depending on the climatic conditions, is controlled separately with a controller, with individual sensors measuring heat, temperature, and the angle of the sun [43, 44]. In addition to the sensors on the facade, a large number of sensors (more than a hundred) are installed on the interior of the building, measuring data such as occupancy and artificial light levels, and generate a distributed intelligent system across the building.

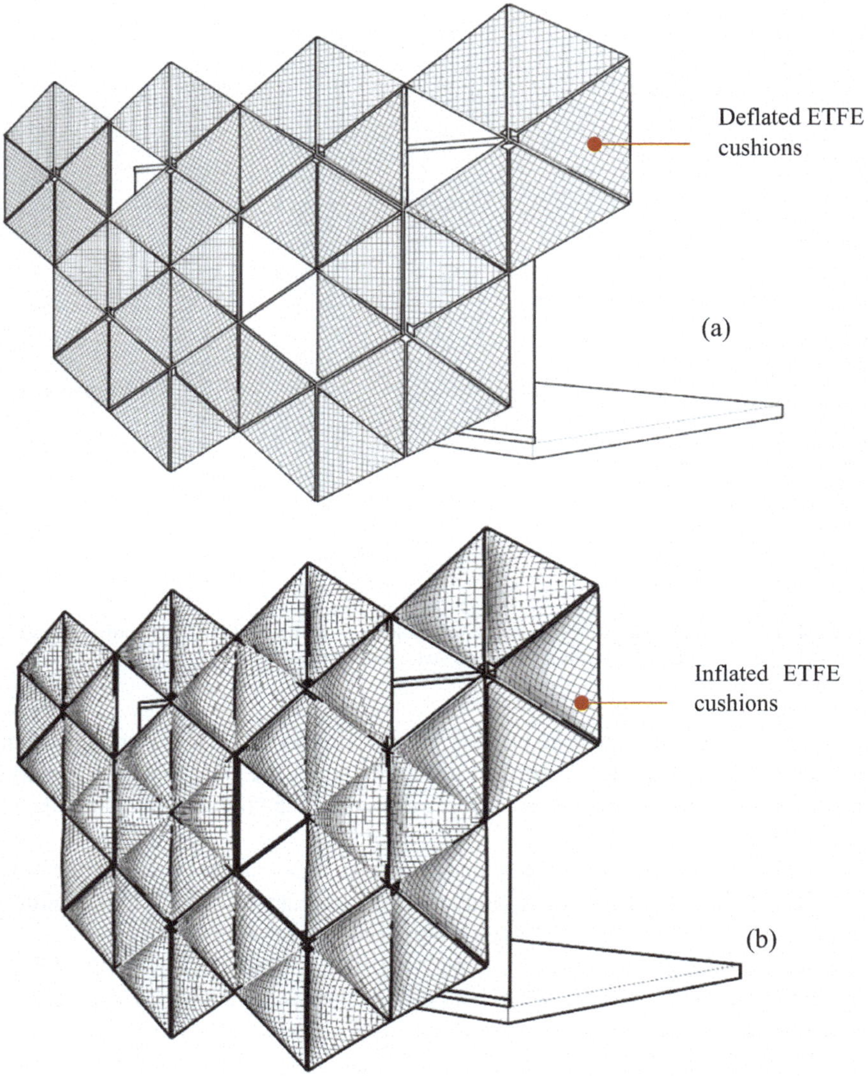

Fig. 3.15 Illustration of the ETFE responsive building envelope applied at the Media-TIC Building, Barcelona

Another promising application involves adaptive fritting. Fritted glass is obtained by applying a ceramic enamel paint to the inner surface of this glass. Adaptive fritting couples the concept of conventional fritting with motorized controls, employing multiple glazing layers with different graphic patterns. The several layers of shifting fritted glass are utilized in the adaptive fritting to induce an overlap or divergence of the graphic pattern, allowing to modify the overall pattern and its degree of transparency (Fig. 3.16).

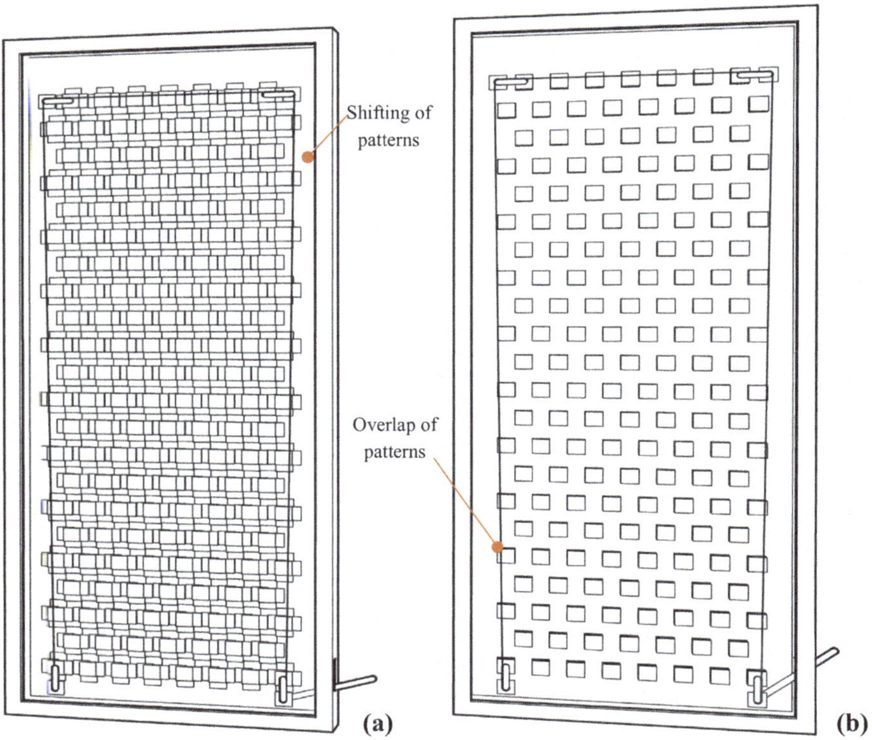

Fig. 3.16 Shifting patterns of the adaptive fritting application of the Harvard University, **a** shifted pattern, **b** aligned patterns

The adaptive fritting mechanically rotates the glass layers in plane. If all fritting patterns of all glass layers are similar and aligned, the overall glazing unit will reveal a single frit pattern, allowing the higher degree of transparency of this system. Once planes are shifted against one another, patterns of various densities are created [45]. Since the movement to shift the patterns is minimal, the required mechanical systems are small and can be self-contained within the unit. Each unit could be controlled individually or through a central control system.

From aesthetic point of view, the density, color, and pattern of an adaptive fritting system can be determined by the designer and the requirements of the specific application. While the implementation of adaptive fritting glass can employ similar techniques and mechanical systems, the fritting patterns could vary greatly, to respond to different requirements.

Above is an example presenting an adaptive fritting design applied at the Graduate School of Design at Harvard University. In this example, the panels were programmed to continuously change, adapting thus to light transmission, views, and overall enclosure appearance. This technique of programming can be coupled with data from sensors to create an adaptive facade that responds to both the interior and exterior environments, and building requirements. Figure 3.16 displays an illustration of the shift of fritting pattern similar to that applied at the Harvard University.

Design Considerations

The concept of climate-adaptive building envelopes is an emerging field, witnessing continuous progress in technology, computational instruments, and materials. Currently, there are still limitations in the application of these systems due to increased initial cost, high replacement and maintenance cost, and difficulties to maintain kinetic elements. Below is a summary of the main design considerations in the implementation of these responsive building systems.

Aesthetics

The aesthetics of the kinetic facade plays an important part in the overall design of the project. The kinetic movement can be used to create visual effects that differentiate the building from other surrounding buildings. While aesthetics is a major consideration in architectural design, it should not be at the expense of practical and environmental criteria, such as construction, maintenance costs, and energy efficiency.

User Control

Similar to all automated systems, the issue of control comes into play when the automated state is not aligned with user preference or the task being performed. Research shows that while an automated approach may provide the best environmental efficiency, occupants sometimes desire a level of control over their environment that may be at odds with the automated system. Some options are designed to allow the occupants some control over their own environments with respect to the amount of light and air that they deem optimum.

The trade-off between full automation of intelligent building envelope systems and user control must be weighed and addressed. Information should be provided back to the typical user of the indoor space to allow better understanding of the impact of certain actions on the overall building performance. In addition, handling of cases of breakdown, to which all automated systems are prone, should be accounted for.

Performance and Maintainability

A major hurdle in the implementation of intelligent building envelope is the acceptance by the developer or client of the additional cost and potential maintenance cost required to upkeep additional mechanical and computational systems. Demonstration projects should be initiated (and subsidized) to demonstrate the long-term advantages in both user comfort and cost savings.

3.2.3 Modularity and Multifunctionality

Multifunctionality is a key characteristic of the majority of advanced facade systems of multistory buildings presented above. Multifunctional facades can be defined as the combination or simultaneous use of building technologies in a single modular integrated unit, with the objective of enhancing the energy performance of buildings [46, 47].

Multifunctional facades are characterized by a number of features. A summary of some of these characteristics is presented in the following.

- *Modularity*: Modularity can be a key strategy to fulfill the multifunctionality aspect of high-performance building envelope, at lower cost. Modular design can enhance the feasibility of multifunctional facades to be mass produced. This allows reducing the cost of production and installation, without compromising the technical aspects of the systems applied [46, 48–50].
- *Flexibility*: Multifunctional facade systems allow a diversity of design solutions and features to fulfill technical, structural, and aesthetic performance. The presence of multiple features allows for adaptability to various environmental conditions and functionalities [46, 47, 51].
- *Prefabrication*: Modularity lends itself to prefabrication, and units are designed to encourage off-site design and on-site assembly [50].
- *Energy Efficiency*: The configuration of technologies is aimed at encouraging energy production and reducing consumption. The assembly of each modular building envelope unit and the materials employed can integrate passive and active energy efficiency strategies [51, 52].
- *Space conservation*: Space on the facade is a limited resource. Multifunctionality enables the accommodation of several technologies in this limited space [53].
- *Quick installation*: Prefabrication allows for quick installation on site, making these systems appropriate for retrofit [46, 51].

The modularity and multifunctionality of double-skin facades and climate-adaptive building envelope systems can be applied to retrofit existing buildings. The portability, ease of site assembly, and modularity are conducive to quick and less invasive renovations.

General considerations

Some general considerations need to be taken into account in the design of multifunctional facade systems. Dimensions, structural elements, accessibility, safety, and the intended occupancy of the building need to be closely considered [48]. The location and orientation of each of the building's facades should be accounted for when studying options of technologies or materials to be implemented. Different locations have inherently different thermal, lighting, and ventilation constraints. Consequently, a building might require different configurations of multifunctional modules, depending on the particular orientations, for optimal performance.

The following parameters should be considered to ensure the feasibility of a particular design:

- Structure includes the resistance to loads, as well as resistance to deformation and shock [54].

- Maintenance may be enhanced by the modularity of the system, enabling one module to be repaired without compromising other parts of the facade.
- Elimination of thermal bridges and proper site assembly must also be considered [53, 54].

The major advantage of this type of advanced facades, including double-skin and climate-adaptive facades, is their capitalization on the benefits of numerous energy efficiency strategies in a single unit, in addition to their function as envelope. The benefits and limitations vary by system design and are dependent on the materials, technologies, and costs associated with each case.

References

1. Aldrich RA, Arena L, Zoeller W (2010) Practical residential wall systems: R-30 and beyond. Res. Rep. Prep. by Steven Winter Assoc. Inc Consort. Adv. Resid. Build
2. Building Science Corporation, "ETW: Wall – Double-Stud Wall Construction." https://buildingscience.com/documents/enclosures-that-work/high-r-value-wall-assemblies/high-r-value-double-stud-wall-construction
3. Wagner R (2012) "Double-stud walls," From Fine Home building Issue 228 June/July 2012. https://www.finehomebuilding.com/2012/05/17/double-stud-walls
4. U.S. Department of Energy, "Double Stud Wall Framing," Updated: 03/14/2016. https://basc.pnnl.gov/resource-guides/double-wall-framing#quicktabs-guides
5. Lepage R, Schumacher C, Lukachko A (2013) Moisture management for high R-Value walls. Building Science Corporation, (November, 2013), p 52
6. Wagner "Double-stud walls"; Bailes, "Is This the End of the Double-Wall, Cold Sheathing Scare?"
7. Ueno K (2015) BA-1501: Monitoring Double-Stud Wall Moisture Conditions in the Northeast. Building Science Corporation, (January 2015), p 4, 5; Lepage et al, Moisture Management for High R-Value Walls, p 52
8. Wagner Double-stud walls; Arena L Building science: Hygrothermal performance of a double-stud wall, Professional Re-modeler, (May, 2014), p 2
9. Mullens Michael A, Arif Mohammed (2006) Structural insulated panels: Impact on the residential construction process. J Constr Eng Manag 132(7):786–794
10. Tracy J (2000) SIPs: overcoming the elements. For Prod J 50(3):12–18
11. Medina Mario A, King Jennifer B, Zhang Meng (2008) On the heat transfer rate reduction of structural insulated panels (SIPs) outfitted with phase change materials (PCMs). Energy 33:667–678
12. Morley M (2000) Building with structural insulated panels (SIPs). The Taunton Press, pp 82
13. Lstiburek J (2008) Builder's guide to structured insulated panels for all climates
14. Andrews S (1992) Foam core panels and buildings systems. Cutter Information Corp, Arlington, Mass
15. Gagnon M, Adams R (1999) A marketing profile of the U.S. structural insulated panel industry. For Prod J 49(7/8)
16. Allen E, Thallon R, Schreyer AC (2017) Fundamentals of residential construction. John Wiley & Sons
17. Oleck RF, Habel AC, Herrit DW (2012) Insulated Concrete Forms (ICF) as blast-resistant barriers. In: Structures congress 2012, pp 35–45
18. Maref W, Armstrong MM, Saber H, Rousseau M, Ganapathy G, Nicholls M, Swinton MC (2012) Field energy performance of an insulating concrete form (ICF) Wall. NRC
19. Boake TM (2003) Understanding the general principles of the double skin facade system

20. Ahmed M, Abel-Rahman A, Ali A, Suzuki M (2016) Double skin facade: the state of art on building energy efficiency. J Clean Energy Technol, 84–89
21. Haase M, da Silva FM, Amato A (2009) Simulation of ventilated façades in hot and humid climates. Energy Build 41(4)
22. Lou W, Huang M, Zhang M, Lin N (2012) Experimental and zonal modeling for wind pressures on double-skin façades of a tall building. Energy Build 54:179–191
23. Oesterle E (2001) Double-skin facades: integrated Planning. Prestel Verlag, New York. Print
24. Knaack U, Klein T, Bilow M, Auer T (2007) Facades: principles of construction. Birkhauser, Boston. Print
25. Joe J, Choi W, Kwak Y, Huh JH (2014) Optimal design of a multi-story double skin facade. Energy Build 76:143–150
26. Heusler W, Lieb R-D, Lutz M, Oesterle E (2001) Double-skin facades integrated planning. Prestel Verlag, New York
27. Beesley P, Hirosue S, Ruxton J (2006) Toward responsive architectures
28. Loonen R, Costola M, Hensen JLM (2013) Climate adaptive building shells: state of the art and future challenges. Renew Sustain Energy Rev 25:483–493
29. Sheikh MM (2011) Intelligent Building Skins: Parametric- based algorithm for kinetic facades design and daylighting performance integration. USC School of Architecture
30. Oldewurtel F, Ulbig A, Morari M, Andersson G (2011) Building control and storage management with dynamic tariffs for shaping demand response. In: 2011 2nd IEEE PES International conference and exhibition on innovative smart grid technologies. IEEE, pp 1–8
31. Krishan A (ed) (2001) Climate responsive architecture: a design handbook for energy efficient buildings. Tata McGraw-Hill Education
32. Schodek D, Addington M (2004) Smart materials in architecture and design
33. Velikov K, Thün G (2013) Responsive building envelopes: characteristics and evolving paradigms. Trubiano F Design and Construction of High Performance Homes, pp 75–92
34. Maragkoudaki A (2013) No-mech kinetic responsive architecture: Kinetic responsive architecture with no mechanical parts. In: 2013 9th International conference on intelligent environments. IEEE, pp 145–150
35. Bullinger HJ (2009) Technology guide: principles-applications-trends. Springer Science & Business Media, Berlin, Germany
36. Sherbini K, Krawczyk R (2004) Overview of intelligent architecture. In: 1st ASCAAD International conference, e-design in architecture Dhahran, Saudi Arabia, pp 137–152
37. Atkin B (1988) Intelligent buildings. Billings & Sons, Worcester
38. Jen L (2006) Terrence Donnelly Centre for cellular and bio-molecular research-the sustainability theme is taken to a whole new level in this sublime new research facility at the University of Toronto designed. Can Archit 51(1):28–35
39. Zaretsky M (2009) Precedents in zero-energy design: architecture and passive design in the 2007 Solar Decathlon. Routledge
40. Derix C, Kimpian J, Karanouh A, Mason J (2011) Feedback architecture. Archit Design 81(6):36–43
41. Fakourian F, Asefi M (2019) Environmentally responsive kinetic façade for educational buildings. J Green Build 14(1):165–186
42. López M, Rubio R, Martín S, Croxford B (2017) How plants inspire façades. From plants to architecture: biomimetic principles for the development of adaptive architectural envelopes. Renew Sustain Energy Rev 67:692–703
43. Dewidar Y, Mohamed N, Ashour Y (2013) Living skins: A new concept of self active building envelope regulating systems. In: Advancing the green agenda; technology, practices and policies conference–BUID, pp 1–8
44. Januszkiewnicz K, Paszkowska N Climate change adopted building envelope for the urban environment
45. Drozdowski Z, Gupta S (2009) Adaptive fritting as case exploration for adaptivity in architecture. In: Proceedings of the 29th annual conference of the association for computer aided design in architecture (ACADIA), pp 105–109

46. Soledad M, Moren P, Korjenic A (2017) Hotter and colder – how do photovoltaics and greening impact exterior facade temperatures: the synergies of a multifunctional system. Energy Build 147:123–141
47. Sarihi S, Derankhshan Z (2018) Advanced Integrated Façades, a New Solution to Energy Concerns. In: Civil Engineering, architecture and urban management, pp 1–9
48. Meno S, Chica JA, Tapia A, Del Portillo LA (2012) New industrialised developments for the envelope energy retrofitting based on heat recovery ventilation. In: VI international congress on architectural envelopes, pp 1–12
49. Vlachokostas A, Madamopoulos N (2017) Daylight and thermal harvesting performance evaluation of a liquid filled prismatic façade using the radiance five-phase method and EnergyPlus. Build Environ 126:396–409
50. Rozanska M (2013) "1 st Year progress report publishable summary on progress: " multifunctional energy efficient façade system for building retrofitting across Europe","
51. Shen J et al (2016) Optimizing the configuration of a compact thermal facade module for solar renovation concept in buildings. Energy Procedia 104:9–14
52. Callegari G, Spinelli A, Bianco L, Serra V, Fantucci S (2015) NATURWALL© - A solar timber façade system for building refurbishment: optimization process through in field measurements. Energy Procedia 78:291–296
53. Ochs F, Siegele D, Dermentzis G, Feist W (2015) Prefabricated timber frame façade with integrated active components for minimal invasive renovations. Energy Procedia 78:61–66
54. Gallo P, Romano R (2017) Adaptive facades, developed with innovative nanomaterials, for a sustainable architecture in the mediterranean area. Procedia Eng 180:1274–1283

Chapter 4
Active Solar Technologies

This chapter presents a summary of active solar technologies employed to convert solar radiation into thermal and electrical energy, to be utilized in various building applications including space heating, domestic hot water, and to meet various electrical requirements. Active solar technologies include various types of photovoltaic (PV) technologies (such as different PV cells, semi-transparent PV, transparent PV, and others), hybrid PV/thermal collectors, and solar thermal collectors. Current advancements in these technologies are summarized. In addition, the methods of integration of these technologies into buildings and especially the building envelope are discussed. The main criteria for successful integration and performance of these technologies are highlighted.

4.1 Introduction to Active Solar Systems

Active solar systems refer to systems that convert solar energy to usable form of thermal or electrical energy. Unlike passive systems, active solar energy technologies require the collection and transport of solar radiation through a medium and then the processing of the collected solar energy into thermal or electrical energy, employing specific components (for each form of energy).

The two main applications of active solar systems in buildings are (1) as a source of electricity and (2) as source of heat for hot water and space heating. Another important solar energy application is as active daylight design. This chapter focuses mainly on electrical and thermal energy generation systems, introducing briefly the utilization of solar energy in active daylighting systems.

The basic solar active systems include solar thermal collectors for domestic hot water (DHW) and space heating, photovoltaics (PV) that generate electricity, and hybrid photovoltaic/thermal (PV/T) systems that can generate thermal and electrical energy simultaneously.

C. Hachem-Vermette, *Solar Buildings and Neighborhoods*, Green Energy and Technology, https://doi.org/10.1007/978-3-030-47016-6_4

4.1.1 Thermal and Electrical Systems

Solar thermal collectors are employed to convert solar radiation into thermal energy while photovoltaic (PV) technologies convert solar radiation directly into electricity [1].

Extensive research exists on various aspects of solar thermal collectors and PV systems. These investigations range from pure technical applications of these solar systems to their integration within the built environment, including their aesthetical impact. Research on social and economic aspects of PV systems is carried out as well, including social acceptance of these technologies, their aesthetics, return on investment, and others.

In contrast to PV panels, whose main function is to generate electricity, thermal collectors are employed for a wide range of applications. These applications include water heating, space heating and cooling, refrigeration, industrial process heat, desalination, thermal power systems, solar furnaces, and others [2]. Figure 4.1 presents an illustration of PV and thermal collectors.

The application of solar collectors is gaining popularity, in view of environmental concerns, on the one hand, and lowering of costs, on the other hand, with production costs regularly decreasing due to developing technologies and mass production. Additionally, these systems can be designed to provide energy for isolated buildings in remote areas, allowing thus to reduce significantly the cost of energy distribution from the main grid, to such areas.

Significant efforts are deployed to integrate these technologies into buildings and building materials, as a method to increase the visual acceptance and functionality

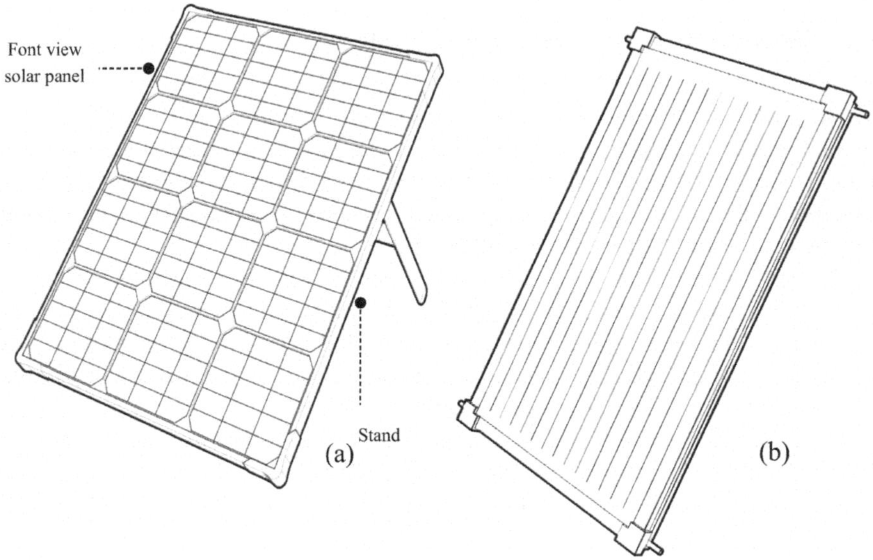

Fig. 4.1 Illustration of **a** PV and **b** thermal collectors

of these systems, and in view of reducing the overall costs. Methods of integration of solar technologies and their benefits are discussed below.

4.1.2 Active Daylighting Systems

The implementation of daylighting in buildings has a number of benefits, including the reduction of energy employed for artificial lighting, in addition to psychological and health benefits for the occupants. In fact, artificial lighting constitutes one of the major energy components in various commercial buildings such as office, retail, and institutional buildings.

Collecting and transferring natural light to specific locations in the building can be achieved through active or passive systems. Traditional daylighting devices rely on passive methods, employing side windows, or top lighting for daylighting. Such devices can provide daylighting for restricted areas, such as the perimeter zones and top floors of a building.

Active daylighting systems rely mainly on the use of devices to convey daylighting into areas within the building that are not directly exposed to natural sunlight. Active daylighting systems incorporate movable devices aiming at tracking the sun to maximize daylight capture. Implementing such methods can enhance the depth of penetration of daylight, increasing the yield and efficiency of daylight collection at an intensity that allows electrical lighting to be turned off. Active daylighting systems involve complex design and maintenance, which increase their cost considerably.

Significant advancement in active daylight systems has been achieved recently due to progress in computer control and optical properties of the materials [3–5]. The technologies for active daylighting applications fall under two main categories: active light guiding systems and active light transport systems. These are briefly described below.

- *Light guiding systems*: Light guiding systems refer to redirecting natural light, both direct and diffuse, into the core of a building. Such systems can provide light penetration depth of about 8–10 m, by means of reflection, refraction, or deflection. An example of active light guiding systems is a moveable light shelf, which can be actively controlled to adapt to the incident sunlight [5].
- *Light transport systems*: Light transport systems are employed to transport light into deeper spaces within a building, such as the core areas. The majority of light transport systems rely on direct sunlight as a light source. The direct sunlight is collected, generally at the roof of the building, and optically delivered by a series of devices including mirrors and tubes to various areas of the building. The selected transmission network may guide the collected solar radiation four to eight storeys downward, with distribution at each floor, and extending as much as 15 m horizontally through the building fabric via reflective or optic devices. Typical optical daylighting systems are mostly based on light tube and mirror systems.

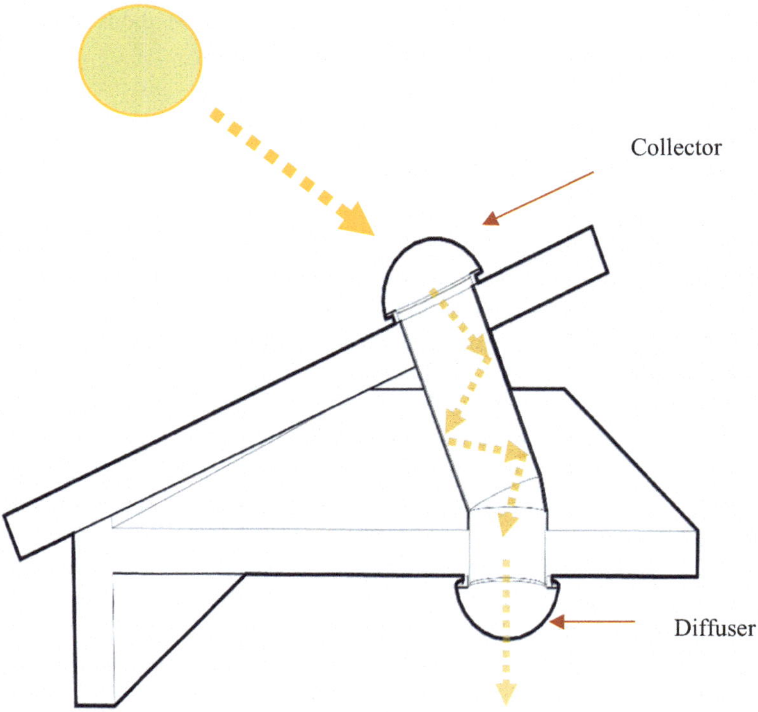

Fig. 4.2 Schematic illustration of a light tube system

Light tubes consist of structures employed for transmitting and distributing daylight to illuminate deep areas with no direct access to natural light (see Fig. 4.2). A light tube comprises a collector, a tubular element for light transmission, and a diffuser at the lower end to distribute natural light within the indoor space [5]. The daylighting mirror system comprises mirror collectors and reflectors [3]. Similar; to the light tube, it is composed of three main parts: a daylight collector; a structure to transmit the collected light, including *optimal anidolic ceilings* or light tubes; and a light distribution element.

The active daylight collection systems have a tracking device that continually orient and adjust the collector system according to the sun's path. This tracking system is coupled with the technologies mentioned above.

4.2 Photovoltaic Systems in Buildings

Photovoltaic technologies are emerging as an important part of the trend toward clean energy source diversification. Despite their potential, the application of photovoltaic (PV) in the built environment is still limited, though increasing at accelerating rate.

The main obstacles that hinder larger deployment of PV applications include cost, electric conversion efficiency, and operating lifetime, as well as the public acceptance of these components as part of the architectural design of buildings [6].

A wide range of new PV technologies are under research and development in pursuit of higher performance and reduced costs.

4.2.1 Basics of Photovoltaics and Methods of Operation

Solar photovoltaics are arrays of cells containing a semiconducting material, such as silicon coated with special additives, which converts solar radiation into electricity. Individual solar cells can be connected in series and in parallel to obtain desired voltage and current.

Standard PV modules are composed of a number of individual cells. The PV module is the basic element of a PV system (see Fig. 4.3). The number of PV modules within a system depends on the amount of electricity required for specific utilization, as well as on available space and costs. Although PV modules may vary in structure, they generally contain the following basic components [7]:

- Transparent glass cover placed over the PV cells for protection;
- Anti-reflective sheet, blocking reflection of the solar radiation;
- The PV cells;

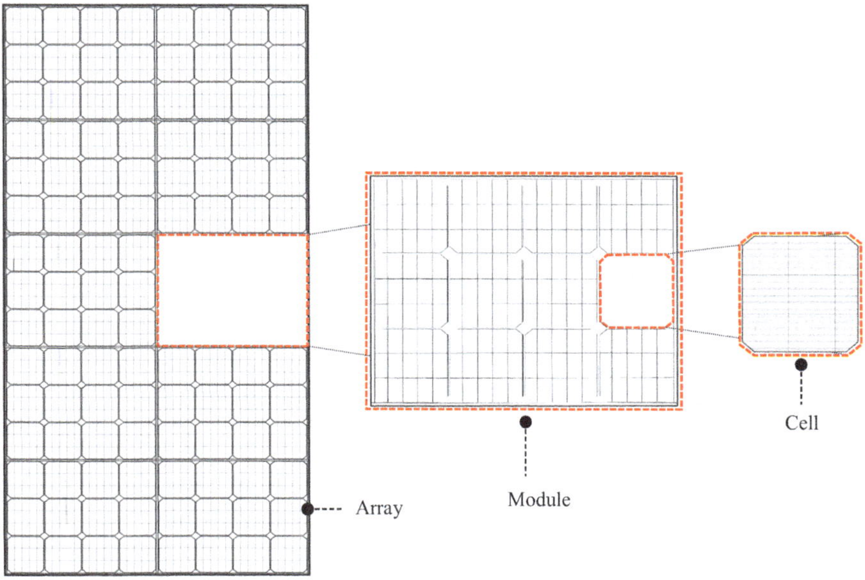

Fig. 4.3 Illustration of PV cells, modules, and array

- The frame and panel back sheet, designed to hold all the above components together.

Different encapsulation technologies are employed in the design of PV panels. Aluminum-framed PV panels are some of the most commonly employed. Modules without a frame, also termed *laminates*, are often employed in building integration applications.

A photovoltaic system consists of a number of components, including PV modules, mechanical and electrical components, in addition to means of regulating and/or modifying the electrical output (conductors, fuses, batteries, and inverters). The components of a PV system vary, depending on the application. PV systems are rated in peak kilowatts (kWp) which refers to the amount of electrical power that the system is projected to generate, under specific lab design conditions [8].

The inherent modularity of PV system allows the system some flexibility to be, for instance, extended, repaired, or completely substituted by another system, if required.

Electricity generation

PV cells produce direct current (DC) when they are exposed to solar radiation. The electricity generated is either consumed directly by DC appliances or converted into alternating current (AC) by means of inverters. The DC current can also be stored in batteries to be used as needed (e.g., during overcast periods, or at night). PV cells respond to both direct and diffuse radiations, and the overall electrical output is directly related to solar radiation intensity and inversely related to the module's temperature.

The electrical energy generated by a PV system constitutes only a fraction of the solar radiation energy incident on the system's surface. This fraction is referred to as the *electrical conversion efficiency* of the PV modules. Existing electrical efficiency currently ranges between 12 and 25%, depending on the type of PV modules (see below). Continuous research efforts are made to increase this efficiency.

Types of PV cells

A number of materials and technologies are employed in PV cells' production. Commonly available PV cells consist of monocrystalline silicon, polycrystalline silicon or thin-film amorphous silicon (A-Si), and other materials [8]. Although most common cell types are based on the same material (silicon), different fabrication techniques lead to the production of PV cells with diverse technical and aesthetic characteristics. The versatility and flexibility in PV appearance and colors are anticipated to increase their potential of integration within buildings, as well as their acceptance by building professionals (e.g., architects) and the general public.

Crystalline silicon PV cells have high production costs, and their dependence on purified silicon as the key raw material creates additional difficulty since there is global shortage of the material [7]. Below is a description of the most commonly employed PV cells, namely, monocrystalline cells, polycrystalline cells, and thin films. These PV technologies are schematically illustrated in Fig. 4.4. In addition, some emerging PV technologies are briefly summarized below.

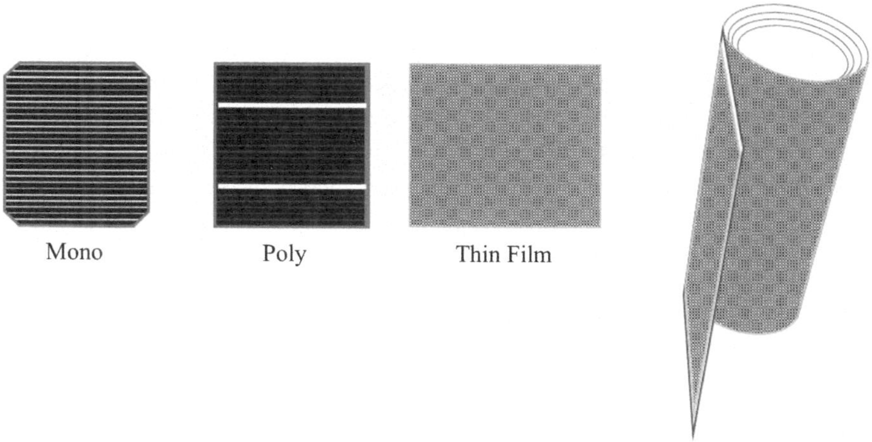

Fig. 4.4 Schematic illustration of the commonly applied PV technologies

Monocrystalline cells

The terminology used to describe this type of cell is based on the method of production, as they are cut into thin wafers from monocrystalline silicon cylinders. Monocrystalline silicon PV cells have a homogenous solid color that ranges generally from blue to black. Monocrystalline cells' dimensions are around 10 × 10 cm by 350 microns in thickness, with an efficiency ranging between 15 and 18% [9].

The form of the monocrystalline cells depends on the methods of cutting, and how much of the cylindrical mono-crystal bar is sliced away. Three cell shapes are obtained, a circular, semicircular, and square. Although circular shape is the least expensive, it is not often implemented in PV modules, due to the space inefficiency when cells are placed next to each other, within PV modules. An interesting application of monocrystalline cells can be achieved within a semi-transparent PV system (see below), for building integration [10].

A wide variety of colors is available but these are of lower efficiency and higher cost. For instance, some colors such as magenta or gold result in a loss of 20% efficiency as compared to the conventional PV modules [11]. Colored PV cells represent one of the solutions toward better integration of PV panels in building facades, as colors can be selected to be homogenous with the colors of various building materials within the same building, as well as with respect to the surrounding buildings. Technical solutions are available (and in development) for all components associated with colored PV, including glass, polymers, and PV-active layers [12].

Polycrystalline cells

Silicon material similar to that employed for mono-crystalline cells serves to produce polycrystalline cells. The main difference is that the raw silicon material is melted and then cast into a mold. This mold is cooled and then cut into square wafers. Controlled heating and cooling are applied, allowing the cast block to cool evenly in one direction. As the material cools, it crystallizes in an imperfect manner, forming random crystal boundaries [13]. The controlled technique of solidification aims at maximizing homogeneous silicon crystals, increasing thus the potential efficiency of this type of cell, which is inherently lower than that of the monocrystalline cells.

The most common color of polycrystalline cells is blue or silver-gray. Polycrystalline cells have an efficiency ranging between 13 and 16% [14]. Although the usual dimensions of the cells are similar to that of the monocrystalline, there is a current trend to increase the size of these cells to improve the efficiency and reduce the cost of production.

Thin-film cells

Thin-film PV modules are produced usually from amorphous-based silicon (a-Si), as well as other base materials such as cadmium telluride (CdTe) and copper indium diselenide (CuISn2, CIS) [15].

The thin-film technology employs a single or multiple thin layers of photoactive semiconductor material, to manufacture solar cells. In this process, the semiconductor photovoltaic materials are deposited onto supporting low-cost substrate, such as a large sheet of glass, metal, or plastic. Typically, less than a micron (0.001 mm) thickness of semiconductor material is sufficient to convert sunlight. This thickness is thus 100–1000 t thinner than the crystalline silicon wafer. The reduced material compared to crystalline solar cells implies reduced costs, forming a key advantage of these types of solar cells. Another advantage is that the unit of production is not restricted to the size of the wafer as for the crystalline cells but can be cut to any required dimension and shape, depending on the size of the substrate [7, 15].

Thin-film PV presents a number of unique advantages, compared to crystalline PV. It better utilizes diffuse solar radiation, works well in low light conditions, and is less sensitive to increase in temperature and less sensitive to shading. From architectural point of view, it can be bent into malleable shapes, offering thus more integration flexibility within the building envelope [14]. A disadvantage of thin-film PV is a toxic manufacturing process, which negatively impacts the environment [16]. This is in addition to the lower electric output. For instance, amorphous silicon-based thin-film cells have efficiencies of 5–7%, and this efficiency diminishes during the first 6–12 months. The copper indium diselenide (CIS) currently has the highest efficiency among all thin-film technologies (9–11%).

Semi-transparent PV

Semi-transparent PV (STPV) can offer an attractive solution to substitute the large extent of glazing in a building. Semi-transparent PV is usually achieved by encapsulating PV modules between two panes of highly transparent glass, or by perforating the silicon wafer. Crystalline silicon PV cells are commonly employed in STPV systems, while ethylene vinyl acetate (EVA) film is widely selected to encapsulate the PV cells [17].

Transparency is obtained by creating specific distance between the PV solar cells. The desired degree of transparency can be achieved by applying an adequate distance between the PV cells. The larger the distance, the higher the transparency [18].

STPV should be designed to balance between electricity generation, glazing transparency, heat gain, daylighting, and view to the outside. Increasing the PV proportion in a semi-transparent PV module as compared to transparent glazing area allows to increase the potential of electricity generation and to reduce solar heat gain, but it will restrict daylighting and available unobstructed view to the exterior [19]. In addition, the opaque cells will create a pattern of shadow in the building interior space that corresponds to the PV cells pattern of the STPV, and thus this consideration should be well integrated within the design of the interior space, and its requirement for solar heat gain and daylighting.

Figure 4.5 presents an illustration of an STPV panel installed within a glazed facade and the shading effect it has on the interior space.

Emerging Solar Harvesting Technologies

In addition to the PV technologies discussed above, additional solar technologies are attracting research efforts and industry interest. Some of these are described below.

Translucent and Transparent PV

Translucent and transparent PV cells are highly desirable for a multitude of applications in buildings. Research on such technologies is attracting a lot of attention. A number of translucent/transparent PV products are being developed by various technologies and manufacturing processes. For instance, thin-film PV (discussed above) can be utilized to achieve transparent PV cells, employing different methods. These include the technique of deposition of the photoactive film on a conductive oxide glass such as fluorine-doped tin oxide (FTO) [20]. The degree of transparency of the thin-film transparent PV cells depends on a number of factors, such as the thickness of the film, the deposition method, and the materials used.

Other types of translucent solar cells are produced by creating evenly distributed microscopic holes through the monocrystalline or polycrystalline silicon wafers, allowing some light to go through them. The holes can be created by a milling process or by laser cut. The efficiency of these cells depends on the methods of producing the holes, varying between 10% for the milled structure and 13% for the laser cut structure. The appearance and colors of these translucent cells derive from the original monocrystalline or polycrystalline wafers [21].

An emerging technology is the transparent luminescent solar concentrator (LSC). This technology consists of transparent polymer such as glass, and narrow PV cells located at the edge of this transparent substrate. Fluorescent dye is embedded in (or painted on) the transparent polymer plate. The fluorescent dye absorbs the incident light which is then propagated by total internal reflection reaching the PV cells (at the edge of the substrate) [20, 22]. Transparent LSC can currently achieve an efficiency of only 1%.

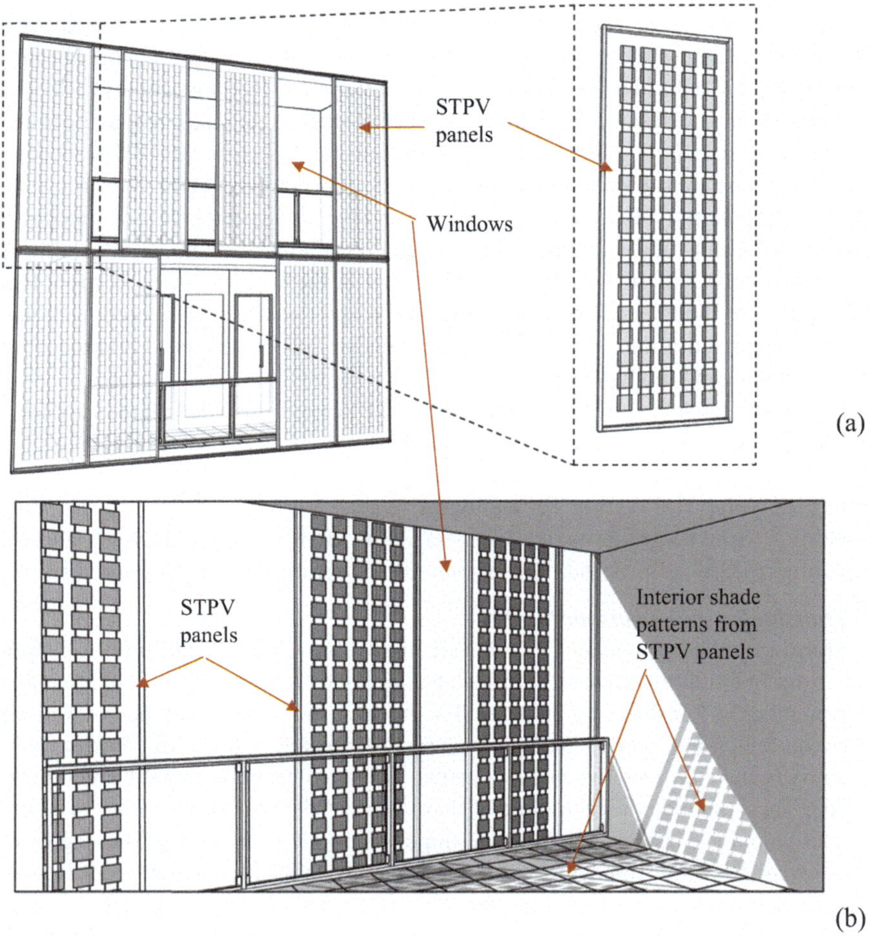

Fig. 4.5 Illustration of STPV panels, **a** in the facade and **b** its shade pattern in the interior space

Biophotovoltaic Panels (BPV)

Another evolving technology for solar harvesting relies on synthetic material, biophotovoltaic panels (BPV). Biophotovoltaics is an emerging field of study in microbial fuel cell research [23]. BPV technology exploits natural photosynthesis to transform light into electrical energy [24, 25]. During this process, incoming light is employed by the oxygenic biomass to split the water molecules, thus releasing electrons which are collected through an anode, to supply electricity [23]. As a nano/bio-material, BPV consists of hybridization of metal nanostructures and photosynthetic biomolecules protein. Figure 4.6 presents an illustration of the BPV technology.

BPV electrical efficiency rate is currently limited, and the associated cost is too high for practical applications. Research on BPV is still in its early stages, and

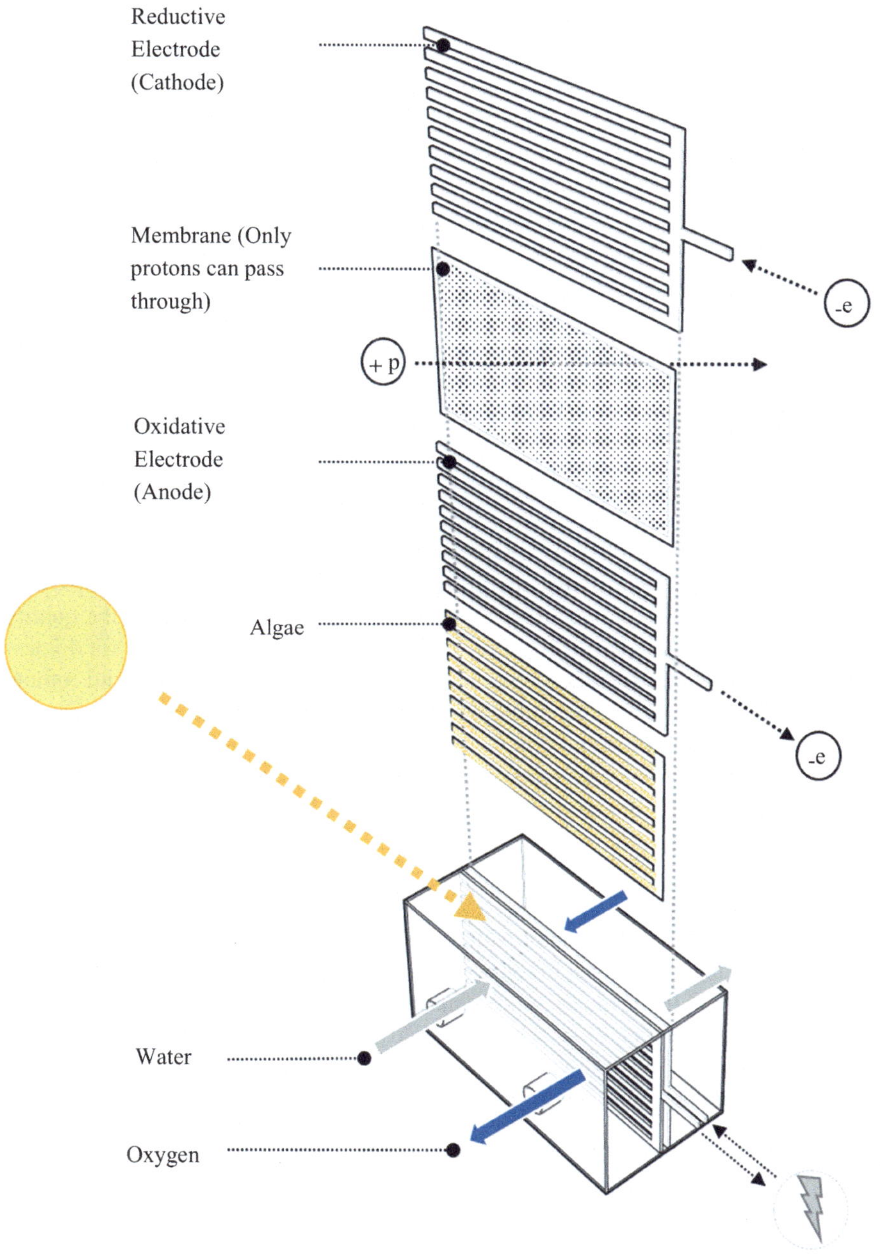

Fig. 4.6 Schematic diagram of biophotovoltaic device

amelioration of various factors may assist in increasing the viability of this technology. A myriad of factors affecting the performance of this technology need to be determined and optimized. For instance, experimental research on BPVs highlights the dependence of these systems on artificial light sources to provide constant illumination. Application into real-life situations is yet to demonstrate the performance of these systems under variable light intensity [23]. Other impactful factors that require more research are the biomass generation, optimal methods of growth of this biomass, and influence of the variation of temperature (diurnal and seasonal) on their growth rate.

4.2.2 Applications of PV Technologies

PV applications are numerous, including terrestrial and outer space applications. The first terrestrial PV applications consisted of rural communications systems. Such PV applications allow to establish communications in developing countries that lack electrical infrastructure, making solar energy an ideal solution for these areas. Employing PV technologies in remote and developing areas improved significantly the quality of life, as it allows a number of essential activities such as pumping water for domestic use, filtering water for drinking, and providing lighting to schools and remote buildings. In addition, PV technologies are increasingly coupled with solar home systems, as well as implemented in megawatt-scale power plants.

Below is a discussion of the types of existing PV systems and their applications in buildings and in the built environment. Advanced applications in building envelope are presented in Chap. 5.

Types of PV systems

There are two basic types of photovoltaic systems: Stand-alone systems and grid-tied systems. Stand-alone systems employ photovoltaics technology only and are not connected to the utility grid. These systems usually have some form of backup (e.g., storage). The second type of system—grid-tied systems—does not necessarily need backup and are tied directly to the utility grid. In this type of system, the excess PV generation is fed back onto the grid.

- **Stand-alone systems and their applications:** The stand-alone system does not supply power to the grid. Applications of stand-alone, off-grid PV systems vary in range and size, extending from small-scale personal applications such as wristwatches or calculators to remote buildings or spacecraft, including communications satellites, terrestrial communication sites, and remote homes and settlements. Storage is, generally, employed in stand-alone systems to provide electricity support during periods when solar radiation is not available [7]. Stand-alone PV systems can be employed to power remote areas, which are too distant to be connected to the conventional utility power grid.

- **Grid-connected PV systems**: These systems are typically connected to an independent electrical grid. The size of these PV systems can vary from restricted rooftop installations to extensive utility-level power production. In some applications, the electricity demand is totally or partially met by the system and the excess, if any, is fed directly to the grid. Most initiatives of net-zero energy houses and buildings use this type of PV system. In other applications, the energy generated is directly supplied to the grid, and the consumption is then supplied by the grid. In this case, the PV power provider gets credit for the power provided.

Inverters are required to convert DC electricity to AC. Figure 4.7 shows an example of PV system connected to the grid and the associated components of the system.

Applications in buildings

Applications of PV technologies in buildings can take two main forms: as complete integration of the PV panels in the building envelope (roofs and facades) and as added elements to the building envelope.

Building added PV–BAPV refers to simply adding PV panels on top of various finished surfaces of the building, without actual integration with the building envelope. In a building-integrated PV system (BIPV), PV modules are integrated into the functional and aesthetic aspects of building envelope. Building-integrated photovoltaic thermal (BIPV/T) is a hybrid system that combines PV and solar thermal collector technologies, simultaneously producing thermal and electrical energy. BIPV and BIPV/T can be implemented in multiple building applications. These are discussed in Sect. 4.3.

Translucent and transparent PV (see Sect. 4.2) are starting to attract considerable attention in view of their application into windows, to allow daylighting and solar radiation into the interior space, while generating some electricity.

Factors affecting performance

PV system efficiency depends mainly on the type of PV cells (see Sect. 4.2.1) as well as on the efficiencies of various components that constitute the system. Weather conditions, such as irradiation level and temperature, influence the efficiency of the system as well. For example, the absence of direct solar radiation under cloudy conditions can cause significant reductions in power generation.

Below are the main factors that affect the performance and the electrical output of the PV system.

- The daily and seasonal variations in solar radiation;
- Geographical location affecting the solar radiation availability at the specific location;
- Tilt angle of the solar panels;
- Azimuth angle, which refers to the orientation with respect to due south (in the northern hemisphere);

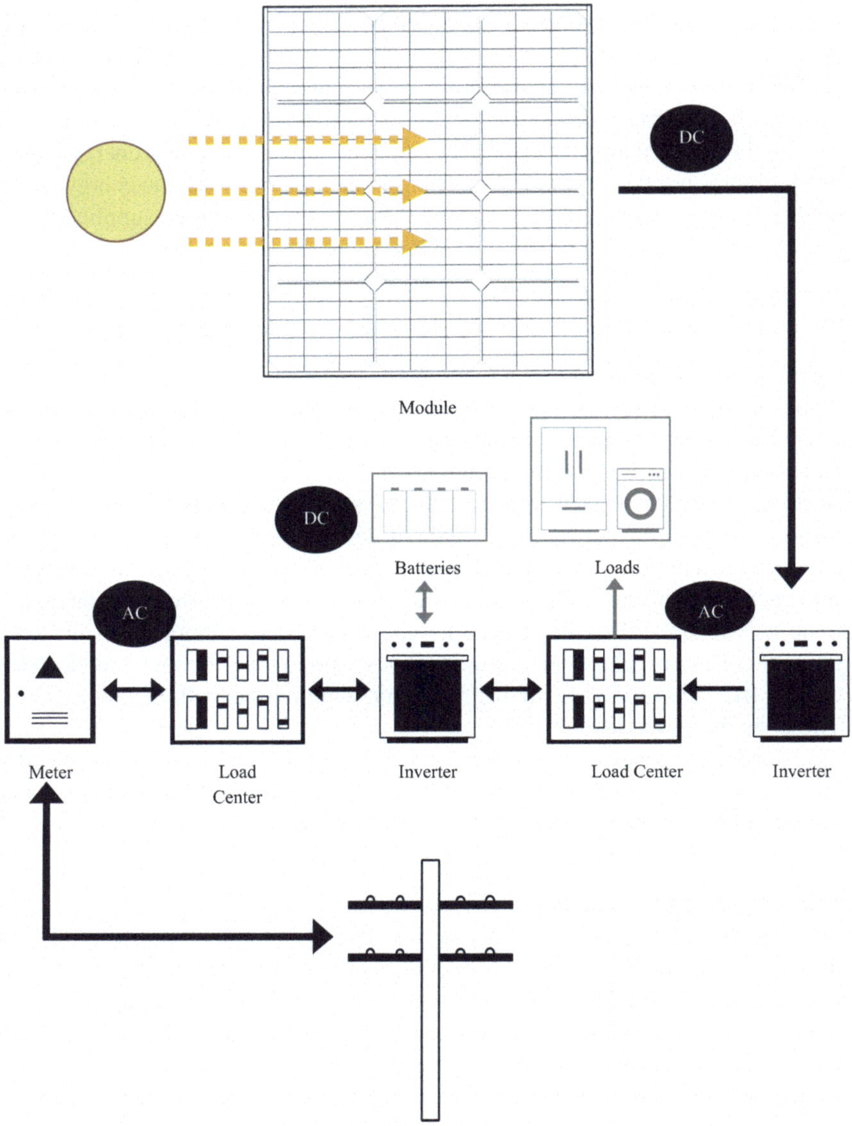

Fig. 4.7 Example of grid-connected PV system, with potential backup batteries

- Shadow that may be cast over the PV panels, from neighboring buildings, trees, or other obstacles;
- Increase in temperature of the PV modules that negatively impact the electrical efficiency of the module.

4.2.3 Integration of PV in the Building Envelope

The main purpose of the integration of PV panels within the building envelope is to establish synergetic relationship between the architectural design, functional aspects, and renewable energy generation [26]. The integration process involves the substitution of a conventional cladding or roofing material by photovoltaic modules. Although this concept has been promoted for a while, it is not currently implemented on a large scale. This is due largely to the extensive planning and architectural challenges involved in the process of integration. New technologies and awareness of both PV and building systems are assisting in the increase of deployment of BIPV.

In principle, BIPV can replace any part of the building envelope. Roof surfaces, with specific tilt angles (the optimal depends on the location, see below), are the ideal medium for installing PV modules due to the high level of irradiation incident on these surfaces year-round. Facades offer wide-ranging application potential, particularly in high-rise buildings in northern regions, where the sun is at relatively low elevation throughout the year, and particularly during the winter. The ratio of facade surface area to roof surface area increases with building height. In addition, the roof area available in commercial buildings (including multistory) is often restricted due to various facilities and installations (such as diverse HVAC equipment). This implies that employing the facades for the integration of PV systems is of particular interest in high-density urban areas. The availability of thin-film PV facilitates practical and efficient integration into facade surfaces.

Architectural integration of PV systems requires a number of considerations to ensure successful implementation. These include color, pattern, size, weather-tightness, wind loading, durability and maintenance, safety (fire resistance, electrical performance, and stability), and cost [27].

Building-integrated photovoltaic (BIPV) systems offer advantages in cost and appearance, as compared to the add-on PV system (BAPV), by incorporating photovoltaic panels into building components, such as roofing, cladding, and glass. When BIPV materials are substituted for conventional materials in new construction, saving involved in purchase and installation of conventional materials are applied to cost of the photovoltaic system. Some advantages of BIPV systems and design considerations that contribute to reducing their cost are presented below (under advantages of BIPV).

Aspects of Integration

A multidisciplinary approach is required to achieve a successful integration of PV systems within the building envelope. Several aspects should be considered including architectural, functional, and technical aspects. A summary of some of these considerations is presented below.

- **Architectural/aesthetic integration:** Several ways of architectural integration have been identified [28]. These include neutral integration, where the system does not contribute to the appearance of the building, or prominent integration, where the BIPV system is distinguished from the total building design. An important criterion of a good architectural integration is the overall coordination with the design of the building [29, 30].

- **Functional integration:** Solar collectors can be engineered to serve multiple functions, in addition to their active energy generation. Examples include integration into passive solar design elements (awnings, light shelves, etc.) and as roof and facade cladding materials [31].
- **Technical integration:** This refers to the integration with the building systems, such as the structural, mechanical, and electrical systems. For instance, the integration of the BIPV/T system with the building HVAC system can contribute to energy savings by preheating fresh air intake. Electrical integration includes voltage and current requirements, wiring methods, in addition to the utility integration.

Building envelope incorporating BIPV systems must be designed to resist water penetration and provide a weather seal and control thermal transfer. In addition, the BIPV systems must be able to withstand the stresses that a conventional building envelope is subjected to, including thermal expansion.

Methods of Integration

Photovoltaic modules are usually integrated into three different parts of the envelope: the roof, facades, or building components, such as balcony railings, sunshades, and sunscreens [32]. BIPV systems can be designed to cover a part or total area of roofs or facades.

Roofs

There is a growing interest in integrating PV systems in roofs, especially in residential or low-rise buildings, since it can provide an ideal exposure to solar radiation. BIPV products that can substitute some types of traditional roof claddings such as tiles, shingles, and slates are becoming commercially available (Fig. 4.8). These BIPV products are developed to match existing building products and are therefore compatible with commonly applied construction methods. For instance, small-scale PV components, such as PV shingles and tiles, are being developed to allow flexible integration within the roof, making them convenient for implementation in different types of designs and buildings (including existing buildings retrofit). Figure 4.8 presents an illustration of such PV roofing products.

Prefabricated roofing systems (insulated panels) with integrated thin-film laminates are starting to penetrate the market as well. These PV sandwiches constitute complete PV systems that comprise PV modules with mounting and interface components. Such products often include dummy elements for aesthetic integration, especially for roofs that are highly visible and when irregular roof shape is involved.

Several methods to integrate PV into the roof can be employed. The most commonly used method consists of designing the PV system as part of the external skin, forming part of the impermeable layer of the construction. Figure 4.9 presents various examples of PV integration within the roof design. PV can be installed in a saw-toothed roof structure (Fig. 4.9a), in a simple gable roof design, covering the total south-facing roof area (Fig. 4.9b) or a part of the roof surface

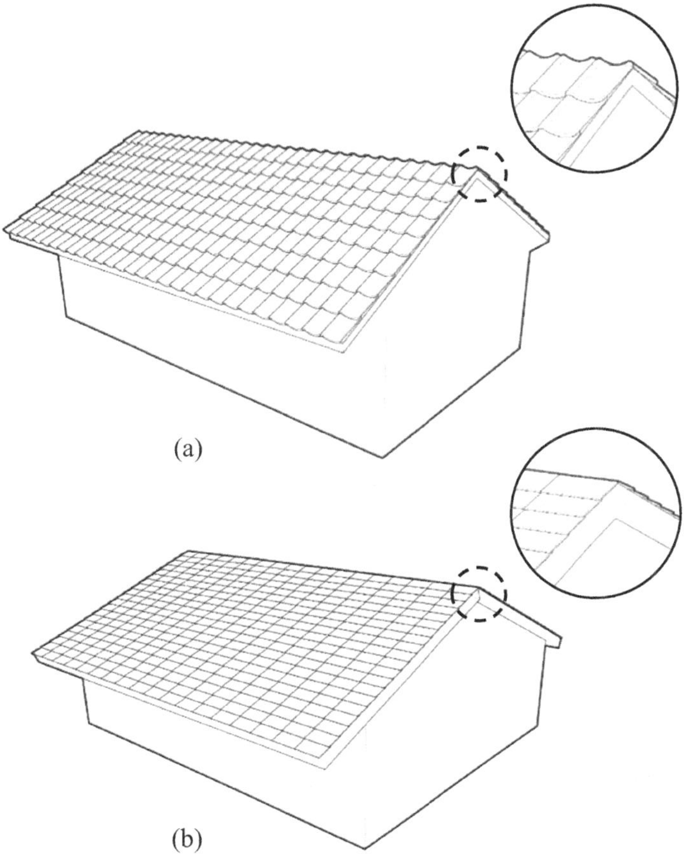

Fig. 4.8 Schematic illustration of types of V roofing products, **a** Shingle, **b** PV tiles

(Fig. 4.9d). Semi-transparent PV panels can be also incorporated within skylight design (Fig. 4.9c).

Using PV modules as roof covering reduces the amount of building materials needed as compared to add-on PV panels, increasing thus buildings' sustainability and reducing the cost associated with additional structural support system. Semi-transparent PV modules applied in areas with extensive glazed surfaces, such as sunrooms and atriums, need to be coupled with shading devices and solar control to avoid overheating.

Advanced integration of PV systems within roofs is further discussed in Chap. 5

Facades

PV systems can substitute or supplement the external layer of facades, as a cladding component, or it can substitute the whole facade system (e.g., curtain walls, opaque, or translucent). Different curtain wall structures can offer a spectrum of architectural effects (see Chap. 5). In the case of PV as external cladding,

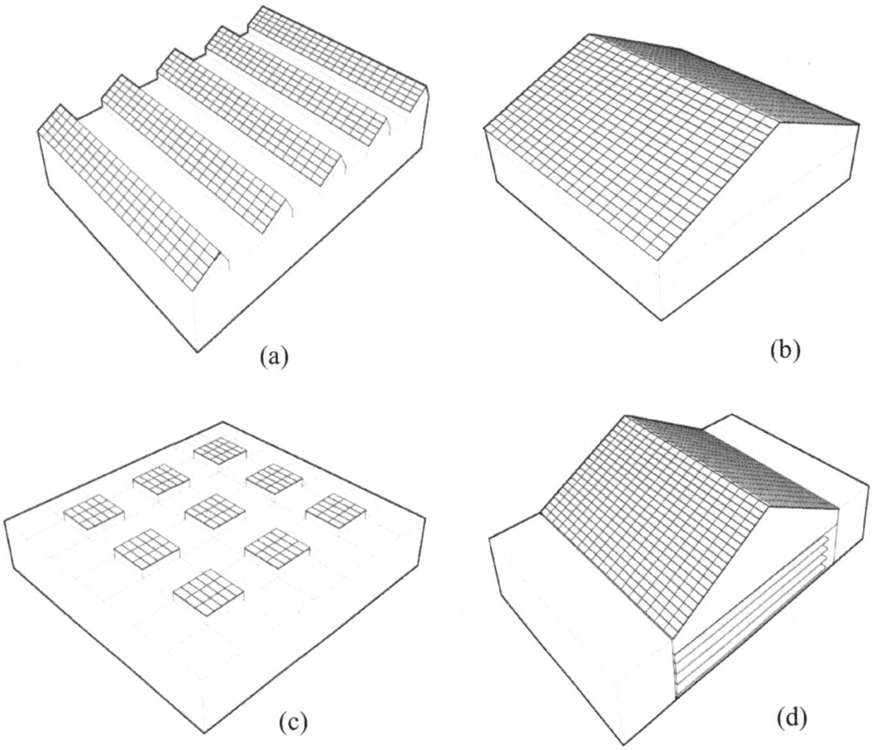

Fig. 4.9 Illustration of the integration of PV in various types of roofs

the back is usually ventilated, to avoid overheating of the panels, which reduces the electrical efficiency of the system. The heated air can be employed for space or water heating (see Sect. 4.3).

In addition, PV can be integrated into the glazing, or replace it under appropriate solar exposure conditions. Semi-transparent PV glazing prevents direct sunlight from entering the interior space, reducing cooling loads and glare. In semi-transparent PV applications, the layout of opaque solar cells on the glass and dimensions of gaps in between them control the amount of solar radiation admitted into the space (see above–semi-transparent PV section). In addition, PV can be integrated within daylighting and shading devices, to provide multiple benefits. Examples of incorporating PV panels within the facade and facade components are presented in Figs. 4.10 and 4.11.

The electricity generation potential of PV-integrated within the facades is less than the generation of PV installed on the roof, especially during summer. This issue can be addressed by designing facades, or facade elements, that are tilted outward as compared to the vertical axis. This is discussed in more detail in Chap. 5. PV panels can be employed as external components to serve various functions such as shading devices, spandrels, or balcony parapets.

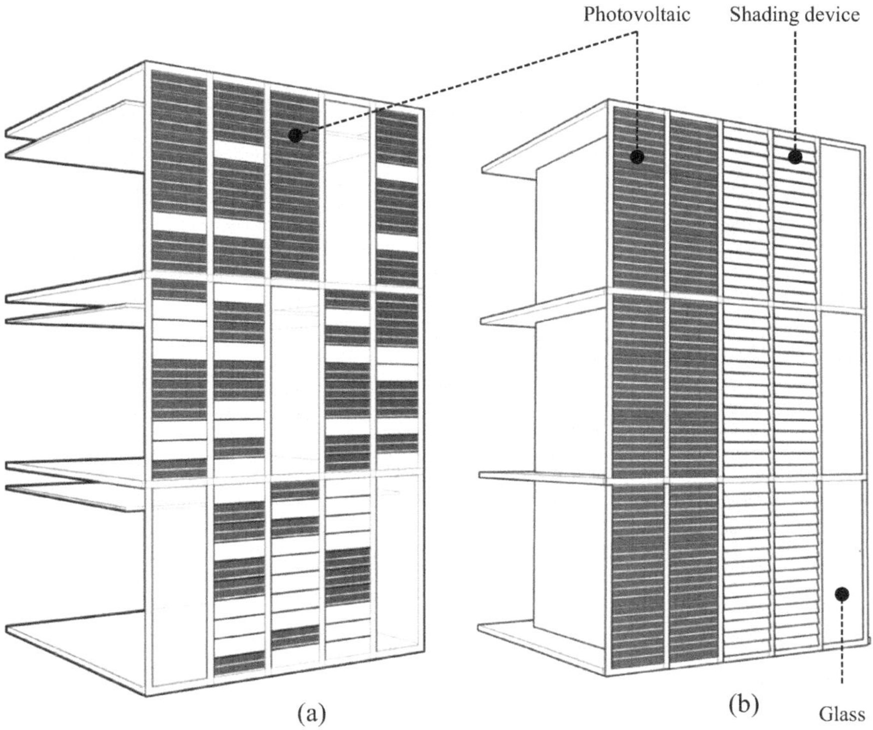

Fig. 4.10 **a** PV panels as cladding elements **b** as window shutters

Figure 4.12 shows examples of BIPV systems as window shutters and awnings.

Advantages of BIPV

There are several advantages associated with the implementation of building-integrated photovoltaic systems, as compared to the add-on PV systems. These include architectural, technical, and financial aspects. Some of these advantages are summarized in the following:

- Elimination of the structural framework required to support free-standing solar collectors. This can assist in offsetting the cost associated with the additional support structure, as well as the cost and potential issues associated with multiple roof penetrations for the supports [33].
- BIPV panels are designed to substitute the external skin of the building envelope (i.e., PV as a cladding), or to substitute the whole technological sandwich (e.g., semi-transparent glass–glass modules as skylights), and therefore it can counterbalance the price of the building materials and systems it replaces [34].
- No additional land area is required, since the building surfaces are used to mount the system, thus allowing its application in dense urban areas [33].

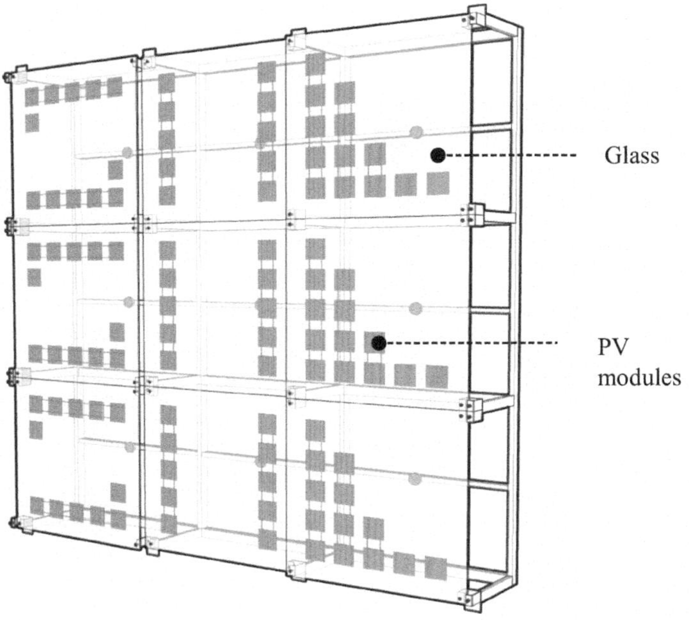

Fig. 4.11 Schematic presentation of an example of integration of PV modules within a glazing system

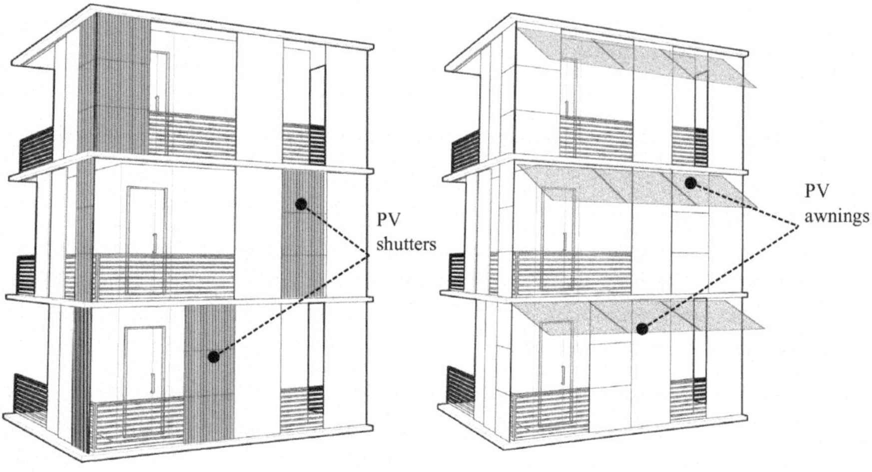

Fig. 4.12 **a** PV panels as window shutters, **b** solar awnings

- PV systems have a life span comparable to other building materials and require no maintenance [35, 36].
- PV systems offer a multitude of architectural design solutions, ranging from urban planning scale to specific building components (e.g., shading devices, spandrels, etc.) [37].

BIPV systems have few disadvantages as compared to add-on PV modules (BAPV); the most significant is its higher cost. This cost, however, is continuously decreasing. In addition, the application of BIPV systems is, in some situations, more suitable for new buildings than for retrofitting applications.

Guidelines for Architectural Integration

Architectural integration of PV systems requires a number of design considerations. These include the material and texture of PV modules, the color of PV cells, shape and size of the modules, and the type of jointing employed [38]. These features of PV modules, relating to all types of PV technologies, should be in harmony with the overall building design to ensure a complete visual, technical, and functional integration. Following are some guidelines for selection of PV products.

- **Materials and surface texture**: The material and surface texture characteristics of PV modules should match those of other cladding and finishing elements of the building envelope. PV panels, with surface finish ranging from shiny polished glass to various types of structured glass, are starting to be commercially available, allowing thus flexibility in the architectural design of buildings [18, 38].
- **Color**: A range of colors of PV cells is available, with anti-reflection layers, which allow a large flexibility in the architectural implementation and integration, albeit at reduced efficiency and increased cost. The range of available colors and textures allows close matching with a desired pattern in a building fabric [39, 40].
- **Shape and size of the modules**: The shapes of the module of the PV modules have to be compatible with the size and shape of other facade elements [38]. The type of PV technology selected for building application affects the basic geometric form of the module. For instance, mono- and polycrystalline PV modules have in general standard sizes and can be bulky, while thin-film PV have diverse shape and sizes, allowing more flexibility.
- **Application**: The type of PV technology implemented has to match the specific application. Although these technologies differ in many aspects (as discussed above), there are some requirements that are common to all of them. Below are some characteristics of size and shape of modules, presented according to the types of application [38]:
  - *Products matching existing building products*: A number of BIPV products, especially for roof applications, are designed to substitute and to be compatible with established products and their mounting systems. Such products include roof components like tiles, shingles, etc. (see Fig. 4.9).
  - *Complete PV systems developed for building integration*: Full integration of PV modules within the building envelope requires an adequate mounting system

and interface components that allow these modules to become an integral part of the building.
  - *Custom-made products developed for specific projects*: Custom-made products offer high flexibility of shape and size to fulfill design requirements. However, this flexibility comes at a higher cost. Additional consideration should be given to producing auxiliary modules that allow replacement in case of damage.

- **Dummies**: In some situation of PV integration into the building envelope, elements with specific dimensions and shapes that do not fit standard PV modules can be required particularly in areas such as edges and corners. Such components, called *dummies,* do not produce electricity but can be useful to emulate PV components, providing the same texture and finish. Dummies can be used as well in areas of the building envelope that are not suitable for the integration of PV, due to cast shade or inadequate solar exposure [41]. Providing non-PV dummy elements permits considerable flexibility in the design of BIPV systems.
- **Joints and Frames**: Joints and frames are prominent components of PV modules, having significant impact on the integration of these modules into the building envelope. The type of jointing and its flexibility depends on the type of product, and type and nature of the application. A number of framing options are available for consideration, as, for example, frameless modules, framing encapsulated in the glazing itself, or conventional framing that covers the edges of modules.

4.3 Solar Thermal Collectors (STC)

Solar thermal collectors absorb solar irradiation as thermal energy, which is then transferred to the solar collector working fluid (air, water, or oil). This heat is either directly utilized in buildings for applications such as providing domestic hot water and space heating or to charge a thermal storage tank. Thermal storage can be used to supply heat when needed (in the absence of solar radiation) as, for example, at night and cloudy days. Various types of collectors exist, depending on the applications and the technologies employed, such as the medium used to transport the absorbed thermal energy.

4.3.1 Air-Based Collector Systems

Air-based thermal collectors serve primary to preheat air for ventilation or for space heating. In this application, the solar thermal collectors are often installed on the facade near the fresh air inlets [42]. Such design considerations allow to reduce the length of air ducts. In addition, the vertical position of the collectors results in higher efficiency during the heating period, since vertical inclination is advantageous at low solar angles.

The basic component of an air-based collector is the absorber that is employed to capture solar radiation. The thermal energy is then transferred to the working fluid (air in this case). The heated air is ducted directly to the indoor space or to the mechanical system for further processing, to bring it to the desirable temperature.

Air-based thermal collectors can be classified into two main types: Glazed collectors and unglazed collectors. These are briefly summarized below.

Glazed Collectors

Glazed collectors consist of a transparent glass sheet, an absorber plate of dark color, and side and back insulated panels, to reduce heat loss to the environment. The recirculating collector type is the most commonly employed for space heating [2]. The indoor air is usually ducted into this system, it collects heat from the absorber (via conduction), and then it is ducted back into the building. The heated air is employed for direct space heating or to supply preheated air for the building mechanical system.

Unglazed Collectors

Unglazed collectors consist mainly of an absorber, without any glass cover (on top of it), and thus it is completely exposed to the outdoor environment. In this collector type, air passes across the absorber or through it (in transpired collectors—see below), picking up heat. Unglazed collectors are employed for preheating ventilation air in different types of buildings, including commercial, and industrial buildings.

A special type of unglazed collectors, termed *Transpired Collectors*, can be obtained by utilizing a perforated plate, to collect heat [43]. This solar collector is typically mounted on building facades, to allow optimal solar exposure during winter months, when the sun is at lower altitude. An air cavity is designed behind the collectors to provide a channel to the outdoor air, passing through the perforated exterior absorber (see Fig. 4.13). A large number of micro-perforations allow the heated layer of ambient air, which is in contact with the exterior surface of the plate to be drawn into the air cavity. This preheated air is then provided to the ventilation system to be distributed to the indoor space. Figure 4.13 presents an example of perforated collectors, linked to air condition system.

Building-Integrated Hybrid Photovoltaic/Thermal Systems

Hybrid photovoltaic/thermal systems (PV/T) combine PV modules and heat extraction devices to produce simultaneously power and heat [44]. Heat extraction from the PV rear surface is usually achieved using the circulation of a fluid (air or water) with low inlet temperature. The extraction of thermal energy serves two main functions. It is exploited for space heating and hot water applications, and it cools the PV modules, thus increasing the total energy output of the system [45].

The total electrical and thermal energy output of the PV/T systems depends on several factors including solar energy input, ambient temperature, wind speed, and heat extraction mode. For locations with large space heating requirements, air-based PV/T systems can be particularly advantageous and cost-effective [44]. The comparison between the performance of hybrid PV/T collectors and more traditional

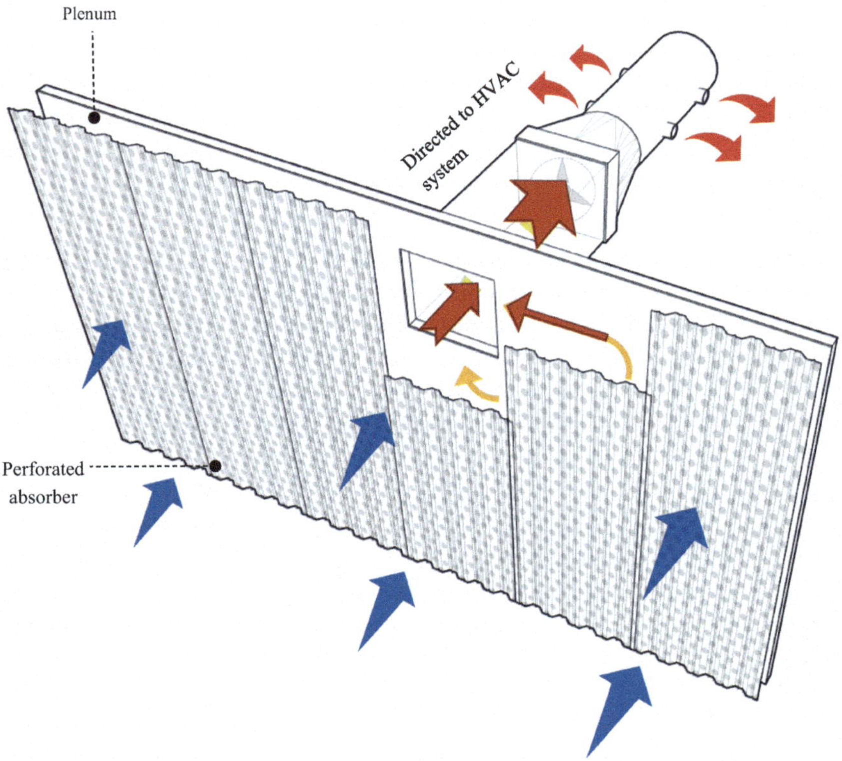

Fig. 4.13 Air-based perforated collector system

PV systems indicates that PV/T systems can achieve increased energy conversion efficiency with potential cost–benefits [46].

Figure 4.14 presents examples of BIPV/T applied to a roof, in a residential building, and to a vertical surface.

4.3.2 Water-Based Collectors

Water-based thermal collectors allow easy storage of solar gains and are suitable both for domestic hot water production and space heating. The medium of this type of collector consists mainly of water charged with glycol in variable percentages to avoid freezing, depending on climatic zones. Since water has a good thermal capacity, this medium is capable of good quality heat exchange with both the heat absorber and the storage. The collected solar thermal energy can be stored in insulated water tanks and used for domestic hot water or for space heating. Four types of water-based solar collectors are distinguished, based on the technology employed: glazed

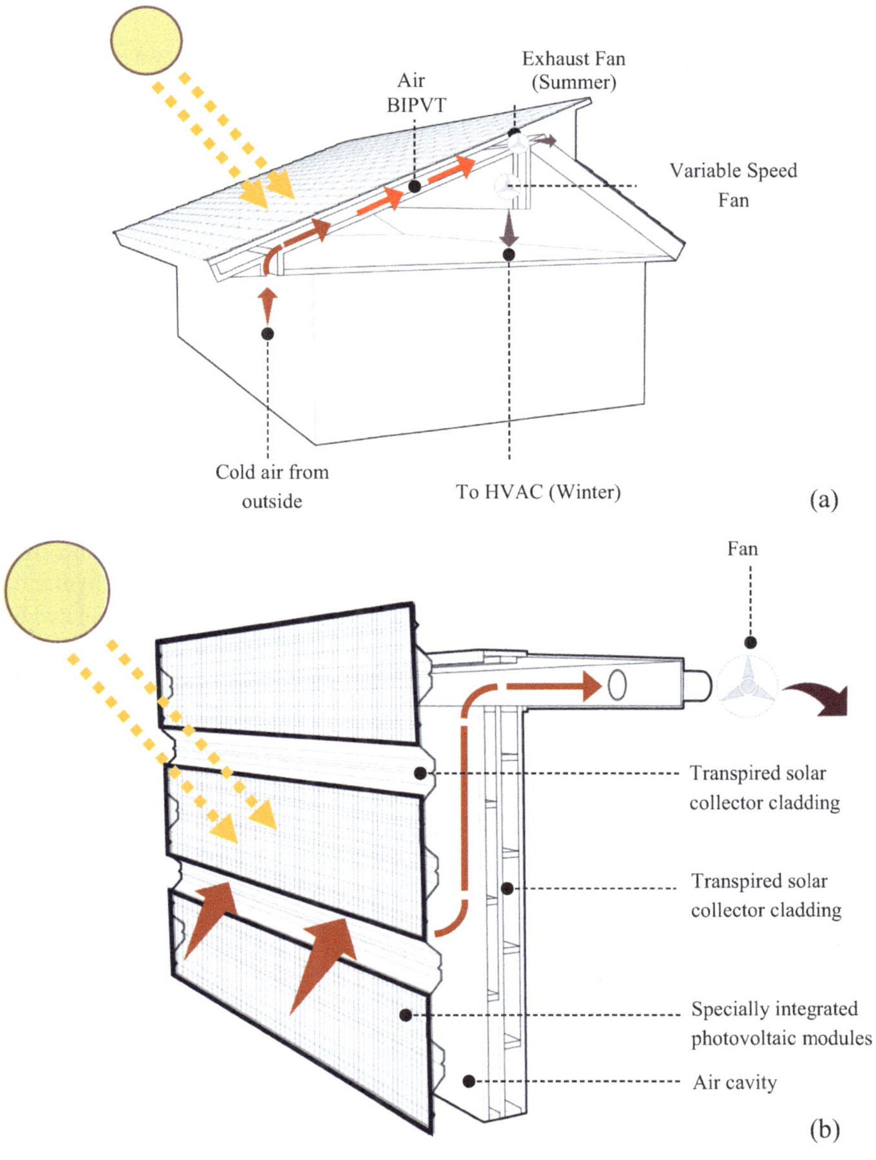

Fig. 4.14 Examples of BIPVT systems applied to **a** roof, **b** vertical facade

flat-plate collectors, unglazed flat-plate collectors, unglazed plastic collectors, and evacuated tubes collectors, as detailed below [47].

Flat-Plate Collectors

Flat-plate solar collectors are usually installed in fixed positions (i.e., tilt and orientation angles). A suitable orientation is crucial for optimal performance of the system. A variety of flat-plate collector can be distinguished, as discussed in the following.

Glazed Flat-Plate Collectors
These are the most commonly used collectors for domestic hot water and space heating. The main components include external glazing, absorber plates, insulation layers, and recuperating tubes filled with heat transfer fluid (water). Glazing is a critical part of the design of this type of solar thermal collector, as it should be selected to maximize the collection of solar irradiation, while reducing heat loss from the absorber plate. Single or multiple sheets of glass can be employed. Although the surface of the absorber plate is usually black, to increase the absorption of thermal solar energy, a range of color-coated absorbers is commercially available, allowing larger flexibility in design and integration with building architecture [48]. To avoid overheating of the solar collector system, a rapid transfer of the absorbed heat to the working fluid should be ensured [49]. Figure 4.15 presents a schematic illustration of a basic flat-plate collector.

Unglazed Flat-:Plate Collectors
Unglazed flat-plate collectors involve less technical complexity than other types of collectors such as glazed or evacuated tubes collectors (see below) but are not as widely used [44]. They are made of fewer layers than the glazed absorber, consisting mainly of the absorber which is a dark color metal plate and a hydraulic circuit that collects heat from the back of the absorber. This circuit is insulated by a back insulation, to reduce heat loss to the environment. This type of collector can be used for swimming pools, for low-temperature space heating systems, and for DHW preheating (Fig. 4.16).

Unglazed Plastic Collectors
Plastic collectors are another kind of flat-plate thermal collectors, consisting usually of black plastic or rubber, especially treated to resist ultraviolet radiation. The collectors are not insulated, and thus a large portion of the heat absorbed is lost to the outdoor environment, particularly under windy, cold conditions. On the

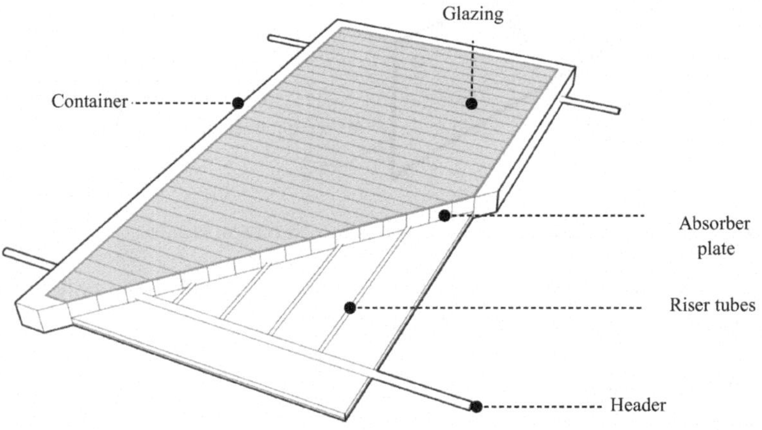

Fig. 4.15 Illustration of a basic, water-based, flat-plate collector

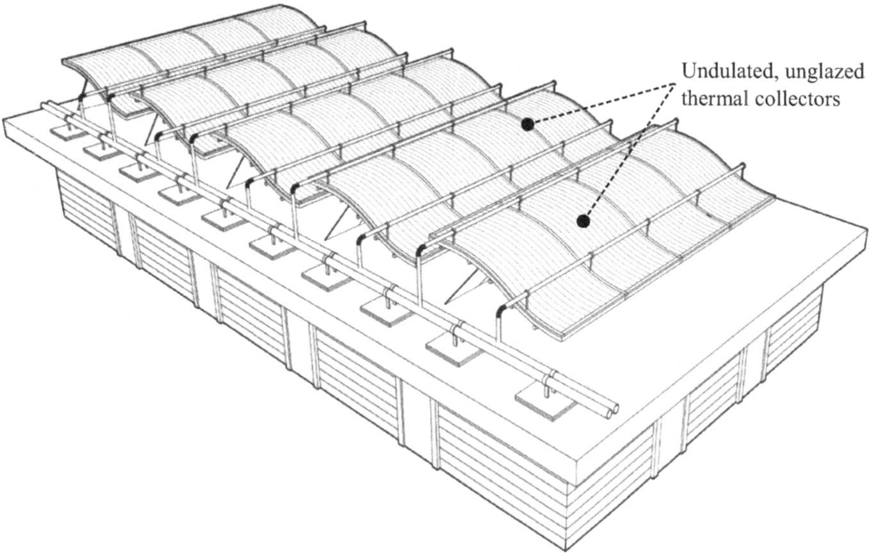

Fig. 4.16 Example of undulated, unglazed thermal collectors (drawn based on an installation for a swimming pool, Freibad Ilanz, Switzerland)

other hand, they can capture heat during the night under hot and windy conditions. These types of collectors are mostly useful to heat swimming pools due to their low working temperatures [50].

Evacuated Tube Collectors

Evacuated tube collectors comprise a number of transparent glass tubes, attached to a header pipe. Each of these tubes is composed of two concentric tubes—an outer tube and an inner tube, with evacuated space between them [51]. The inner and outer tubes are sealed together at their extremities. The vacuum acts as a highly effective insulated layer, to reduce heat loss from the inner tube, increasing thus the energy conversion efficiency [52]. The inner tube is coated with high absorptivity and low emissivity selective coating material, which allows to absorb incident solar radiation. The collected heat is then transferred to the medium inside the inner tube [53], and then to a heat exchanger, located transversely at the edge of the evacuated tubes, through various methods (e.g., employing heat pipe, see Fig. 4.17).

The tubular form of the individual evacuated tubes maximizes the capture of solar radiation, at different sun angles. This increases the overall efficiency of this thermal collector type, as compared to flat-plate collectors. Evacuated tubes collectors can be employed for various applications, including domestic hot water for residential use, space heating, as well as in different industrial applications.

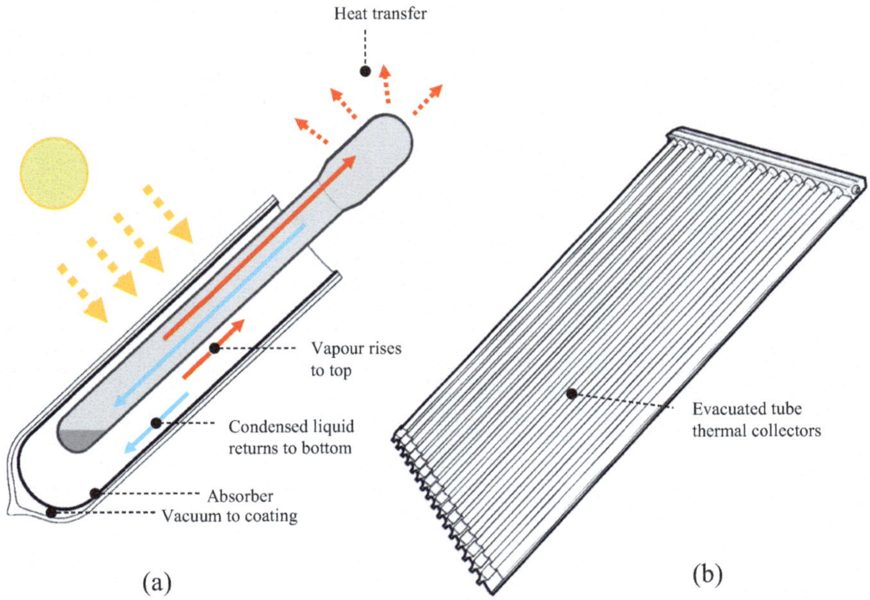

Fig. 4.17 Evacuated tube thermal collectors, **a** section of an individual tube, **b** sketch of the thermal collector

4.3.3 Integration of STC in Buildings

Integration criteria of solar thermal collectors (STC) in buildings hold some similarities with those employed for the integration of PV systems, as well as some differences. Similar to BIPV, integration of solar thermal collector systems in buildings requires addressing technical constraints related to the specific solar thermal technology, while fulfilling various architectural functions (such as cladding). Additional similarities include considerations regarding the modules size and shape, the jointing method, color of the collectors, and texture and finish.

Due to the large dimensions and complexity of their modules, thermal collectors are less flexible than PV systems, and the options of available products are restricted, making the integration of STC in building envelope face a number of hurdles. Unlike PV modules, the integration of different types of STC modules is affected by the large variations in their characteristics [47]:

- The energy transfer medium (air, water…);
- The materials employed for the collector (plastic, metal, glass…);
- The form of the collector (flat plate, multilayer flat plates, vacuum tubes).

The main methods of integration of STC in buildings are summarized below.

- **Roof integration**. Flat-plate collectors can be obtained in various sizes, which presents an opportunity for optimal integration in roof surfaces.

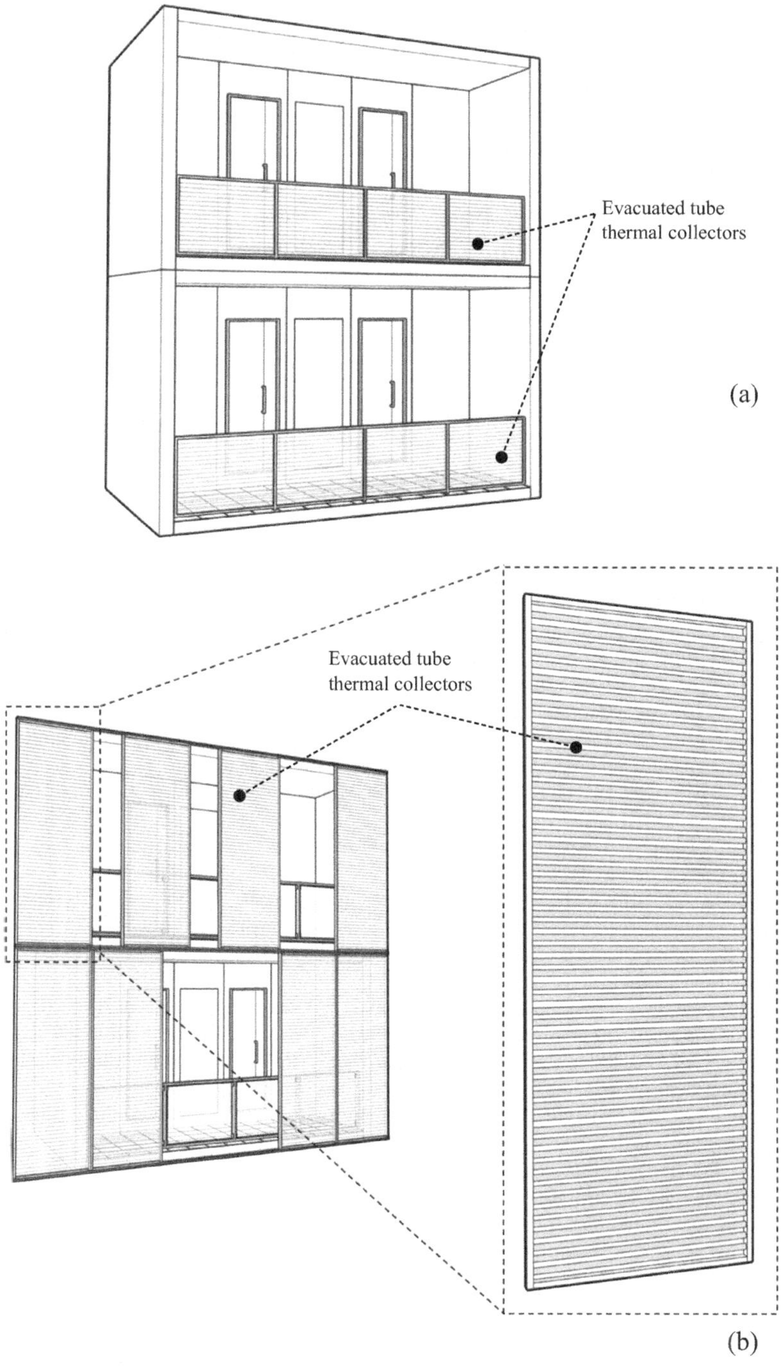

Fig. 4.18 Illustration of the application of evacuated tube thermal collectors, **a** as balcony railings, **b** as window shutters

- **Facade integration**. Facade integrated solar thermal collectors are gaining more popularity, as solar irradiation can be exploited even during winter due to lower sun altitude, corresponding thus to higher space heating requirement. Installing STC on facades can be especially beneficial when roof space is insufficient or unsuitably oriented. While the dark color of thermal collectors may be a drawback for facade integration, collectors' color can be modified through coloring of the glass sheet covering the absorbers.
- **Other integration possibilities**. Solar thermal collectors can be employed to fulfill different other functions, as, for example, sun screening components or balcony railing. Evacuated tubes can particularly present an interesting design for balcony railings and vertical sunscreen. Figure 4.18 presents a schematic illustration of such application.

References

1. Singh GK (2013) Solar power generation by PV (photovoltaic) technology: a review. Energy 53:1–13
2. Kalogirou SA (2004) Solar thermal collectors and applications. Prog Energy Combust Sci 30(3):231–295
3. Kim JT, Kim G (2010) Overview and new developments in optical daylighting systems for building a healthy indoor environment. Build Environ 45(2):256–269
4. Mirkovich DN (1993) Assessment of beam lighting systems for interior core illumination in multi-story commercial buildings. ASHRAE Trans 99(1):1106–1016
5. Mayhoub MS, Carter DJ (2010) Towards hybrid lighting systems: a review. Light Res Technol 42(1):51–71
6. Rehman S, Bader MA, Al-Moallem SA (2007). Cost of solar energy generated using PV panels. Renew Sustain Energy Rev 11:1843–1857,
7. Poullikkas A (2010) Technology and market future prospects of photovoltaic systems. Int J Energy Environ (4)
8. Parida B, Iniyan S, Goic R (2011) A review of solar photovoltaic technologies. Renew Sustain Energy Rev 15(3):1625–1636
9. Razykov TM, Ferekides CS, Morel D, Stefanakos E, Ullal HS, Upadhyaya HM (2011) Solar photovoltaic electricity: current status and future prospects. Sol Energy 85(8):1580–1608
10. Pearsall N (ed) (2016) The performance of photovoltaic (PV) systems: modelling, measurement and assessment. Woodhead Publishing
11. Reijenga TH, Kaan HF (2011) PV in architecture. Handbook of photovoltaic science and engineering, 2nd ed. Wiley, Chichester, UK, pp 1043–1077
12. Eder G, Peharz G, Trattnig R, Bonomo P, Saretta E, Frontini F, Polo López, CS, Wilson HR, Eisenlohr J, Chivelet NM, Karlsson S (2019) Coloured BIPV: market, research and development
13. Ramalingam K, Indulkar C (2017) Solar energy and photovoltaic technology. Distributed generation systems: design, operation and grid integration, p 69
14. Jelle BP, Breivik C, Røkenes HD (2012) Building integrated photovoltaic products: a state-of-the-art review and future research opportunities. Sol Energy Mater Sol Cells 100:69–96
15. Aberle AG (2009) Thin-film solar cells. Thin solid Films 517(17):4706-4710
16. Shah AV, Schade H, Vanecek M, Meier J, Vallat-Sauvain E, Wyrsch N, Kroll U, Droz C, Bailat J (2004) Thin-film silicon solar cell technology. Prog Photovolt Res Appl 12(2–3):113–142
17. Skandalos N, Karamanis D (2015) PV glazing technologies. Renew Sustain Energy Rev 49:306–322

18. Heinstein P, Ballif C, Perret-Aebi LE (2013) Building integrated photovoltaics (BIPV): review, potentials, barriers and myths. Green 3(2):125–156
19. Miyazaki T, Akisawa A, Kashiwagi T (2005) Energy savings of office buildings by the use of semi-transparent solar cells for windows. Renew Energy 30:281–304
20. Husain AA, Hasan WZW, Shafie S, Hamidon MN, Pandey SS (2018) A review of transparent solar photovoltaic technologies. Renew Sustain Energy Rev 94:779–791
21. Boxwell M (2010) Solar electricity handbook: a simple, practical guide to solar energy-designing and installing photovoltaic solar electric systems. Greenstream Publishing
22. Zhao Y, Meek GA, Levine BG, Lunt RR (2014) Near-infrared harvesting transparent luminescent solar concentrators. Adv Opt Mater 2:606–611. https://doi.org/10.1002/adom.201400103
23. Tschörtner J, Lai B, Krömer JO (2019) Biophotovoltaics: green power generation from sunlight and water. Front Microbiol 10:866
24. Mccormick AJ, Bombelli P, Scott AM, Philips AJ, Smith AG, Fisher AC et al (2011) Photosynthetic biofilms in pure culture harness solar energy in a mediatorless bio-photovoltaic cell (BPV) system. Energy Environ Sci 4:4699–4709. https://doi.org/10.1039/C1EE01965A
25. Bradley RW, Bombelli P, Rowden SJ, Howe CJ (2012) Biological photovoltaics: intra- and extra-cellular electron transport by cyanobacteria. Biochem Soc Trans 40:1302–1307. https://doi.org/10.1042/bst20120118
26. Skea J, van Diemen R, Hannon M, Gazis E, Rhodes A (2019) Building integrated photovoltaics. In Energy innovation for the twenty-first century. Edward Elgar Publishing
27. Roberts S, Guariento N (2009) Building integrated photovoltaics: a handbook. Walter de Gruyter
28. Schoen T, Prasad D, Ruoss D, Eiffert P, Sørensen H (2001) Task 7 of the IEA PV power systems program–achievements and outlook. In Proceedings of the 17th European photovoltaic solar conference
29. Wall M, Probst MCM, Roecker C, Dubois MC, Horvat M, Jørgensen OB, Kappel K (2012) Achieving solar energy in architecture-IEA SHC Task 41. Energy Procedia 30:1250–1260
30. Probst MM, Roecker C (2012) Criteria for architectural integration of active solar systems IEA task 41, subtask. A Energy Procedia 30:1195–1204
31. Keoleian GA, Lewis GM (2003) Modeling the life cycle energy and environmental performance of amorphous silicon BIPV roofing in the US. Renew Energy 28(2):271–293
32. Reijenga TH, Kaan HF (2011) PV in architecture. Handbook of photovoltaic science and engineering, 2nd edn. Wiley, Chichester, UK, pp 1043–1077
33. Pearsall NM, Hill R (2001) Photovoltaic modules, systems and applications. Clean Electr Photovolt 1:1–42
34. Strong S (2010) Building integrated photovoltaics (BIPV). Whole Buil Des Guide 9
35. Wambach K, Muller A, Alsema EA (2005) Life cycle analysis of a solar module recycling process. In: European photovoltaic solar energy conference, p 8AV. 3. 1
36. Lamnatou C, Notton G, Chemisana D, Cristofari C (2015) The environmental performance of a building-integrated solar thermal collector, based on multiple approaches and life-cycle impact assessment methodologies. Build Environ 87:45–58
37. Kaan H, Reijenga T (2004) Photovoltaics in an architectural context. Prog Photovolt Res Appl 12(6):395–408
38. Farkas K, Frontini F, Maturi L, Munar Probst MC, Roecker C, Scognamiglio A (2013) Designing photovoltaic systems for architectural integration (No. REP_WORK). Farkas, Klaudia pour international energy agency solar heating and cooling programme
39. Zhang W, Anaya M, Lozano G, Calvo ME, Johnston MB, Míguez H, Snaith HJ (2015) Highly efficient perovskite solar cells with tunable structural color. Nano Lett 15(3):1698–1702
40. Lee KT, Fukuda M, Joglekar S, Guo LJ (2015) Colored, see-through perovskite solar cells employing an optical cavity. J Mater Chem C 3(21):5377–5382
41. Scognamiglio A, Farkas K, Frontini F, Maturi L (2012) Architectural quality and photovoltaic products. In Proceedings of the 27th European photovoltaic solar energy conference and exhibition (EU PVSEC), Frankfurt, Germany, pp 24–28

42. Buker MS, Riffat SB (2015) Building integrated solar thermal collectors–a review. Renew Sustain Energy Rev 51:327–346
43. Shukla A, Nkwetta DN, Cho YJ, Stevenson V, Jones P (2012) A state of art review on the performance of transpired solar collector. Renew Sustain Energy Rev 16(6):3975–3985
44. Tripanagnostopoulos Y, Tzavellas D, Zoulia I, Chortatou M. (2001) Hybrid PV/T systems with dual heat extraction operation. In Proceedings of the 17th PV solar energy conference, Munich (pp 22–26)
45. Charron R, Athienitis AK (2006) Optimization of the performance of double-facades with integrated photovoltaic panels and motorized blinds. Sol Energy 80(5):482–491
46. Tian Y, Zhao CY (2013) A review of solar collectors and thermal energy storage in solar thermal applications. Appl Energy 104:538–553
47. Probst MCM, Roecker C (2011) Architectural integration and design of solar thermal systems. EPFL Press
48. Mills D (2004) Advances in solar thermal electricity technology. Sol Energy 76(1):19–31
49. Slaman M, Griessen R (2009) Solar collector overheating protection. Sol Energy 83:982–987
50. Munari Probst MC (2009) Architectural integration and design of solar thermal systems (No THESIS). EPFL
51. Abd-Elhady MS, Nasreldin M, Elsheikh MN (2017) Improving the performance of evacuated tube heat pipe collectors using oil and foamed metals. Ain Shams Eng J
52. Tang R, Li Z, Zhong H, Lan Q (2006) Assessment of uncertainty in mean heat loss coefficient of all glass evacuated solar collector tube testing energy conver. Manage 47:60–67
53. Shah LJ, Furbo S (2004) Vertical evacuated tubular-collectors utilizing solar radiation, from all directions. Appl Energy 78:371–395

Chapter 5
Advanced Solar Envelope Design

This chapter presents advanced geometrical designs of roofs and facades, for increased thermal and electrical solar energy generation. Roofs are designed to accommodate solar technologies for low-rise buildings, while facade designs are mostly associated with multistorey buildings. A number of designs of roofs and facades are presented, ranging from simple and conventional designs, to more sophisticated designs, based on multifaceted geometries. The impact of these designs on the heating and cooling loads of these buildings is discussed. In addition, the impact of buildings' layout, on the design of roof and facade surfaces for solar collectors' integration and on the overall energy performance is analyzed. Applications of some of these designs are presented.

To demonstrate various design impacts, the chapter relies on hypothetical examples that are designed, and systematically simulated and analyzed. While simulations are conducted for specific geographic locations, modifications of these designs in view of their applications to other locations are discussed.

5.1 Simple and Advanced Roof Design

Roofs can be ideal for integration of PV modules, particularly for low-rise buildings, as it provides an area that is similar to the occupied floor area. Roof surfaces can be designed with optimal characteristics to integrate solar thermal collectors and maximize solar energy capture by these collectors. Some of these characteristics are detailed below.

5.1.1 Basic Surface Parameters and Their Effect

This section presents the effect of roof surface parameters, namely, tilt angle and orientation relative to south, on the energy performance of PV modules integrated

C. Hachem-Vermette, *Solar Buildings and Neighborhoods*, Green Energy and Technology, https://doi.org/10.1007/978-3-030-47016-6_5

within the roof. The main response variables in assessing performance are the annual and monthly electricity generation. The summary presented below is based on simulations conducted for a building design located in northern mid-latitude cold climate (45° N).

Effect of Tilt Angle

The electricity generation per unit area of BIPV system varies widely over the months of the year, following the sun's elevation, as well as the total number of hours of solar radiations per day. The generation is high during March to September, and drops (by up to 50%) during November to February. This variation in performance is independent of tilt angles (see Fig. 5.1). Tilt angle of roof is the angle between the normal to the surface and the vertical direction [1].

Tilt angle has an impact on the amount of solar radiation captured by the PV modules, and therefore on the total energy generation. Steeper tilt angles, relative to the geographic latitude, lead to better performance in winter months, and reduced performance during summer months [1, 2].

The annual electricity generation is not significantly affected by a tilt angle that ranges between 30° and 50°, for this specific mid-latitude location. A 60° tilt angle, as compared to 45°, indicates a reduction of electricity generation by some 7% annually and by 16% in June, while an increase of generation by some 6% is observed in December.

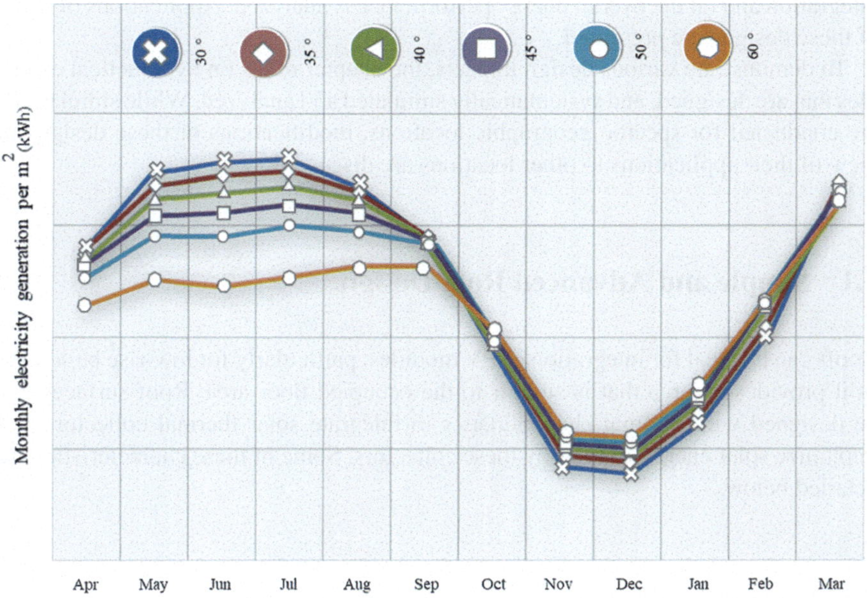

Fig. 5.1 Monthly electricity generation for different tilt angles, for mid-latitude cold climate location

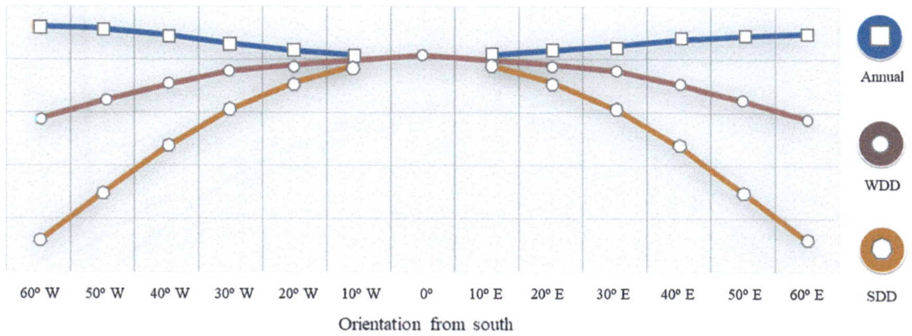

Fig. 5.2 Effect of the angle of orientation on annual electricity generation ratio to south-facing orientation

The optimal tilt angle forming a compromise between summer and winter months ranges within latitude angle ±15°. Figure 5.1 presents the average monthly electricity production of various tilt angles.

Effect of Orientation Angle

Orientation is the angle between the south and the horizontal projection of the normal to the surface. Orientation affects the solar potential in two ways: the amount of generation and the time of peak generation [3, 4].

Annually, the highest energy yield is associated with a south (equatorial) facing system. Deviation of the orientation of the system from the south by up to 30° west or east leads to an approximate reduction of up to 5% of annual electricity generation, while rotation by 60°, west or east of south, results in a reduction of some 12%. Figure 5.2 presents the effect of the orientation on electricity generation, for selected winter and summer days and the total annual generation, associated with a 45° tilt angle BIPV system.

The average monthly electricity generation indicates that in the summer months, orientation of the BIPV system toward west (°W) or east (°E) results in electricity generation that is equal to, or slightly higher than, the south orientation. Figure 5.3 presents the effect of orientation on monthly electricity generation [1].

The orientation of the BIPV system affects not only the value of the electricity generation, but also the time of peak generation. For a south-facing system, the peak generation is at solar noon. Rotation of the BIPV system results in shifting the peak electricity generation of this system to the afternoon for west rotation and to the morning for east rotation. A 30° orientation (east or west) enables a shift of the time of peak generation by up to 2 h, relative to solar noon. An orientation of 60°–70° enables a 3 h shift of peak. Roofs that combine both east and west orientations can lead to a spread of peak generation time, reaching six hours over the day.

In some cases, return on hourly energy produced during the course of a year may be a more important objective than the total energy produced, particularly in locations where prices of electricity vary with time of day. Optimizing return on electricity production involves consideration of orienting the BIPV systems to obtain peaks at

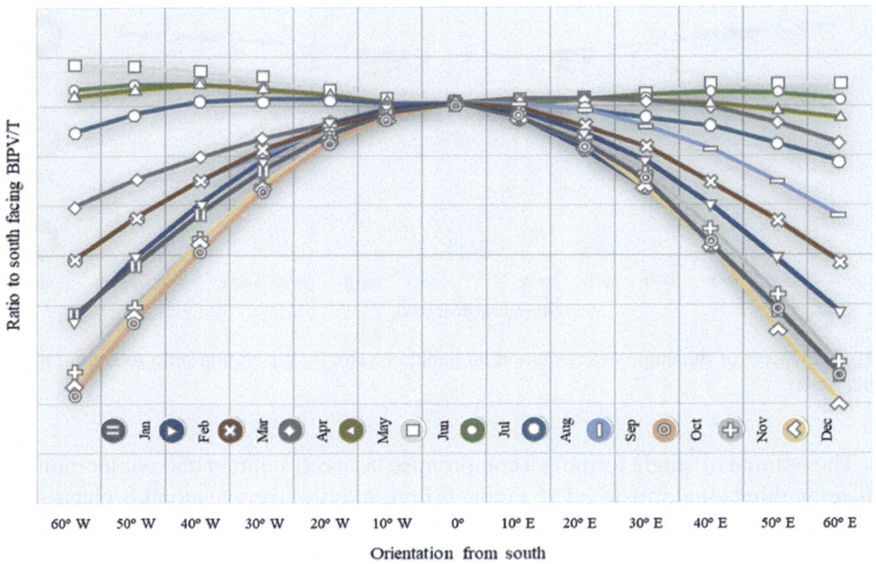

Fig. 5.3 Effect of the angle of orientation on the monthly electricity generation of the BIPV systems

time of high electricity demand, enabling thus larger annual income from selling the excess electricity to the grid and cost saving for consumption at high demand time.

Figure 5.4 presents electricity generation profiles for selected summer and winter days, at selected orientation angles, relative to south.

5.1.2 Design of Solar Optimized Roofs

This section presents some design options of roof shapes for a hypothetical rectangular residential building. The designs discussed below are conceptual, aiming at exploring the potential of different roof designs of enhancing electricity production by BIPV. Technical considerations relating to PV technologies are not addressed in this discussion. Since the purpose of the employed examples is to present the comparative performance of various design options, a fixed energy efficiency of PV systems is assumed.

The photovoltaic system is assumed to cover the total area of all south and near south-facing roof surfaces. In practice, some percentage of the roof area is used for various technical considerations such as the mounting structure, framing, and others [5, 6]. In addition, it is assumed that, due to the continuous development in PV technologies and associated components, future technologies may be capable of accommodating specific requirements raised by proposed roof designs, such as PV modules of varying shape, size, and color, structural requirements, as well as inverters for different BIPV orientations [4].

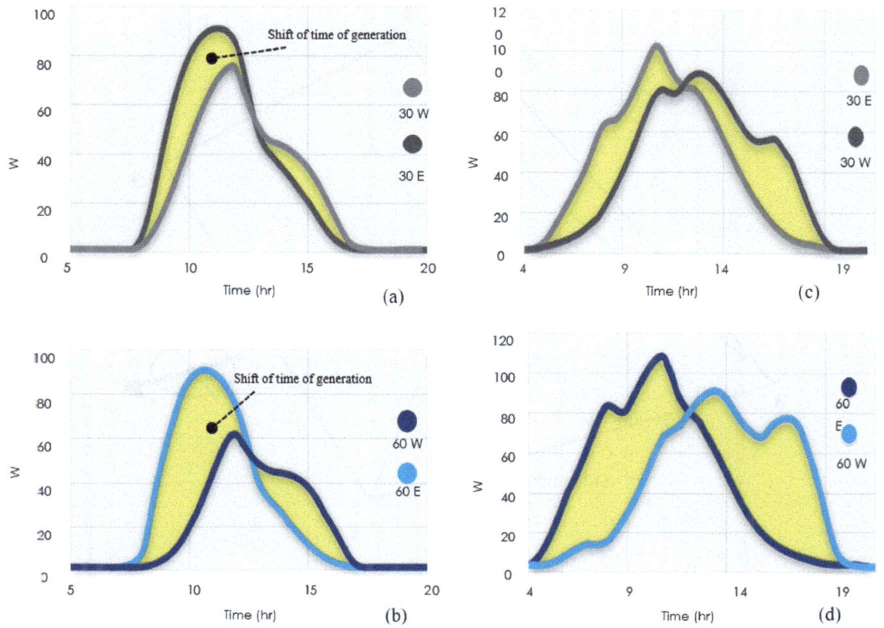

Fig. 5.4 Effect of the angle of orientation on the electricity generation (W per m^2 of PV), **a** 30° for the winter day, **b** 60° for the winter day, **c** 30° for the summer day, **d** 60° for the summer day

Three basic roof geometries are discussed. The first geometry is a commonly used hip roof with a fixed tilt angle (within the optimal range) and varying side angles (Fig. 5.5). The second and third types of roofs are relatively independent of the shape of the building, employing a multifaceted roof surface combining a range of tilt and orientation angles [1, 4].

Hip and Gable Roofs

This is the simplest roof design that permits a good integration of PV panels in the south-facing surface with the ridge running east–west (E–W) (Fig. 5.5). The ridge is assumed at the center for practical and aesthetic considerations, such as limiting the height of the roof.

Variations of hip roof can be obtained by changing the side angle, which is the angle between the hip plane and the horizontal plane, as shown in Fig. 5.5. This angle affects mainly the area of south (and north) facing surfaces. To highlight this impact, three values of the side angles—45°, 60° and 90° are discussed. A side angle of 90° leads to a gable roof. The tilt angle is kept constant at 45°.

In addition to the effects of tilt and orientation angles discussed above, the main design consideration is the roof surface area available for PV integration. In general, electricity generation is proportional to the roof surface area available for PV integration. A hip roof surface area can be manipulated through various design strategies such as:

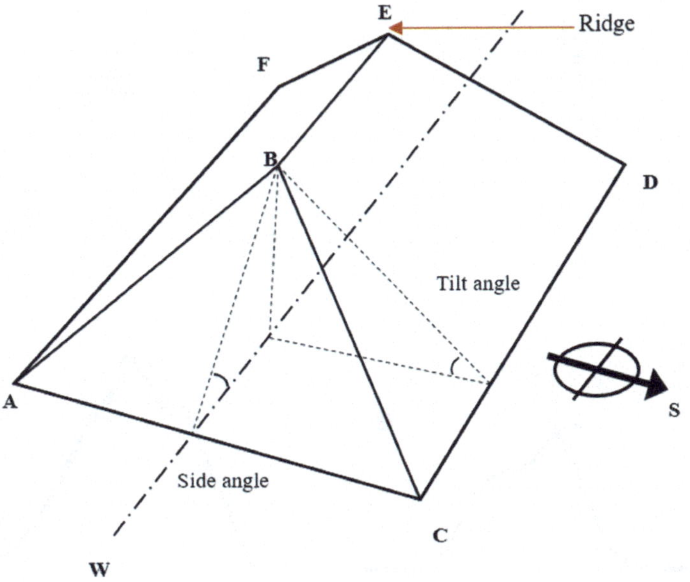

Fig. 5.5 Illustration of a hip roof and associated tilt and side angles

- **Combination of side and tilt angles**. For instance, a rectangular layout with side angle of 45° and tilt angle of 30° possesses a larger south-facing surface area than with a tilt angle of 45°. Consequently, the yearly electricity generation is larger with the 30° tilt, even though a tilt angle of 45° allows higher, yearly average radiation per unit area [1, 4].
- **Optimizing side angle**. A larger side angle allows a larger south-facing roof area. For instance, the optimal roof is characterized by a side angle of 90°, corresponding to a gable roof. The electricity generation of a hip roof with a 45° side angle is 40% lower than for the corresponding gable roof (for the studied example).
- **Impact on building performance**. The effect of different designs of hip roof side angle on heating and cooling loads of the corresponding 2-storey house is not significant. For instance, the heating load required for the housing unit with gable roof is approximately 5% larger than for the unit with hip roof of 45° side angle.

Advanced Roof Design

This section presents roof systems that combine surfaces of various tilt and orientation angles. Although the plan of the proposed and analyzed roofs is not rectangular, a rectangular building layout is assumed, which results in roof overhangs over the facades of the building. Change of building layout together with the roof systems is discussed in Sect. 5.3. To ensure that the presence of overhang is not impacting the generation potential for these advanced roofs, a similar depth of overhangs is assumed for the rectangular hip roof, serving as a reference for the study.

The main goal of these roofs is to offer design options that increase the solar potential of the BIPV systems. Orientation and tilt angles employed in these roof designs that are optimal for the summer and winter months are selected and combined in such roof systems [1, 4].

Split-Surface Roof

The south-facing roof surface can be divided into a number of surfaces with different tilt and orientation angles, to benefit from seasonal and daily variations of solar radiation [7, 8]. In the example presented below (Fig. 5.6), the south-facing portion of the roof is divided into three plates of differing orientations and tilt angles. A BIPV system is assumed to cover the total area of each of these plates.

Two variants are considered, as well as some variations of these options. The mid plate with 45^0 tilt angle is south oriented, while the side plates are rotated by equal angles toward the east and toward the west. In the first option, the orientation angle

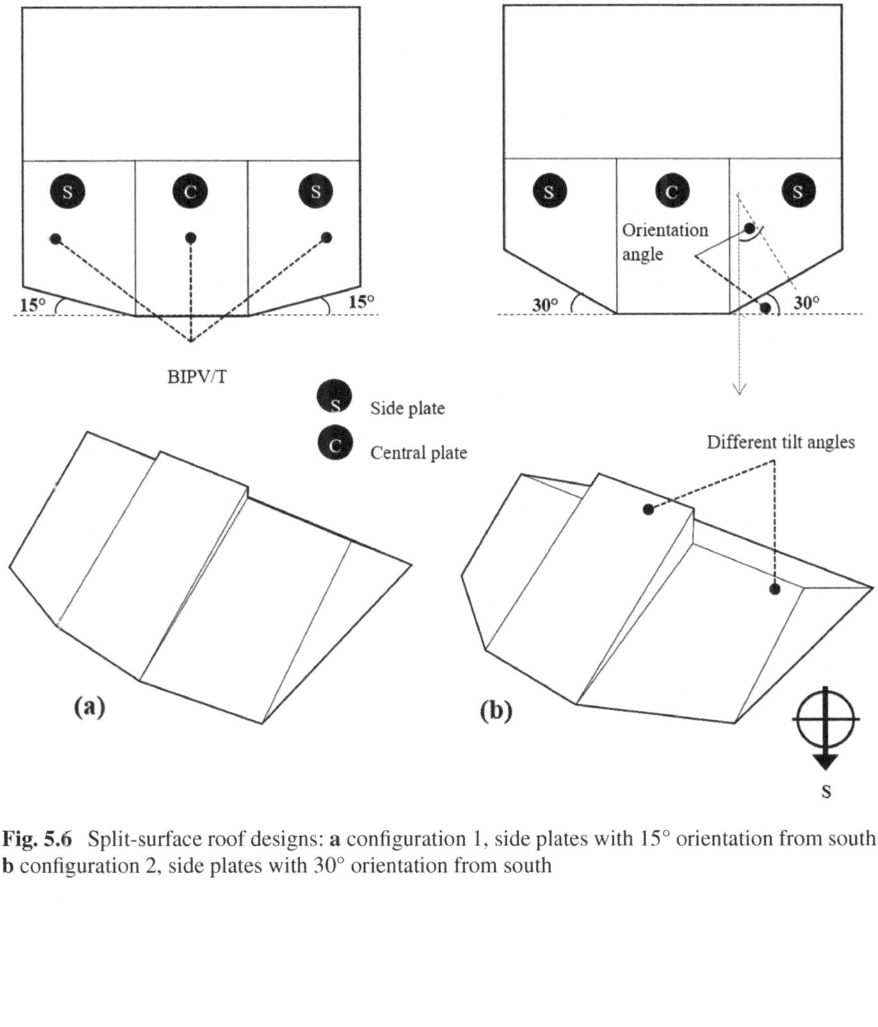

Fig. 5.6 Split-surface roof designs: **a** configuration 1, side plates with 15° orientation from south; **b** configuration 2, side plates with 30° orientation from south

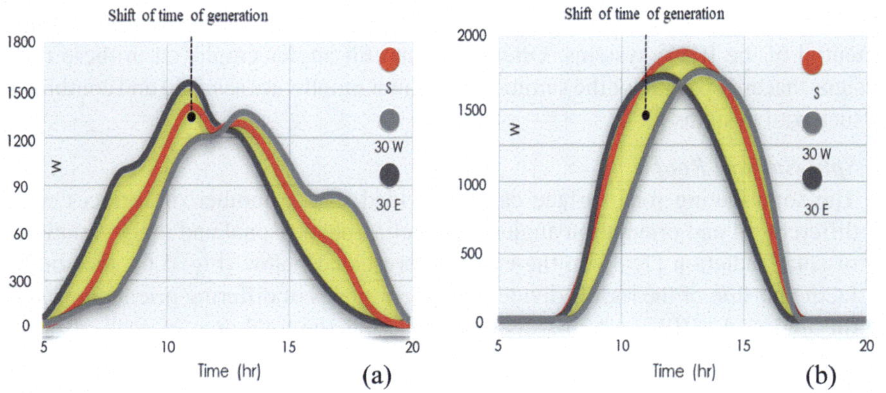

Fig. 5.7 Electricity generation on design days for the plates of the 30° (E, W) split-surface roof option, **a** Summer day, **b** Winter

of the side plates toward east and west is 15°, while in the second option this angle is 30°. The two configurations of the split-surface roofs are presented in Fig. 5.6. The main characteristic of a split-surface roof design is that it facilitates combinations of orientation and tilt angles, which allows obtaining spread of peak electricity generation time of up to 3 h (Fig. 5.7).

Folded Plates

Another example to maximize solar energy capture and available surfaces for PV integration is provided below. This example relies on folded plates roof design, which refers in this particular case to the geometrical shape of the roof, not necessarily to the structural system.

The folded plate roof geometry is composed of triangular plates with various orientations. Two basic shapes are presented in this section. The first configuration is composed of four plates, with the two side plates facing south (Fig. 5.8). The second basic shape consists of three plates with the central plate facing south (Fig. 5.9b) [1, 4]. All configurations are designed with similar geometry on the north-facing slope, for architectural visual effect, but with no BIPV.

More complex designs can be derived by joining together two or more units of the basic shape. Figure 5.9 presents configurations of folded plates that join two basic units. Configuration 2 (Fig. 5.9a) is composed of two basic four-plate units, with central plates rotated 30° east and west. Configuration 3 (Fig. 5.9b) is composed of two three-plate basic shapes with side plates rotated 30° east and west.

Folded plate roof design enables obtaining various orientations for the same rectangular building roof. Furthermore, this roof shape has significantly higher south-facing surface area than the gable roof. The difference in the BIPV potential of the different configurations analyzed is not significant (maximum difference of 4%).

The roof options with 15° orientation allow a spread of time of peak generation of approximately 2 h while the 30° enables 3 h difference. The electricity generation

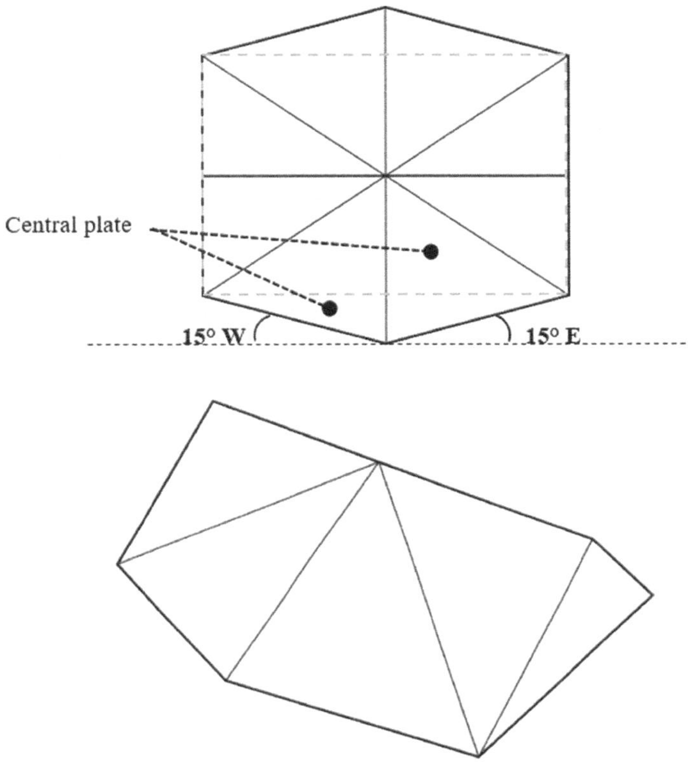

Fig. 5.8 Illustration of the four plates basic unit (Configuration 1)

of the six-plate folded roof (Fig. 5.9b) exceeds the generation of the gable roof by approximately 30%.

Overall Design Summary

Optimization of roof design for solar capture and for the integration of solar technologies requires optimal choice of orientations and tilt angles of roof surfaces. Both orientation and tilt angle of the BIPV system affect its overall energy generation. Deviation of the surface azimuth angle of the BIPV system from due south by up to 40° west or east does not lead to a significant reduction in generation. The orientation of a roof affects the time of peak generation. This can be of particular advantage in cases where the value of electricity varies with the time of day. A tilt angle that approximates the latitude of a specific location is optimal for PV systems, as demonstrated in this study (see Sect. 5.1.1).

Multifaceted roofs, such as folded plate and split-roof configurations, can significantly increase electricity production, primarily through increased effective surface area. For instance, dividing the roof surface into three plates with varying tilt/orientation angles can increase electricity generation (by up to 17% as compared to a gable roof, as in the presented example). Replacing the gable roof with a

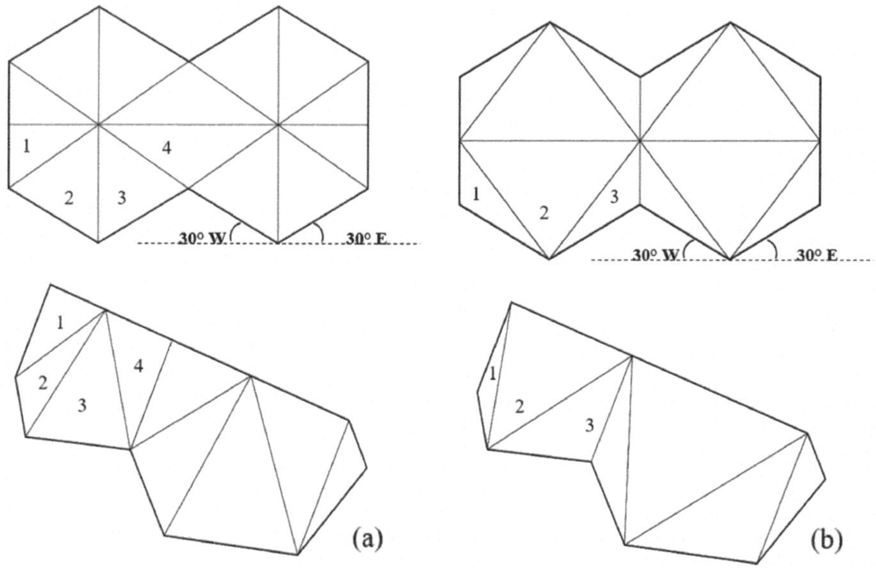

Fig. 5.9 Folded plate roof designs, **a** Configuration 2–two basic 4-plate units with 30° orientation, **b** Configuration 3–two basic 3-plate roof with 30° orientation

folded plate surface increases the electricity generation potential by up to 30% in the example. Varying surface orientations in such roof designs enable the spread of peak electricity generation over up to 3 h, potentially reducing the impact on the electric grid.

Although the presented examples relate to a northern mid-latitude climatic zone, the methodology implemented in the study can be applied to different locations and climates, with few modifications. Such design modifications include varying the range of the tilt angles of various roof configurations. Other considerations should include reducing snow accumulation in northern cold climate, which favors increased tilt angles.

Complex roof systems, such as folded plates, are associated with increased cost due to larger BIPV system and increased manufacturing complexity. However, the cost rise can be offset by increased potential of BIPV system which results in increased electricity generation coupled to diversification in the timing of the PV electricity generation. Roof systems such as folded plates can be designed as structural systems, providing extra functional space, while offsetting the cost associated with additional structural elements [1, 4]. The additional structural elements (e.g., trusses) are not used for the installation of the PV system, but to support the roof itself (e.g., in the conventional gable roof).

The examples presented above demonstrate an approach to increasing solar potential of buildings, while maintaining maximum design flexibility that can accommodate functional and other considerations not related to energy efficiency.

5.1.3 Roof Applied BIPV/T System

Hybrid building-integrated PV/thermal systems (BIPV/T) have considerable potential to supply space heating and domestic hot water, in addition to generate electricity, as discussed in Chap. 4. The potential of heat generation of such roof-integrated system is demonstrated in the example below.

An open loop air-based BIPV/T system is assumed to be implemented in a residential building (single-family house) located in northern mid-latitude climate, similar to the examples presented above. Correlation between thermal and electrical efficiency of the BIPV/T system is derived based on a simple numerical model [1, 4]. This model employs a transient quasi-two-dimensional finite difference method to determine the thermal energy generation potential of the BIPV/T system [9, 10] and to establish a relationship between electricity and useful heat generation. Environmental parameters, including outside temperature, solar radiation, and wind speed and sky temperature, are provided by the weather data files of EnergyPlus (the simulation program employed to assess the energy performance of the presented examples).

The air circulated behind the PV panels increases the PV electrical efficiency by cooling the PV panels, while recovering heat that can be used for space and/or water heating [11]. The useful fraction of thermal energy generation depends on the temperature of the solar-heated air and on the end use. For instance, a BIPV/T system coupled with a heat pump enables exploiting lower temperature outlet air to offset heating load or for direct space heating, while higher outlet air temperature can be employed to heat domestic hot water or sent to storage [12].

Case Study

A gable roof of a rectangular housing unit, with a tilt angle of 45°, is employed to develop the numerical model, employed in this case study. The BIPV/T system, illustrated in Fig. 5.10, is divided into five control volumes along the direction of the ridge [10]. The PV panels are assumed to have negligible thermal resistance and thermal capacity [13].

The bottom surface of the air cavity is assumed to be well insulated. The model is applied to roofs with different tilt angles, ranging from 30° to 60°, at 5° increments. An electrical efficiency of 12.5% is assumed for the BIPV system. The model is applied to design days—representing a cold and sunny winter day and hot and sunny summer day, as well as to selected sunny days representing each month of the year.

Based on the numerical simulations of the model described above, a ratio (Q_u/Q_e) is established to determine the correlation between the heat and electricity generation of the BIPV/T systems. Q_u/Q_e for selected sunny days, of each month of the year for a 45° tilt angle, is presented in Fig. 5.11a. Figure 5.11b presents the results of the simulations of the BIPV/T systems with various tilt angles, for the winter design day (WDD).

The ratio of solar thermal production to the electricity production (Q_u/Q_e) varies between 3 and 3.5 (mean value of 3.1 and standard deviation of 0.2). A value of $Q_u = 3Q_e$ is adopted in this case study for an air speed of 2 m/s in the BIPV/T system, selected to ensure high efficiency of the BIPV/T system.

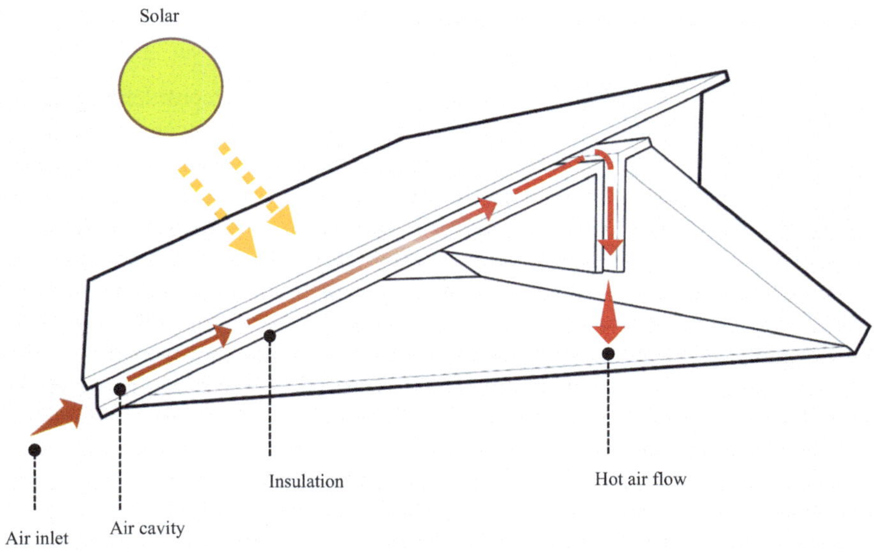

Fig. 5.10 Illustration an open loop BIPV/T system

The main goal of this approach is to explore the potential of the BIPV/T system, allowing thus to determine approximately the amount of thermal and electrical energy that can be expected from a hybrid PV/T system. In practice, for design purpose, detailed models involving yearly simulations would be advisable to regulate outlet temperature according to desired applications and therefore to maximize useful heat produced by the BIPV/T system.

Variable speed fan enables controlling the outlet air temperature, in order to be used for different applications. Figure 5.12 presents the relation between air velocity and the average air temperature change (ΔT- the difference between inlet and outlet air temperature of the BIPV/T system), over the winter design day. This relation is associated with the gable roof of the selected case study.

The outlet air temperature decreases exponentially with increasing flow rate. Flow rate and outlet air temperature should be determined to fit the intended application. For instance, for heat pump lower outlet air temperature can be used, while for space heating and hot water higher temperature is needed.

5.2 Advanced Facades for Multistorey Buildings

Multistorey residential buildings enable achieving high urban density. When properly designed, such buildings can offer numerous benefits, including high-energy

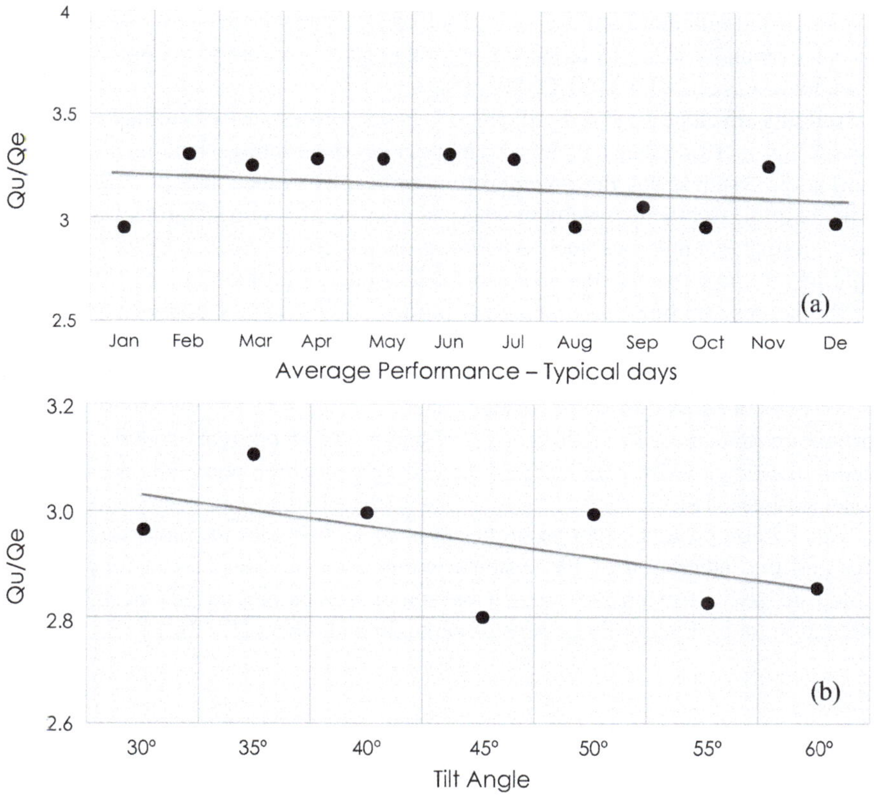

Fig. 5.11 **a** Qu/Qe for 45° tilt angle roof for one sunny day of each month, over a year; **b** Qu/Qe for WDD of roofs with different tilt angles

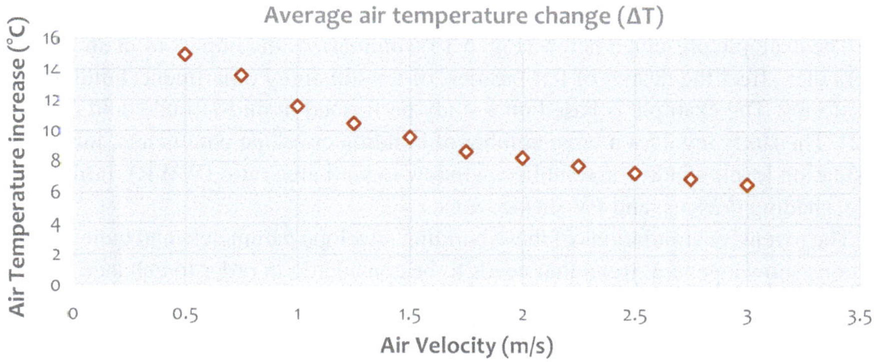

Fig. 5.12 Relation between air velocity and the average air temperature change in the cavity (ΔT), on a WDD

efficiency and efficient land use, and promote efficient infrastructure. These benefits can eventually lead to a reduced per capita energy consumption, and associated greenhouse gas emissions (GHG) [14, 15].

Building envelope plays a significant role in the performance of multistorey buildings. A high-performance facade, which integrates daylighting, shading, and natural ventilation systems, has the potential to significantly reduce energy consumption of the building, while increasing the thermal comfort of occupants. Recent developments in solar technologies, such as building-integrated photovoltaic systems, enable building envelopes to play the additional role of energy generators [16, 17].

Multistorey buildings have inherently limited roof area per overall occupied floor area. This implies a reduced potential for solar electric and thermal generation from roof surfaces, relative to energy demand [18]. Added to this limitation is the fact that PV systems integrated in south facades generate about 40% less electricity per unit surface area than those in south-facing roof surfaces (with optimal tilt angle). Manipulating facade geometry can increase the energy generation potential, as compared to flat surfaces [18].

This section presents some examples of advanced facades for multistorey buildings and the integration of PV systems within these facades. The examples presented are based on extensive research and simulations employing advanced building simulation tools (EnergyPlus, openstudio plugin and others [19–21]).

5.2.1 Flat Facades

Flat facades are the most common design in multistorey buildings. It often presents a large amount of glazing to achieve some aesthetic and architectural effects, resulting, when low efficiency glazing system is employed, in poor energy performance of the building.

The example presented below (Fig. 5.13) summarizes the impact of main parameters that affect the design of flat facades, of a multistorey (apartment) building, of 12 stories. The example is based on a study performed in mid-latitude cold climate [22]. The study involves a large number of building envelope parameters, including insulation level, window assemblies, window to wall area ratio (WWR), infiltration rate, shading devices, and PV surface area.

The extensive simulations of these building envelope parameters and their combinations show a general trend that needs to be considered in order to enhance energy performance.

Areas available for PV integration depend on the amount of glazing implemented in the facade. Assuming that the opaque area of the facade incorporates BIPV systems, this area is increased with the decrease of WWR, and therefore electricity generation potential is increased.

The parameters that allow optimizing net energy consumption include wall insulation, airtightness (in terms of Air Change per Hour (ACH)), shading devices (overhangs, fins—Fig. 5.14), glazing—dimensions (WWR), conductance (U-value) and

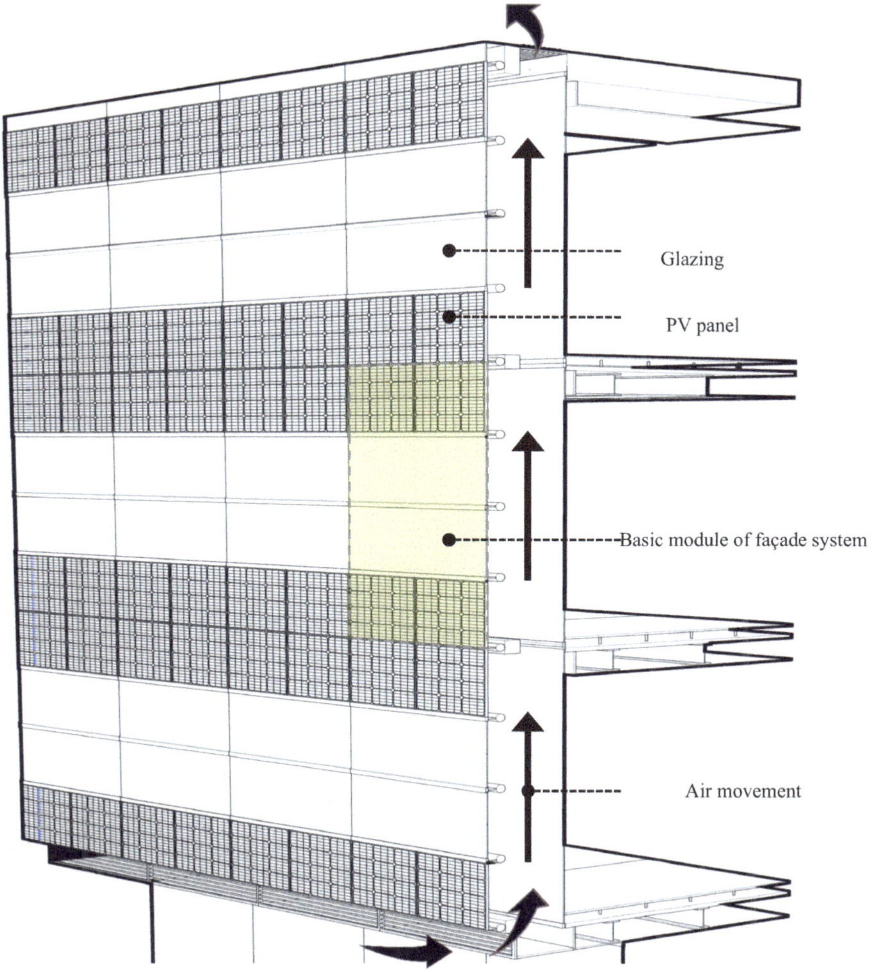

Fig. 5.13 Illustration of an advanced double skin facade, based on modular design. Each module includes two PV parts and a vision section

solar heat gain coefficient (SHGC). Some flexibility in the values of these parameters can be achieved, where negative effects in some parameters can be compensated for by other parameters [17, 22].

Considering that the general tendency in current design of multistorey buildings is in favor of highly glazed buildings, it is important to understand that an increase in glazing results in significant reduction in potential generation of electricity from PV integrated in the opaque area of the facade, while it increases the energy required for heating and cooling the building, due to reduced insulation of the glazed area (as compared to the opaque area). The effect on thermal loads is strongly dependent on

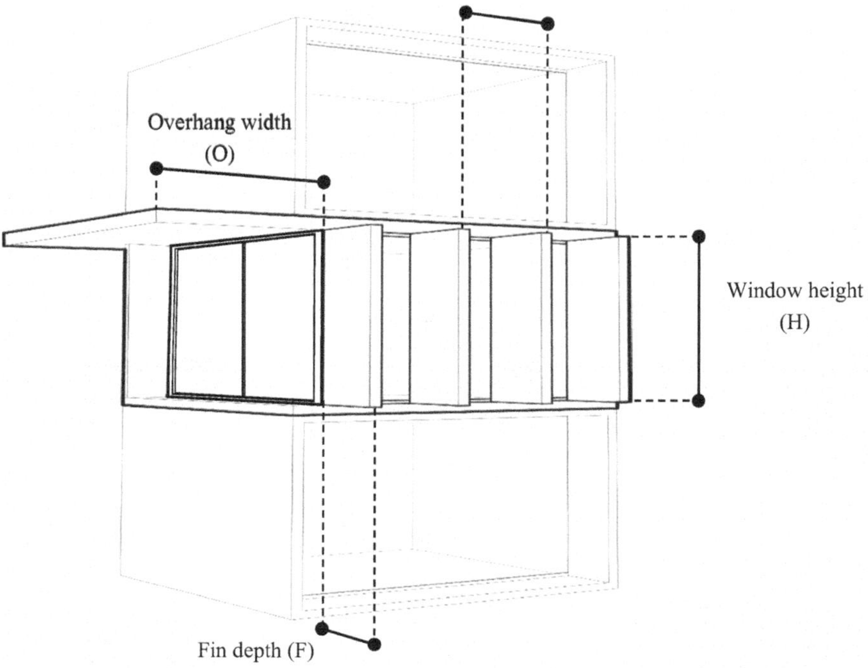

Fig. 5.14 Illustration of overhangs and fins. Overhang is mostly applied to the south facade, while fin is applied to east and west facades

the apartment units' orientations. The combined heating and cooling loads for south-facing perimeter zone are not highly sensitive to window area, due to the balance between cooling reduction and heating increase. Perimeter zone is defined as the building interior space adjacent to the facade, with a maximum depth of 5 m, from the outside boundary.

The graphs presented in Fig. 5.15 illustrate the impact of change in WWR on the thermal load as well as potential PV generation, of 2 apartments of an apartment building of 12 floors. These two apartments, a south-facing apartment and a corner apartment (situated on the southwest corner of the building), are assumed to be located on the 6th floor (of 12-floors building), to reduce the impact of upper and lower floors on the heating and cooling load, since the heat transfer between floors in mid-sections is negligible [18]. The S apartment shows an increase of PV electricity generation of about 30% and increase of heating load of about 24% when reducing the window area from 80 to 20% of the south wall. The SW corner apartment, featuring the same WWR on the south and the west facades, shows that the combined thermal load is reduced by up to 35% when the WWR is reduced from 80 to 20%. This is associated with an increase of the PV generation of about 25%.

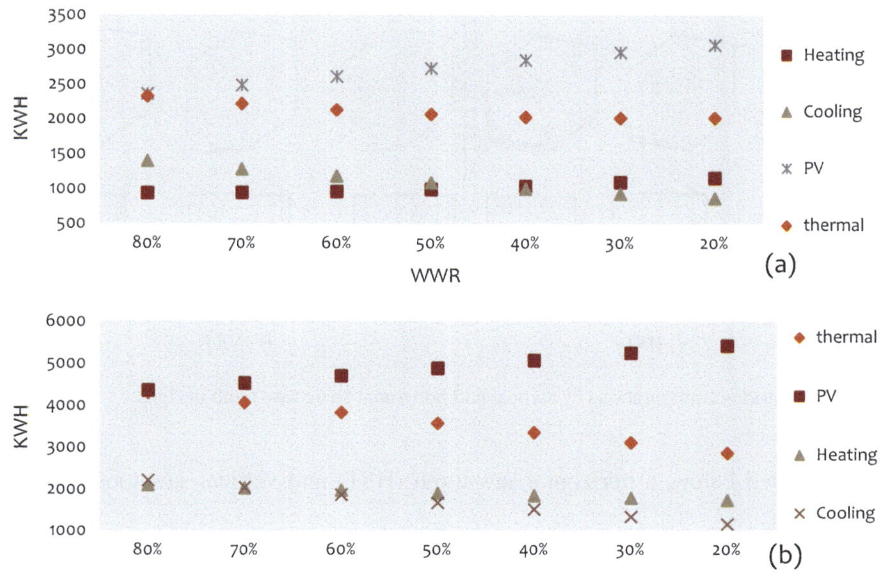

Fig. 5.15 Impact of WWR on heating, cooling, combined thermal load and PV generation for **a** S apartment, **b** SW corner apartment

5.2.2 *Multifaceted Facade Design*

This section presents examples of manipulation of the geometry of the facades of multistorey buildings in order to increase their potential of capture and utilization of solar energy.

Altering facade geometry provides the opportunity to enhance solar potential for south, west, and east facades. Optimizing facades for solar capture involve including multiple folded planar surfaces with varying tilt and/or orientation angles. Folded plate geometries can take a variety of patterns, such as saw-tooth or polyhedral facets. Below are examples of such geometries, followed by energy performance in northern mid- to high-latitude climate of selected designs, based on extensive simulations.

The developed facade patterns are applied to a double skin *box window* category, where the two skins are continuous over the building face and divided both horizontally at each floor plate and vertically between specific facade modules (see Chap. 4). The folded plate modules constitute the outer skin of these DSF modules.

While the examples relate to an apartment building, similar designs can be applied to other types of multistorey buildings, including office buildings.

Saw-tooth

The saw-tooth configuration is the simplest folded plate geometry. It consists of units with a single fold, the plate on one side of which consists of glazing, and the plate on the other side is designed to integrate PV modules. Two main geometries

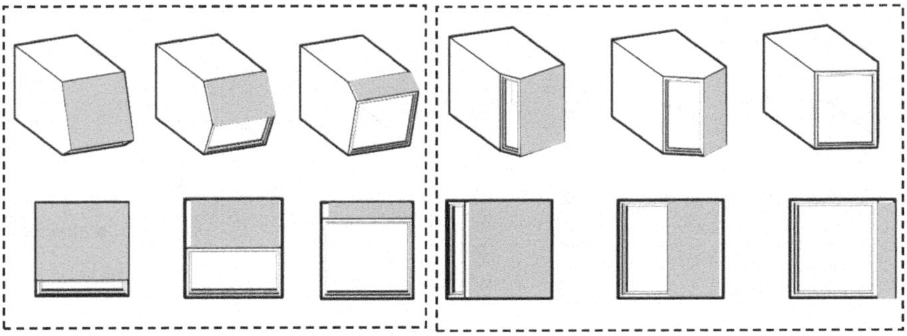

Fig. 5.16 Various configurations of vertical and horizontal basic saw-tooth designs

are presented below, a horizontal saw-tooth (HST) and vertical saw-tooth (VST) (Fig. 5.16).

These basic geometries can be designed in various patterns as illustrated in Fig. 5.17. The four configurations of the basic HST designs (Fig. 5.17) have an inclined opaque surface and an inclined window (with a single fold per storey), The difference between the configurations is the location of the window relative to floor level bottom (HST1, 2) or top (HST 3, 4)–Fig. 5.17.

The VST geometries have vertical fold, and therefore the main variable that affects the design is the orientation angle (rather than the tilt angle of HST). Two main variations are studied–VST 1 and VST2 (see Fig. 5.17). While in VST1 the glazed area is installed at specific orientation, inclined to the facade surface, the glazed area in VST2 is parallel to the facade. The PV modules can be integrated in the west-oriented plates or in east-oriented plates or can be alternated between east and west orientation. The plate orientation–east or west–does not affect significantly the total energy production, but it affects the timing of peak production.

A modification of the HST with split orientation (MHST) is presented in Fig. 5.18.

Pyramidal Units

More complex folded plate units are based on pyramids. A large number of facade patterns can be obtained by manipulating the base geometry as well as the dimensions of the folded units. Triangular pyramid unit (TP), rectangular pyramid unit (RP), and hexagonal pyramid unit (HP) are shown in Fig. 5.19a, b, d, respectively. Variations of selected geometries are also designed to study the effect of truncation on the performance (Fig. 5.19c, e).

Glazing, in all variations, is assumed on the down facing plates (see Fig. 5.19). In the truncated design, glazing is integrated, in addition to the downward-oriented plates, on the truncated faces.

More complex configurations are obtained by dividing the facade module into multiple folded plate units–Fig. 5.19d, e.

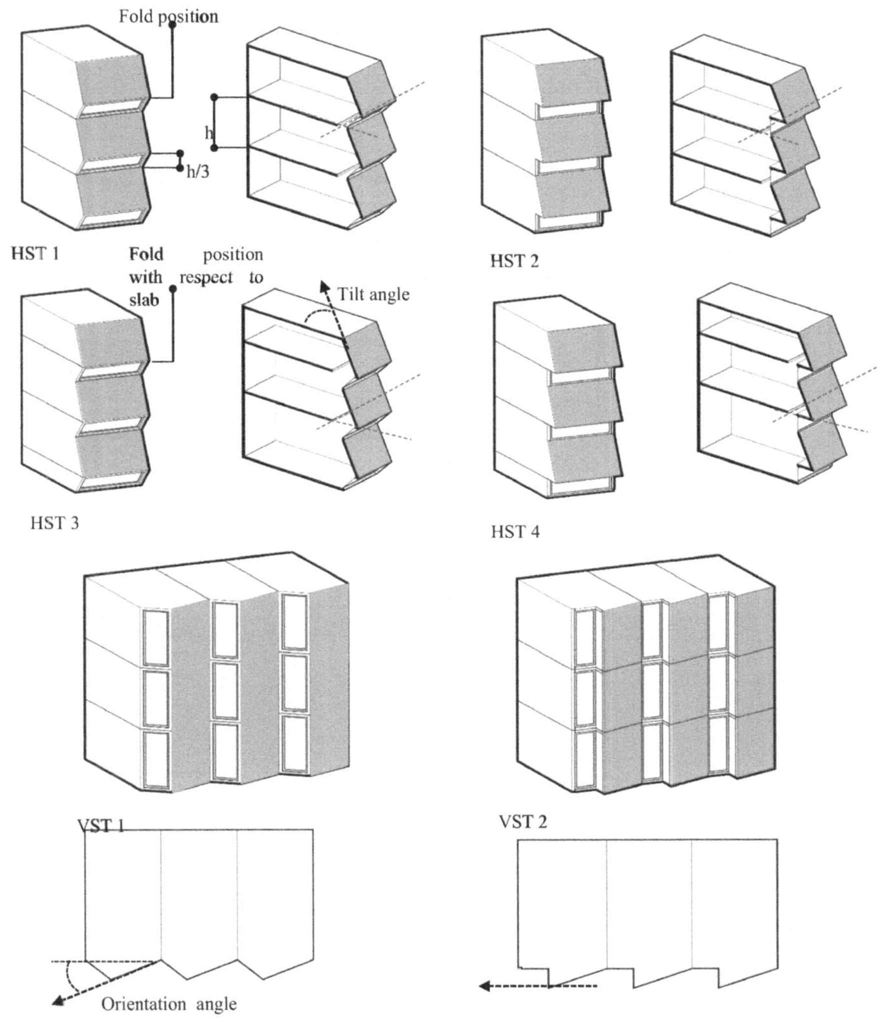

Fig. 5.17 Variations of saw-tooth facade designs

5.2.3 Performance of Multifaceted Facades

This section present results of a sample analysis of the impact of facade design. This analysis, which forms a basis for some general observations, is carried out for selected facade patterns, applied within a double skin facade system (see above). All geometries are studied for the south-facing apartment and for the two corner apartments, in an intermediate storey of a sample apartment building (Fig. 5.20). In the corner apartments, the studied geometry would replace the two flat facade design

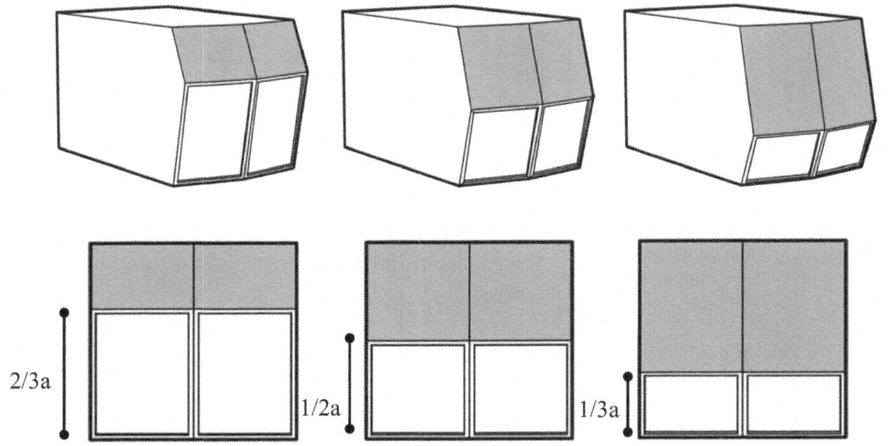

Fig. 5.18 Modifications of HST

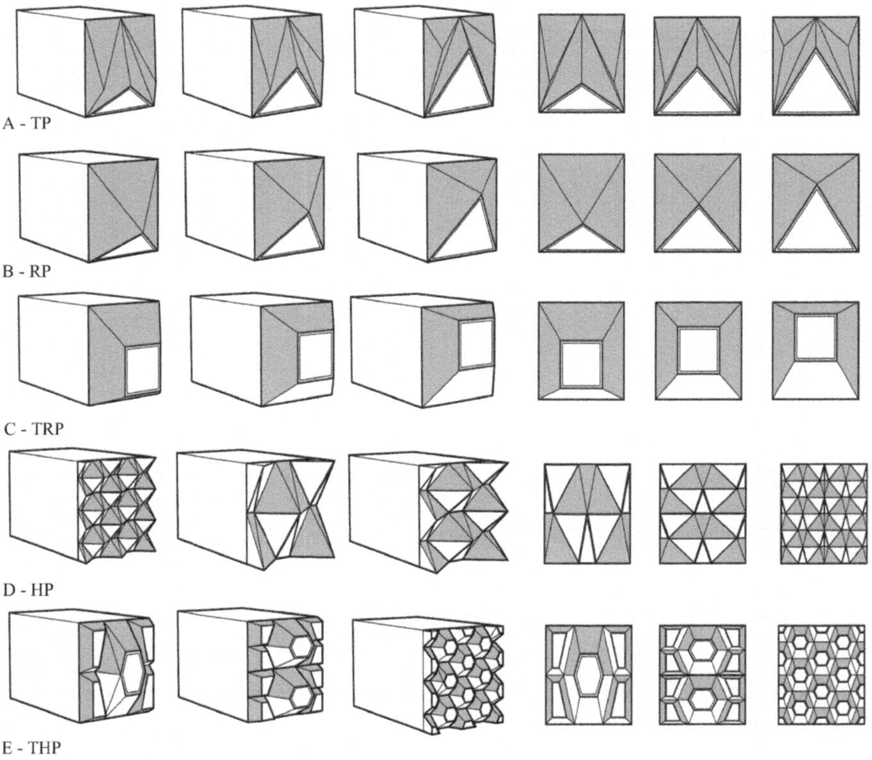

Fig. 5.19 Pyramid-based folded plates

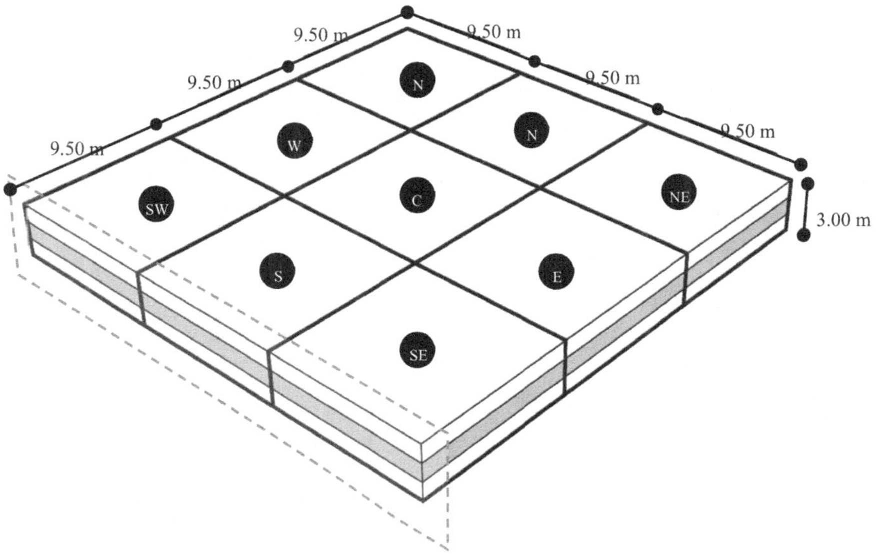

Fig. 5.20 Illustration showing the studied apartments

for both the south and east/west walls. The two corner apartments (SW and SE) have very similar performance.

The performance of folded plate geometry is governed by two main parameters: the position of the ridge fold, or apex, of the folded unit, relative to its edges, and the unit's depth–namely, the maximum cavity depth. The ridge/apex position and the unit's depth (depth hereafter) determine the angles of the plates and the size of the window, which can occupy any of the pyramid's faces. In horizontal saw-tooth units further parameters determine the position of the window (relative to floor) and its size–Fig. 5.17.

The performance of selected pyramidal and saw-tooth patterns, as compared to flat facade, is presented below, as an example to highlight various effects. In addition, discussion of the impact of various design parameters on the performance is briefly presented. A WWR of 30% is assumed for all folded plate designs, as well as for the flat facade, which serves as reference.

Overall Performance

Solar Radiation

Figure 5.21 shows the average solar radiation (W/m^2) incident on various facade systems, on four days of the year, representing the four seasons. The facade systems selected are the flat facade, the VST, HST, and the rectangular pyramid folded plates. The figures depict the south and west facades of a SW corner apartment. Rectangular pyramid performs better than the other two configurations in capturing solar radiation during various times of the year, for both west and

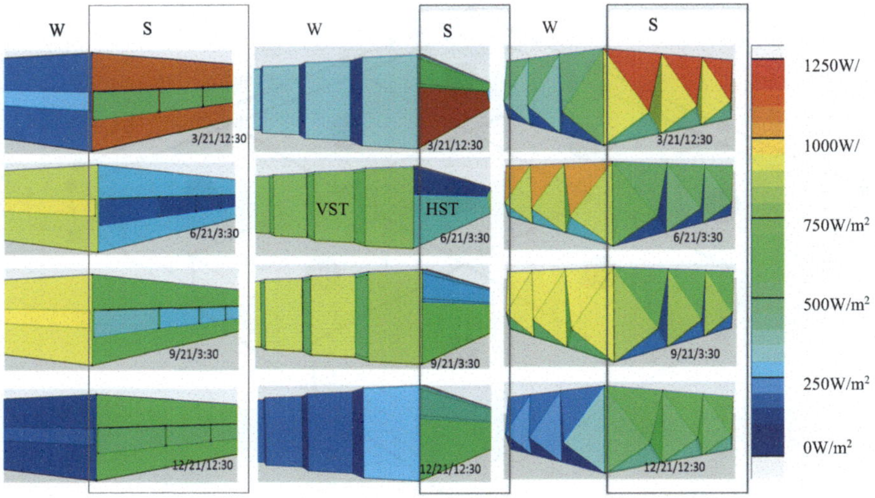

Fig. 5.21 Solar radiation potential in W/m^2 of 3 facades patterns, during 4 days of the year. The south facade (S) is in the right of each figure, and the west facade (W) is on the left [18]

south orientations. HST on the south facade has a slightly better solar potential than similar performance of the flat facade, while the VST on west performs significantly better than the corresponding flat facade, during the spring and winter days.

Energy Performance

The energy performance in terms of heating, cooling, and potential electricity generation of a whole floor of the apartment building, featuring 8 apartments (see Fig. 5.20), is presented in Fig. 5.22 for the selected facades–flat, saw-tooth, triangular pyramids (TP), and rectangular pyramids (RP). In addition, the ratio of potential energy generation to total energy consumption (ratio of performance–ROP) is included in the graphs. Total energy consumption includes heating and

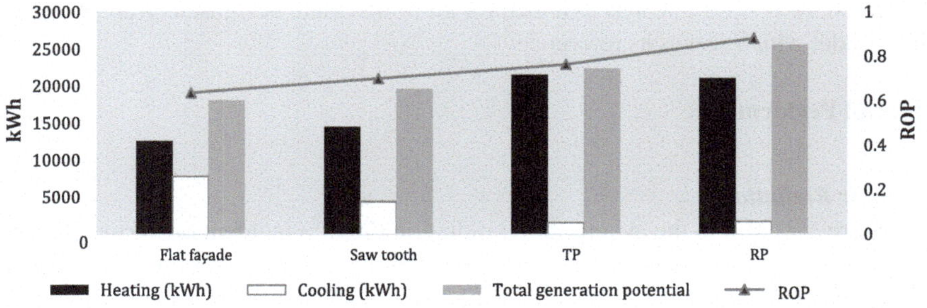

Fig. 5.22 Energy performance of a whole floor (8 apartments), featuring the various facade designs

cooling energy, energy used by appliances, lighting, and domestic hot water (DHW).

The results indicate that heating load is increased for multifaceted facade, while cooling load is reduced. Simultaneously, energy generation potential is increased for the multifaceted facades. The overall energy performance ratio indicates that while flat facade (FF) scenario can supply about 63% of the total energy consumption from PV integrated in the south, east, and west facades, the RP scenario can supply about 88% of the energy consumption, especially when energy-efficient mechanical system such as heat pump is adopted.

Impact of Folded Facade Parameters

The effects of the governing parameters–the position of the ridge/apex in the folded plate unit and the tilt angle/depth (maximum cavity depth)–are briefly discussed below for the south-facing apartment unit of the studied apartment building. The analysis presented below applies to HST units with window facing down (HST1, HST3) in Fig. 5.15 and modified HST, as well as to VST configurations, displaying similar trend of results.

Effect of Fold/Apex Position

The fold position refers to the distance (in plan view) of the fold or apex from the base of the glazing. Three different positions are analyzed in this example: low position (narrow position for VST), where the fold occurs at the third of the overall module height (Width for VST) (3 m); in mid-position the fold is at the mid-point of the height (width for VST); and in high-position, the fold is at a distance equal to 2/3 of the total height/width (see Fig. 5.18).

In all folded plate configurations, the impact of fold/apex positions on both heating and cooling loads follows the same trend: heating load is higher for lower fold positions, due to the reduced area of glazing, and of the resulting reduction in passive heat gain. Cooling load displays an increase with a higher fold position, due to increase in passive heat gain. These trends are illustrated in Fig. 5.23 for the rectangular pyramid configurations. The results are presented as function of the cavity depth/tilt angle (m/°), as discussed below.

PV generation exceeds total heating and cooling consumption for all configurations. PV generation potential is increased with a low position (narrow position for VST) of fold were the PV area constitutes more than 2/3 of the facade surface area (about ¾ in the case of the RP configurations). This increased generation is further accentuated by a larger tilt angle associated with larger cavity (see below).

Effect of Cavity Depth/Tilt Angle

Increasing the tilt angle/cavity depth leads to an increase in heating load, across all configurations, while cooling load is slightly reduced. Electricity generation is increased with increased tilt angles, as well as with the associated increased surface area for PV integration. Increase of the tilt angle/cavity depth should, however, be applied in moderation in order to avoid self-shading among facade modules. The study shows that shade becomes more significant with a larger

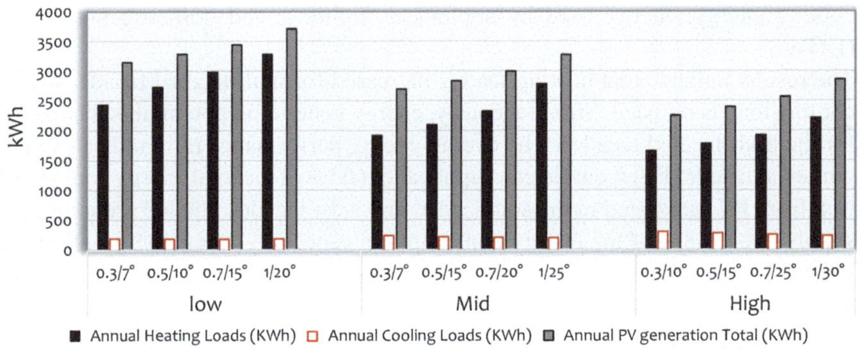

Fig. 5.23 Impact of apex position (low, Mid, or high) of RP units associated with different cavity depth (m)/tilt angle combinations on average loads and electricity generation per apartment (studied on the basis of the 8 apartments' floor)

cavity (>0.5 m). Figure 5.23 presents the impact of tilt and cavity depth on the performance of RP units, associated with different apex positions.

5.3 Impact of Building Layout

5.3.1 Low Rise, Small Buildings

This section summarizes the impact of small-scale buildings and their shape on solar energy capture and utilization. The design of roofs associated with these building layouts and their impact on the integration of PV systems, and electricity generation are discussed.

Optimization of Building Form for Solar Energy Capture

Rectangular layout with specific dimensions (see below) is generally considered the optimal shape for energy efficiency. In urban context, however, this shape may not be optimal (e.g., around curved roads). Non-rectangular and particularly self-shading shapes, like L-shapes, offer flexibility in architectural as well as solar design, but their energy efficiency is influenced by several design factors. To design energy-efficient shape, with an optimized solar potential a number of parameters should be considered at early stages of the design. These parameters are discussed below, for rectangular shapes and for some self-shading shapes, and illustrated in Fig. 5.24.

Rectangular Shapes

Aspect ratio (AR), which is the ratio of south (north in southern hemisphere) facing facade to the perpendicular facade, is a key parameter in the design of a rectangular shape for increased solar potential, on roof and facades. Heating demand decreases with larger aspect ratio while cooling demand increases. AR

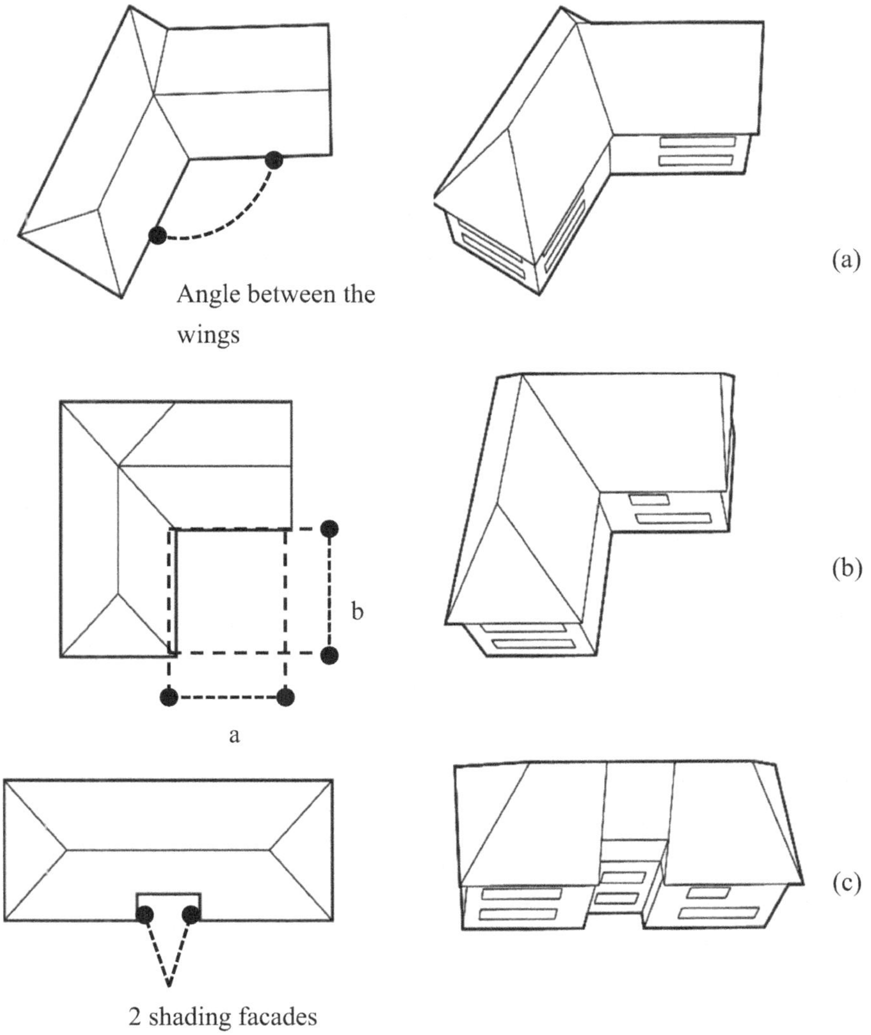

Fig. 5.24 **a** Illustration of shading facades, **b** Depth ratio, **c** L-variants with obtuse angle between shading facades

of 1.2– 1.3, in mid-latitudes offers a balance between heating and cooling loads. For AR less than 1.2 heating demand is significantly increased, due to reduction in south-facing surface areas. For a cold climate, a ratio between 1.3 and 1.7 can offer a reasonable choice, allowing to increase south-facing areas and thus passive heat gain, while maintaining the functionality of the plan.

Self-Shading Geometries

There are some situations where it is not feasible or advantageous to design simple rectangular building shapes, due to the site shape, proximity of various road layouts, or simply for aesthetic and functional considerations. In this case, the design should take into consideration the following aspects:

- **Shading depth ratio**: Solar radiation incident on non-convex shapes (relative to south) is significantly affected by the shading depth ratio (SDR)–the ratio of shading to shaded facade lengths (Fig. 5.24 b). Assuming a south-facing shaded facade, a ratio of ½ is suggested in cold climate design to reduce the effect of shading on heat gain by the windows of the shaded facade. The difference in heating load between L-shape with SDR of 1 and ½ can reach about 10%.
- **Number of shading facades**: Number of shading facades (Fig. 5.24c) has significant effect on heating load. It reduces heat gain due to shade from a number of facades as compared to a single one as in L-shape. It is suggested, therefore, especially in the design of houses, to reduce the number of shading facades on the south and near south orientation. A larger number of wings, or their length, can be accommodated when the angle between the wings is larger than 120°. Shading depth ratio (SDR) is particularly critical with a larger number of shading facades.
- **Shading angle**: By increasing the angle enclosed between shading and shaded facades, self-shading can be controlled and manipulated by variations of the basic geometry (Fig. 5.24a).
- **Building envelope**: The envelope in buildings of non-convex shapes has, generally, significantly larger surface area than in a rectangular shape of the same plan area. The envelope size is directly correlated with increase of heating load. Using optimal south-facing windows (as percentage of south facade) can assist in reducing this effect.

Roof Design and Building Layout

It is advisable to design roof geometry for potential installation of solar collectors. The basic roof design, consisting of hip and gable roofs, applied to rectangular building is discussed in Sect. 5.1 above. This roof design, and its variations obtained by changing tilt and side angles, can be applied to various building layouts. The following example illustrates the application of this roof design to a number of basic building layouts, such as square, trapezoid (with aspect ratio of 1.3 and base angle of 60°), L (shading depth ratio 1), T-, U-, and H-shapes (with SDR = 1/2). The tilt/side angle combinations presented in the example below are 45°/45°; 45°/60°; 30°/45° and 30°/60°. L- and T-shapes have gable ends for the main wings (where side angle is 90°).

Figure 5.25 presents an illustration of these roofs, corresponding to the same floor plan area.

The main design consideration for hip/gable roofs presented above (Sect. 5.1) applies for various plan layout. Different building layout implies different south-facing roof area and thus available surface area for PV integration. For instance, building layouts like a trapezoid, with the larger facade facing south, can maximize the area of south-facing roof, and thus potential PV electricity production.

Figure 5.26 presents the expected annual electricity generation of roof designs of the layouts of Fig. 5.25, with various tilt and side angles.

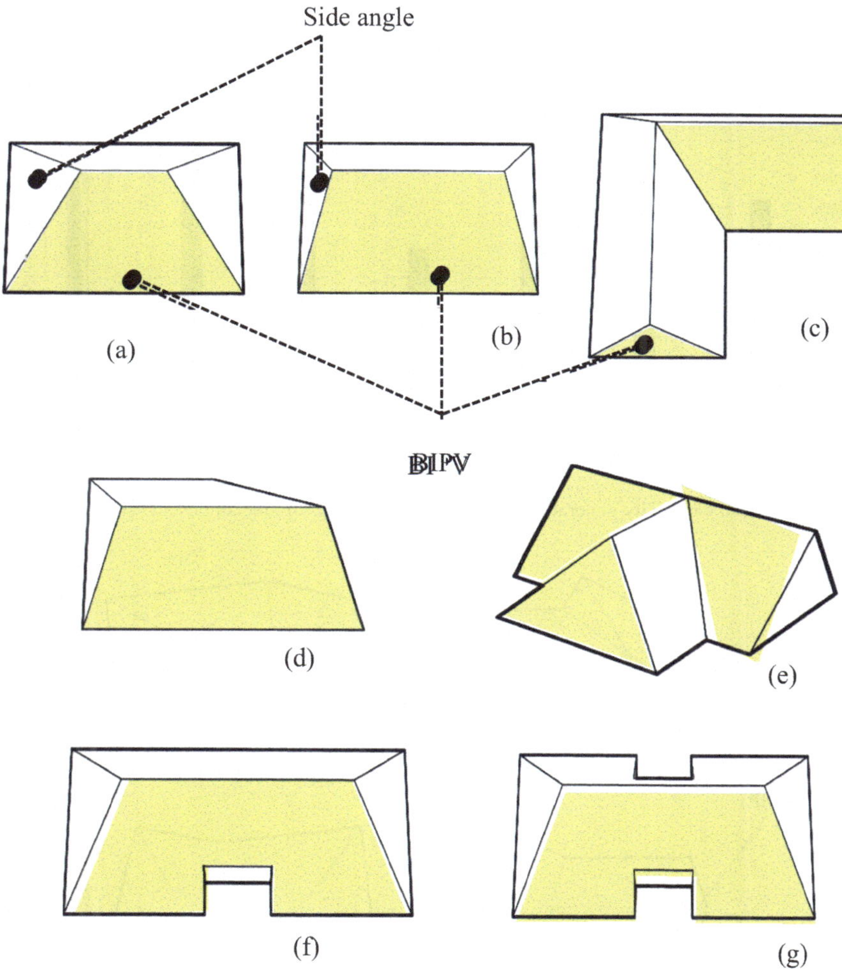

Fig. 5.25 Illustration of roofs of basic shapes, **a** rectangular shape with hip roof, **b** rectangular shape with gable roof, **c** L-shape roof, **d** trapezoid shape roof, **e** T-shape roof, **f** U-shape roof, **g** H-shape roof

South-facing roof area for a given floor area can be managed by manipulation of the aspect ratio and/or the shape design, as demonstrated in the examples of L variations (Fig. 5.27). In such example, rotation of one wing leads to enlargement of the south-facing roof surface, and consequently the area available for the integration of solar technologies.

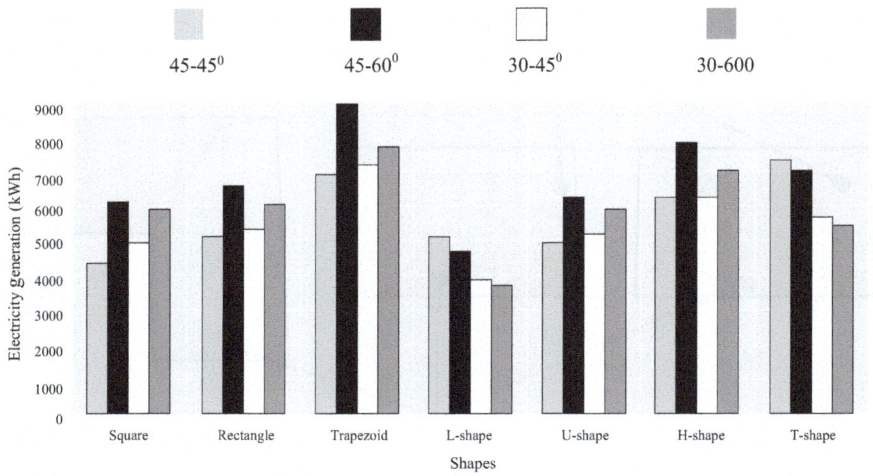

Fig. 5.26 Annual electricity generation of roofs with differing tilt-side angles and shapes

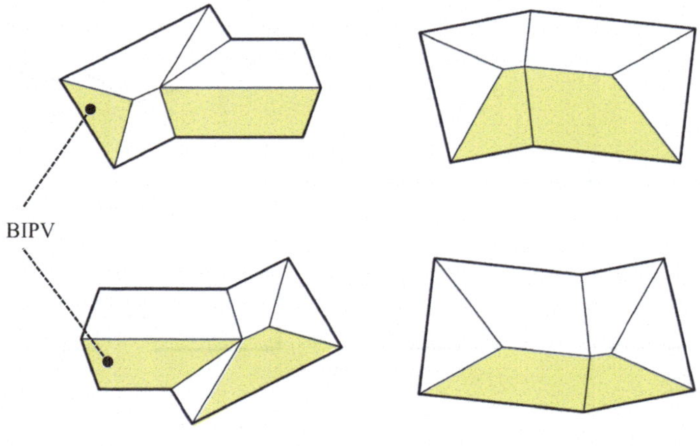

Fig. 5.27 Variant L-shapes and PV integration. PV integrated surfaces are highlighted

5.3.2 High-Rise Multistorey Buildings

The layout of a multistorey building can be manipulated to maximize the surface area of the most advantageous facades for solar exposure and PV integration. This section presents examples of some plan layouts and the analysis of their performance in terms of their impact on building thermal loads and PV electricity generation. The building envelope geometries considered in this section are flat or folded plates, similar to those of Sect. 5.2 (HST/VST (Fig. 5.16), RP and TP, (Fig. 5.19a, b). HST is applied to south and near south surfaces and VST to east and west surfaces.

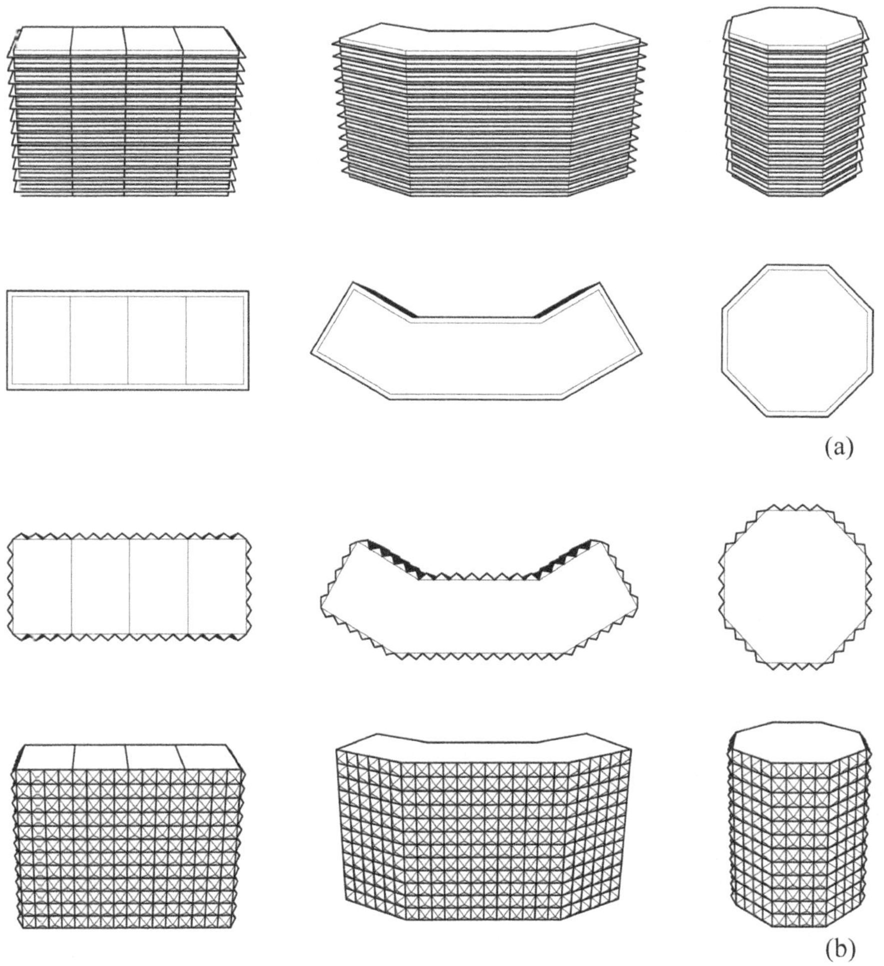

Fig. 5.28 Illustration of all studied building layouts, **a** with flat facade, **b** with RP folded plates facades

The analyzed layouts are square, rectangular, octagonal, and an irregular U-shape. In all configurations window to wall ratio is assumed as 30%. These layouts are as follows:

- The rectangular layout assumes 4 apartments on each of the south and North (2 middle apartments and 2 corners). This results in an elongated plan on the east–west axis.
- The octagonal layout contains one apartment facing each of the S, SW, W, NW, N, NE, and E directions.
- The obtuse U-shape contains a single row of flats along its three branches, facing south, with the service core located at the center, and a corridor along the north

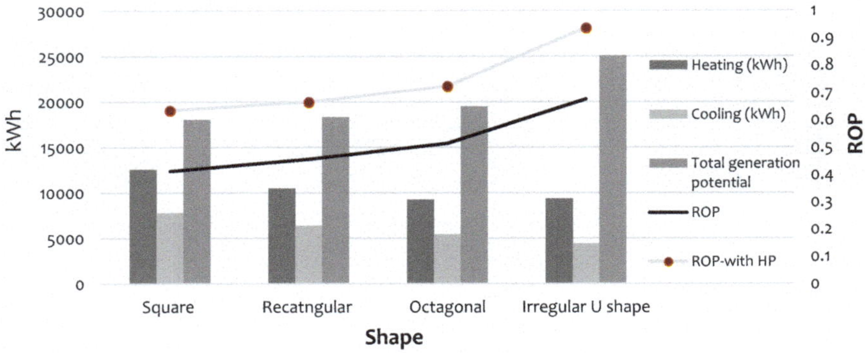

Fig. 5.29 Performance of various building layouts, with flat envelope assumes a heat pump (HP)

facade. The wings of the U layout face toward north at obtuse angles (120°), such that all apartments have a south or near south exposure (30° SE or SW).

Figure 5.28 presents schematics of the studied layouts, with flat facade and rectangular pyramid (RP) folded plates.

The performance of all layouts is studied in terms of heating and cooling loads and energy generation potential. In addition, the ratio of energy generation to total energy consumption (ROP) is included in the analysis. This is performed for the flat facade, and for the folded plate geometries presented above (ST, RP, and TP). The same assumptions such as apartment size, number of apartments per floor, electrical loads, and domestic hot water (DHW) are maintained for all layouts.

The results associated with the flat facade are presented in Fig. 5.29, while those associated with the various folded plate configurations are presented in Fig. 5.30. For the flat facade, the rectangular shape presents a slight improvement as compared to the square layout. The octagonal shape performs slightly better than both rectangular and square layouts. The U-shape performs significantly better due to its higher potential for capturing solar radiation from south and near south facades. Under the assumptions of this study, this irregular shape can generate between 67 and 94% of its energy consumption, depending on the efficiency of the heating and cooling system. The scenario with higher energy performance mechanical systems assumes a heat pump (HP).

The folded plate geometries increase further the potential of some of the layouts presented above. The saw-tooth facade patterns, where HST are implemented on south and near south facades and VST, oriented to the south, on east and west facades do not result in significant change in the performance of various layouts as compared to the flat facades. The folded plate patterns, especially those that do not have significant self-shading, such as the RP, increase significantly the overall performance of various layouts, enabling near net-zero energy and energy positive status for the octagonal and U-shape layouts, respectively. This higher performance is dependent on implementing energy-efficient HVAC systems, employing for instance heat pump (see Fig. 5.30).

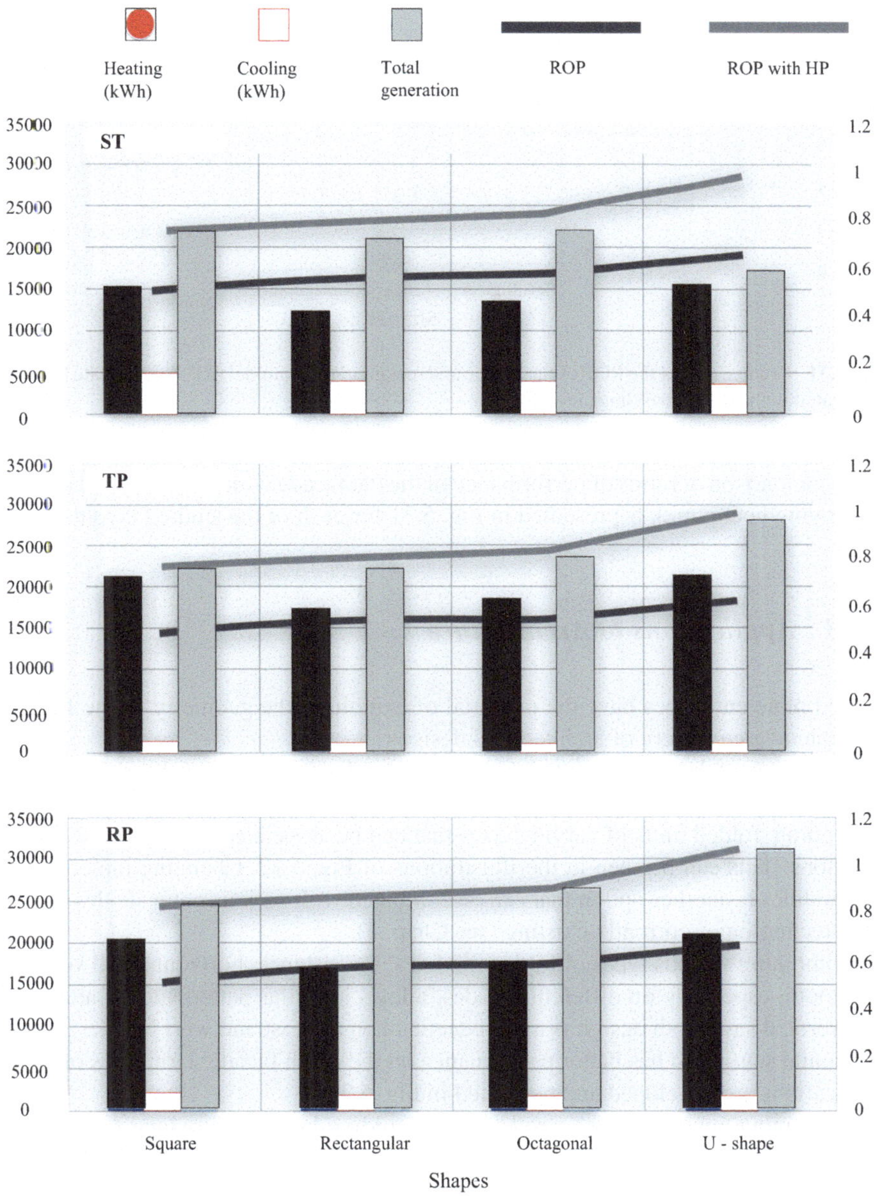

Fig. 5.30 Performance of various building layouts, with folded plates envelope

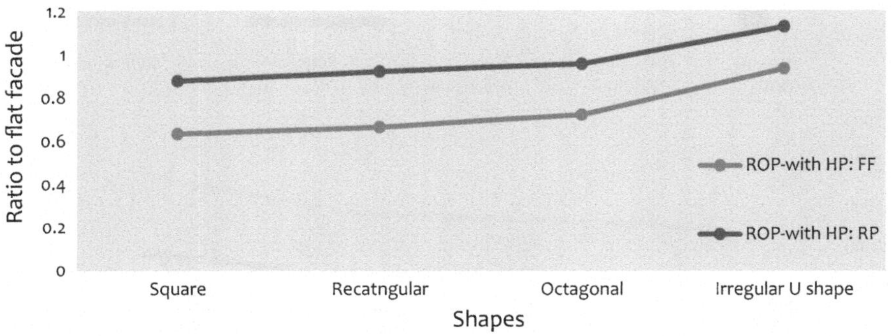

Fig. 5.31 Comparison of the ROP of flat facade and rectangular pyramidal (RP) folded plate facade, associated with studied buildings layouts

A comparison of ratios of performance of the flat facade with the RP folded plates, employing heat pump, is presented in Fig. 5.31 for each of the studied layouts.

5.3.3 Applications to Architecture

Folded plate envelopes have the potential of exploiting the geometry of the facade to create a large variety of architectural designs.

Flexibility of design associated with some visual effects can be achieved by varying some geometrical features within these patterns. The design of the facade can incorporate folded units of varying sizes that can be associated with various indoor functions. This can be seen in the illustrations of Fig. 5.32. Changing the color of the modules is another option that can be combined with various color shades of PV solar technologies currently existing (see Chap. 4).

Combining various types of folded modules, for instance, horizontal and vertical saw-tooth, especially on different facades, allows irregular patterns to be attained. The vertical saw-tooth design is more beneficial on the east and west facades, while horizontal saw-tooth has better performance on the south facade. Examples of such applications are developed and presented in Fig. 5.32.

Combining envelope design with overall building layout offers increased architectural flexibility with potentially increased performance. Layouts that allow south, near south, and east and west orientation (such as the obtuse U-shape and octagonal buildings analyzed in this study) have higher performance than a building of square or rectangular layout. This study demonstrates that implementing multifaceted facades with some of these shapes, together with the implementation of energy-efficient mechanical systems can achieve net energy positive multistorey buildings.

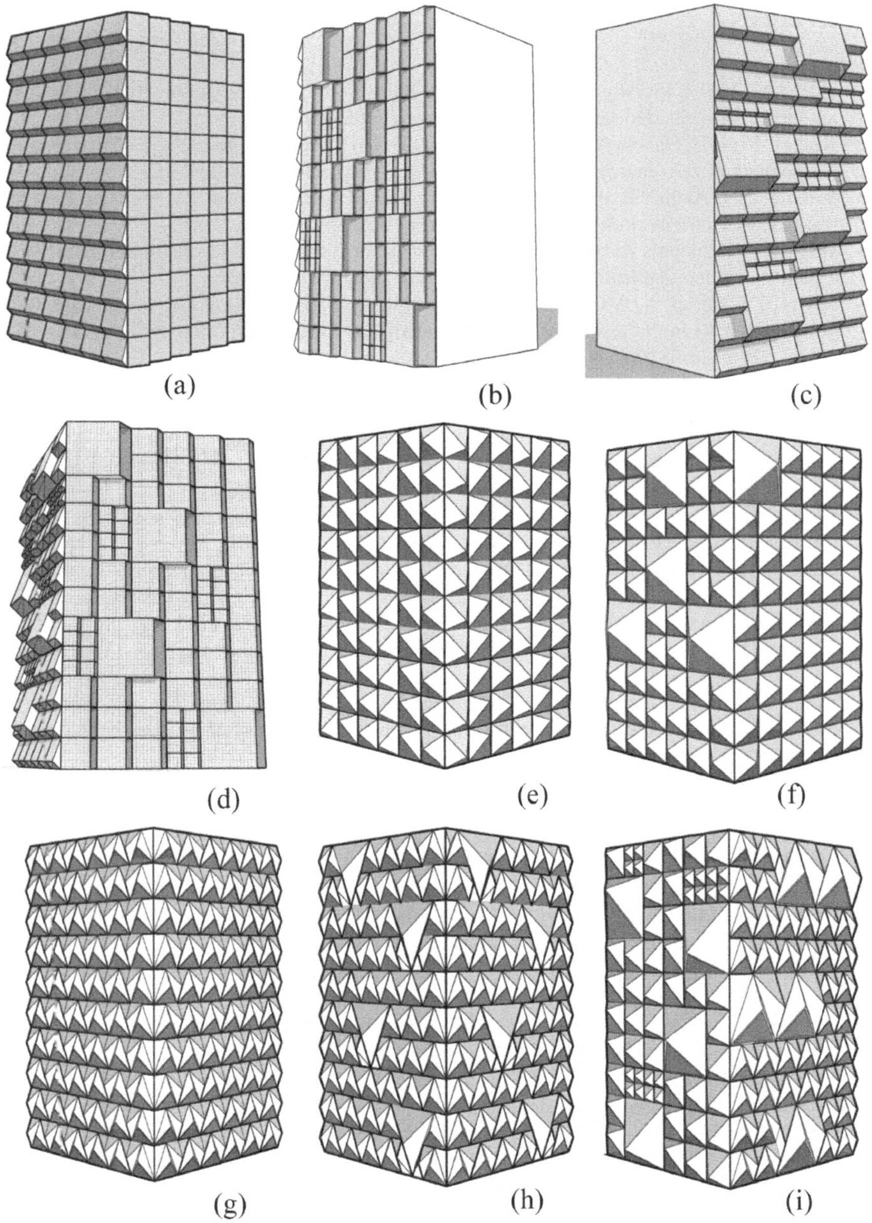

Fig. 5.32 Implementation of various design variations to folded plated patterns to demonstrate flexibility of design **a** Saw-tooth with HST and VST on E/W facades **b** Saw-tooth with\VST, **c** HST with different size of modules, **d** Saw-tooth with HST and VST; **e** RP, **f** RP with different size of modules, **g** TP; **h** TP with different size, **i** RP and TP with various size of modules

References

1. Hachem C (2012) Investigation of design parameters for increased solar potential of dwellings and neighborhoods (Doctoral dissertation, Concordia University)
2. Kim JT, Kim JH, Hassan A, Hachem C, Boafo FE (2016) Active systems. In: ZEMCH: toward the delivery of zero energy mass custom homes. Springer, Cham, pp 237–274
3. Sadineni SB, Atallah F, Boehm RF (2012) Impact of roof integrated PV orientation on the residential electricity peak demand. Appl Energy 92:204–210
4. Hachem C, Athienitis A, Fazio P (2012) Design of roofs for increased solar potential BIPV/T systems and their applications to housing units. Ashrae Trans 118(2)
5. James T, Goodrich A, Woodhouse M, Margolis R, Ong S (2011) Building-Integrated photovoltaics (BIPV) in the residential sector: an analysis of installed rooftop system prices (No. NREL/TP-6A20-53103). National Renewable Energy Lab (NREL), Golden, CO (US)
6. Shukla AK, Sudhakar K, Baredar P (2016) A comprehensive review on design of building integrated photovoltaic system. Energy Build 128:99–110
7. Duffie JA, Beckman WA (2006) Solar engineering of thermal processes. Wiley
8. Li DH, Lam TN (2007) Determining the optimum tilt angle and orientation for solar energy collection based on measured solar radiance data. Int J Photoenergy
9. Candanedo LM, Candanedo JA, O'Brien W, Chen Y (2010) Transient and steady state models for open-loop air-based BIPV/T systems. ASHRAE Trans 116:600
10. Chen Y, Athienitis AK, Galal K (2010) Modeling, design and thermal performance of a BIPV/T system thermally coupled with a ventilated concrete slab in a low energy solar house: Part 1, BIPV/T system and house energy concept. Sol Energy 84(11):1892–1907
11. Chow TT (2010) A review on photovoltaic/thermal hybrid solar technology. Appl Energy 87(2):365–379
12. Hailu G, Dash P, Fung AS (2015) Performance evaluation of an air source heat pump coupled with a building-integrated photovoltaic/thermal (BIPV/T) system under cold climatic conditions. Energy Procedia 78:1913–1918
13. Athienitis AK, Bambara J, O'Neill B, Faille J (2011) A prototype photovoltaic/thermal system integrated with transpired collector. Sol Energy 85(1):139–153
14. Mir A, El-Kodmany. K (2012) Tall buildings and urban habitat of the 21st century: a global perspective. Build-Open Access J 384–423
15. Parker D, Wood A (2013) The tall buildings reference book. Routledge, Abingdon
16. Hachem C, Elsayed M (2016) Patterns of façade system design for enhanced energy performance of multistory buildings. Energy Build 130:366–377
17. Hachem-Vermette C (2018) Multistory building envelope: Creative design and enhanced performance. Sol Energy 159:710–721
18. Hachem C, Athienitis A, Fazio P (2014) Energy performance enhancement in multistory residential buildings. Appl Energy 116:9–19
19. US Department of Energy (2016) Energy plus. Lawrence Berkeley National Laboratory, Berkely, CA. https://energyplus.net/downloads
20. Google SketchUp Plugins (2011). http://sketchup.google.com/intl/en/download/plugins.html
21. University of Wisconsin et al (2014) TRNSYS 17. http://sel.me.wisc.edu/trnsys/index.html
22. Hachem C, Beckett R J Archit Environ Struct Eng Res

Chapter 6
Introduction to Solar Neighborhoods

This chapter introduces the general concept of solar neighborhoods. The characteristics of a neighborhood are presented, followed by an overview of energy performance on an urban scale, including energy consumption and potential of renewable energy generation. Impact of neighborhood design on solar energy potential is briefly discussed in this chapter. Specific parameters that influence solar energy access and utilization, such as building design, density and community layout, are briefly introduced. These factors are presented in more detail in Chaps. 7 and 8, with relation to two types of neighborhoods: small-scale residential neighborhood, and mixed-use larger scale neighborhood.

Advanced neighborhood design trends which pose challenges both for research and application are presented. These include N*et-zero Energy Neighborhoods* and *Zero Carbon Emissions Neighborhoods*. These are neighborhoods that are closely affected by the potential of their buildings and urban surfaces, to capture and utilize solar energy, while reducing the dependence on fossil-based fuel.

The chapter discusses modeling and simulation methods, as means to study various energy efficiency and solar potential strategies, applied on neighborhood scale.

6.1 Energy and Neighborhoods

Many factors need to be considered in the design of an energy-efficient neighborhood, especially a neighborhood that is aiming at producing on-site renewable energy.

The scale of a neighborhood and its general layout (including streets, built area, public open space), building types included within this neighborhood and their density, constitute important part of the overall energy performance of the neighborhood. This section presents the characteristics of a neighborhood, from energy analysis perspective, in terms of energy consumption and potential renewable energy generation, within the physical boundary of the neighborhood.

C. Hachem-Vermette, *Solar Buildings and Neighborhoods*, Green Energy and Technology, https://doi.org/10.1007/978-3-030-47016-6_6

6.1.1 Neighborhood Characteristics

Neighborhoods can be classified according to a number of criteria. For example, a neighborhood can be only residential or a mixed-use neighborhood, composed of various types of buildings, including commercial and residential. Neighborhoods also can be classified by their locations, for instance, urban neighborhoods or rural isolated neighborhoods. Other characteristics of neighborhoods include such features as transport modes–e.g., walkable or vehicle-oriented neighborhoods, density (low, medium, and high), and others.

Figure 6.1 summarizes various factors that determine the characteristics of a neighborhood. These characteristics are determined based on their potential impact on the energy performance of the neighborhood. Six main categories of factors are distinguished: building design, neighborhood design, energy, transport, policy and human factors. This manuscript is mainly dealing with the design of buildings, neighborhood and energy systems. The impact of neighborhood design on modes of transportation and the resulting environmental impact is briefly discussed (see Chap. 8).

In this manuscript, neighborhoods are classified mainly by their buildings' composition and associated representative size. Two main neighborhood types are studied: a small scale, residential neighborhood, and larger scale mixed-use neighborhood (Chaps. 7 and 8, respectively). The term *spatial scale* is used in this document for describing the size of a cluster area for energy planning/simulation purpose.

Neighborhood size and scale is hard to define. Britter and Hanna's [1] distinguish four spatial scales of urban areas. These are the regional scale (less than 100–200 km across), the city scale (less than 10–20 km), the neighborhood scale (less than 1–2 km), and the street scale (less than 100–200 m). A number of research studies indicate that urban-scale energy modeling should be conducted within 1 km, to achieve meaningful analysis of urban energy systems, while taking into account various building and urban physical characteristics (e.g., [2, 3]).

The small-scale neighborhoods presented in this study assume a street scale neighborhood, with two-storey residential buildings (Chap. 7), while the larger scale mixed-use neighborhoods are within 800 m across (see Chap. 8).

The reference to neighborhood size and scale indicates a rough cross-dimension and does not restrict the shape and layout of the neighborhood. Figure 6.2 presents an illustration of a building cluster and the assembly of clusters into a larger scale neighborhood.

6.1.2 Neighborhood-Scale Energy

Multiple and diverse factors influence the overall energy profile at urban scale. These factors include urban morphology parameters, such as density and spatial urban structures (including street layouts and green public areas), geometry of buildings, their types, energy demand, and available energy resources [4–6]. The energy scenario

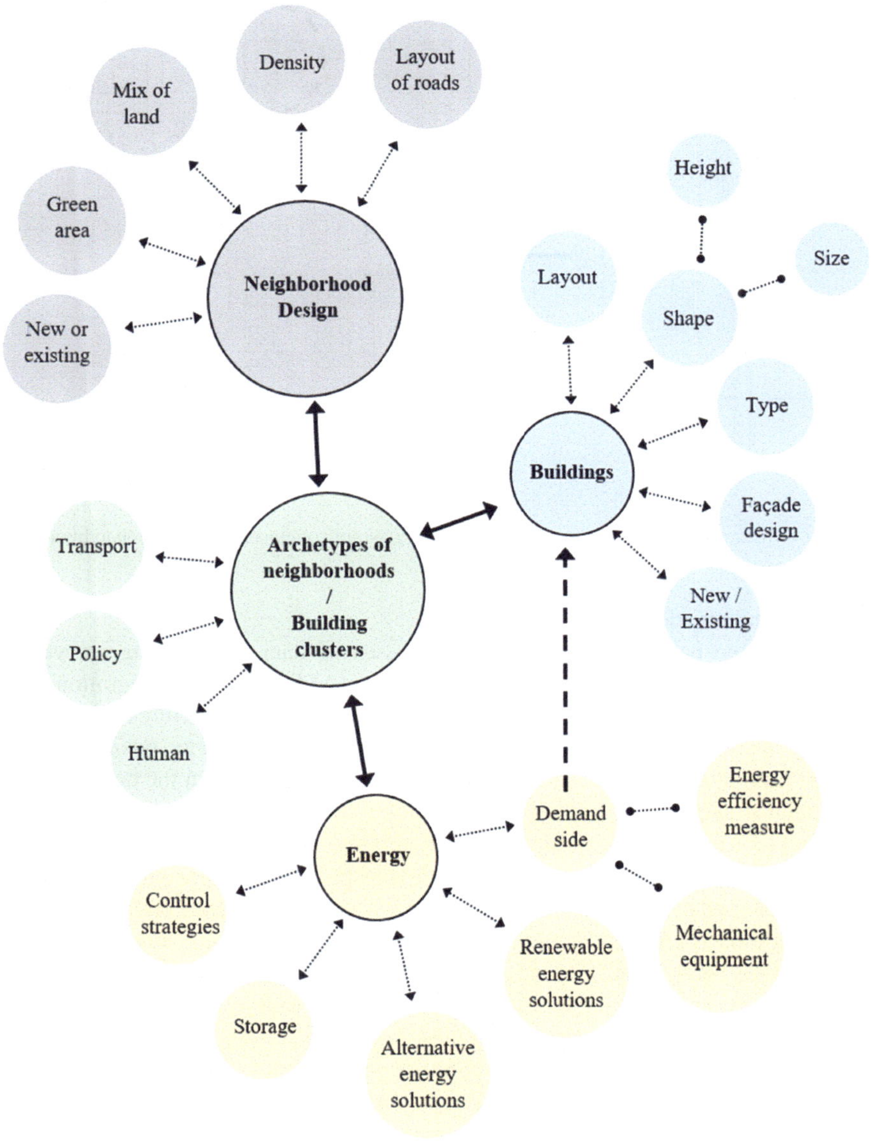

Fig. 6.1 Factors influencing neighborhoods' characteristics

of a neighborhood is affected by the climate zone, the neighborhood location, and the buildings' construction period [7]. In addition to buildings related energy consumption, modes of transportation are becoming an urgent issue, from the point of view of energy demand and GHG emissions. Transportation modes are significantly affected by urban planning [6]. In many residential suburban areas, a personal vehicle is required to fulfill various daily needs.

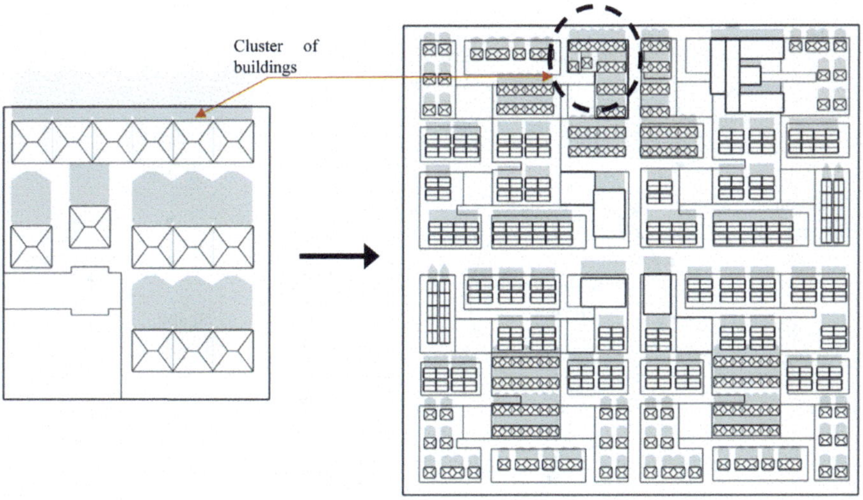

Fig. 6.2 Representation of a cluster of buildings, and grouping of clusters into larger neighborhood

Energy Consumption

Energy demand patterns at urban scale are essential data for planning renewable energy sources (RES), alternative energy solutions (AES), and their integration with the conventional utility grid to match capacity of energy infrastructures [5]. Understanding the interaction between various energy systems, energy resources, and demand profile influences decision-making at various levels, from the development of regional strategies to the detailed design of buildings.

Energy consumption of urban areas and neighborhoods is composed mainly of building energy operations and energy used in transportation. This section focused on energy consumed by building operations. Energy required to operate buildings within a neighborhood is affected by multiple factors, including climatic conditions, solar radiation (see below), building design (including shape and materials employed), building envelope design and construction method, as well as mechanical systems, appliances and equipment employed in these buildings.

The estimation of energy consumption of urban areas is a target of extensive research efforts. A number of studies focus on analyzing the impact of various factors, including urban design, on energy consumption of urban developments of different scales (ranging from a block-level to whole cities). Various methods are developed to determine the effect of urban form on energy demand of buildings. These methods include experimental work, modeling and simulations, and statistical analysis [8]. For instance, the impact of building typology and morphology, including their 3-dimensional elevations, is thoroughly studied, in several regions of the world, suggesting various building types to reduce thermal energy consumption [8–10]. Other research areas focus on developing tools to analyze the impact of building typology as well as urban textures, including street layout, on energy performance and solar access of urban areas [e.g., 11, 12]. This is presented in more detail in Sect. 6.3 below.

Table 6.1 Strategies for reducing energy consumption

Architecture/Design	Technology	Infrastructure
Thermal mass heating/cooling	Triple glazing	Tree shading
Verandas/porches	Low-E glass	Pond/water feature
Solar chimney	Increased Insulation	Green roof
Daylighting	Building-integrated PV	Green wall
Reflective surfaces	Solar thermal collectors	
Earth coupling	Heat pumps	
Screens/Louvers/Overhangs	Thermal storage	

Increasing energy efficiency in buildings constitutes a major consideration in reducing the overall energy demand of a neighborhood. A number of approaches are adopted to achieve these objectives, such as the employment of passive design strategies and deploying various efficiency measures (e.g., mechanical and lighting systems). These approaches can be categorized under the following: (1) architectural design and features of buildings and open space structures; (2) technology on the level of systems (equipment) and their operations; and (3) Neighborhood infrastructure. Some of these strategies straddle more than one approach. Some of the strategies employed to reduce energy consumption in buildings and the built environment are outlined below and listed in Table 6.1.

- **Architectural Design and Features**: Designing buildings for passive solar heating or with features that exploit local conditions (wind, geothermal access, etc.) can significantly mitigate energy demand for heating and cooling (as discussed in Chaps. 1 and 2). Adding indoor or outdoor features to block unwanted solar radiation, obstruct or redirect wind, can increase the passive energy performance of buildings.
- **Technology**: Decreasing demand for energy by building operations by employing the following technologies: high-performance building envelope system, integrating advanced window materials and technologies, efficient mechanical systems, efficient lighting, and energy efficiency-rated appliances.
- **Natural Infrastructure**: Strategically placed outdoor vegetation/water features can reduce cooling load demand in summer months by lowering ambient air temperatures with shadow casting and evapotranspiration.

In addition, urban energy systems such as district heating and cooling can assist in increasing the efficiency of neighborhood energy operations. Urban energy systems are discussed below.

Energy Efficiency for Equipment and Operations

HVAC Systems

Energy use for space heating constitutes a large portion of energy demand in cold northern climate (about 60% of the total energy consumption in houses, in Canada) [13, 14]. Considerable energy, cost savings, and emission reduction can be achieved by the implementation of efficiency measures for heating, ventilation, and air-conditioning (HVAC) systems, in conjunction with integrated building design [15, 16]. These measures include the installation of energy-efficient equipment, such as heat pumps, solar collectors, heat recovery ventilators, solar air-conditioning, geothermal energy, and effective distribution and controls [16, 17].

For instance, ground source heat pumps (GSHPs) can supply heat of up to quadruple the energy of the electricity they consume by using ground extracted heat [18]. Heating or cooling of fresh air supply can be minimized by employing a heat recovery ventilator (HRV), which further reduces energy consumption for space heating and cooling [19].

Smart control management systems enable preheating or precooling buildings before the peak hours, being especially useful in residential buildings [20]. Preheating and precooling can be readily applied through strategic exploitation of thermal mass, highly efficient building envelope, and controllable mechanical ventilation [21].

Domestic Hot Water, Lighting, and Appliances

Solar thermal collectors can be designed to provide significant portion of the DHW demand for residential applications [22]. A typical solar thermal collector hot water system consists of a solar collector, a circulating system to transfer heat from the collector to the preheated insulated storage water tank, and a backup water heating system (as discussed in Chap. 4). Insulated storage tanks should be used to eliminate heat losses.

Low-energy appliances can reduce electricity demand in the range of 10–50% [23]. In addition, application of energy-efficient lighting such as LED can reduce energy consumption for lighting significantly. This can have a large impact especially in commercial and institutional buildings including offices.

Energy Supply

Fuel source including renewable energy sources (RES) and other low-carbon alternative energy solutions (AES) play an important role in the design of high-energy performance, low environmental impact neighborhoods. RES include a broad range of technologies, such as electricity generating (e.g., PV, small wind turbine, fuel cells), heat generating (e.g., PV/T, thermal collectors), and energy-efficient technologies [24]. AES include various technologies, such as waste to energy (WtE), biomass, and other co-generation and tri-generation combinations. Research on such systems has increased significantly in recent years [e.g., 25, 26], concentrating primarily on specific technologies (e.g., BIPV and BIPV/T [27, 28], solar facades [29–32], district heating [33, 34]), and on issues related to their application.

On a neighborhood scale, the emergence of RES and AES should be supported by an evolution in urban energy planning, modeling techniques, assessment of synergies between energy supply systems, and management schemes for matching energy supply and demand. Energy balancing and congestion issues are expected to occur with the integration of increased levels of distributed energy resources within existing energy distribution systems [35]. This entails flexibility of various components of the energy system including demand, generation, and storage [36]. Integrated energy systems, such as *hybrid energy hubs* (handling multiple energy carriers), can be designed to combine an amalgam of individual energy consumers and producers, allowing to take into account variable loads, energy sources, control and storing components [37]. The benefits of such integrated systems include increased reliability, load flexibility, and efficiency gained through synergetic effects [38]. The proper grouping of buildings with diverse energy characteristics to maximize energy sharing can offer a multitude of benefits for increased and efficient energy sharing potential [39]. Example of energy hub is discussed below (in Sect. 6.2.2).

Distributed Energy
Distributed energy generation system (DG) relies on multiple small generation sites instead of a central generation system. A hypothetical DG system within cities with decentralized units integrates several technological advancements: a wider variety of renewable generating methods, such as solar, wind, geothermal, biomass, tidal, hydrogen generation, increase in storage capacity, and higher efficiencies in end-use products [40, 41].

There is extensive research on urban DG systems, examining, not only their potential and feasibility, but also the associated technologies, as well as sociological and economic aspects of their implementation. Deploying distributed energy resources in urban centers presents an effective way to lower overall urban carbon emission rates. Although this concept is gaining a lot of attention, there is yet to be an actual implementation on a large scale, and demonstration of the synergy between various decentralized systems on such scale. A major obstacle in the implementation of such systems in existing neighborhoods and urban complexes is the expected disruption to existing systems and associated costs.

Studying synergy between the various technologies can be beneficial, since different energy resources have various energy generation profiles, which may be in synergy or conflict with each other. For instance, the waste to energy (WtE) combined heat and power generation (CHP), operated using community waste, has varying energy production levels, due to seasonal fluctuations in waste disposal, as shown in Fig. 6.3. Similarly, wind and solar-based RES have daily and monthly generation limitations depending upon weather conditions (Fig. 6.3). The difference in the generation model of these technologies need not be a handicap, at least on a daily basis, given appropriate electrical energy storage facilities, which can be designed as part of such systems.

Considerations in the deployment of DG should encompass other factors than technological aspects, such as social, governance, and finance issues. The financial issue is significant due to many factors, such as long timeframe for expected return on investment, lack of industry incentives, and on-going subsidies for the fossil fuel industry [42, 43].

Other barriers impeding the implementation of a DG system are socio-logistical, requiring human, and technical interaction. For instance, the installation of renewable technologies such as solar, wind, and geothermal can be complicated. Combination

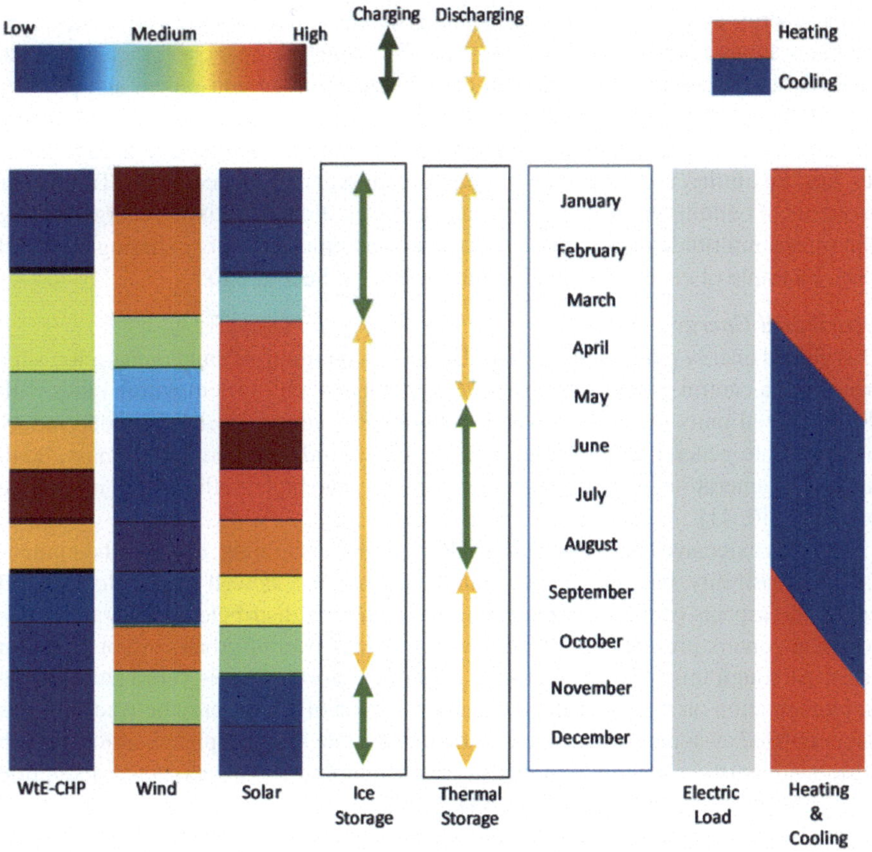

Fig. 6.3 Energy generation and storage by various renewable energy sources

with on-going maintenance and engagement with the existing system can also work against implementation [44].

Urban Energy Systems

The integration of different energy sources in an urban area requires designing an advanced urban level, energy management, which allows to capture the benefits of the diverse sources and their integration options.

Integrated urban energy systems are thus developed to accommodate the shift toward distributed energy resources. These urban systems effectively integrate a variety of local generation of thermal and electric energy, energy demand, as well as energy storage. The main purpose of integrated urban energy systems is to fulfill the energy requirements of local communities through better synergies among different energy carriers, integrating smart-grid technologies and demand side management, leading to increase in reliability and efficiency of such local energy systems.

Employing integrated urban energy systems, buildings of a neighborhood become interconnected to the same energy infrastructure. A change in the energy performance of one of these buildings can have synergetic or disruptive impact on the energy infrastructure and on other buildings [45].

6.2 Solar Neighborhoods

Incorporation of solar access principles and passive solar design in the design of new neighborhoods results in multiple benefits. Such benefits are associated with increased passive heating, cooling, and daylighting potential, which reduce the overall energy consumption of the neighborhood. In addition, ensuring an adequate solar access to buildings and open public areas allows the incorporation of solar collector technologies for thermal and electrical energy generation.

Solar neighborhood design should employ a holistic approach taking into account the shapes of buildings and their design, the neighborhood layout, as well as the density of the neighborhood. Figure 6.4 presents the parameters involved in the design of residential neighborhoods.

In mixed-use neighborhoods, the nature of the mixture (commercial and public buildings) should also be considered.

6.2.1 Impact of Urban Design

Urban design factors and their implementations can have significant impact on the microclimates of urban areas, including access to daylight, availability and intensity of solar radiation, wind flow characteristics (for potential implementation in natural ventilation), and local temperature. These microclimates in turn can affect energy demand and consumption in buildings.

Some of the main factors that govern the design of a neighborhood include building types and size, density of development, and layout of streets. These parameters are usually prescribed for a given site. Solar neighborhood, designed for exploitation of useful solar radiation for heating, daylight and electricity generation, requires consideration of additional parameters, such as building geometry, roof shapes, the arrangement of buildings along streets, and the configuration to match the required density (e.g., in residential neighborhoods-attached units, rows, apartments). These design parameters have substantial impact on passive solar gain, daylighting, and the feasibility and performance of photovoltaic systems. Other parameters identified in research, which allow to quantify the urban layout design and its effect on solar access of buildings within this layout, include distance between buildings, building site coverage, complexity of building form, and variations in building height [46].

Each of the identified neighborhood parameters has its own potential benefits and drawbacks. For example, high density can reduce energy use per capita [47],

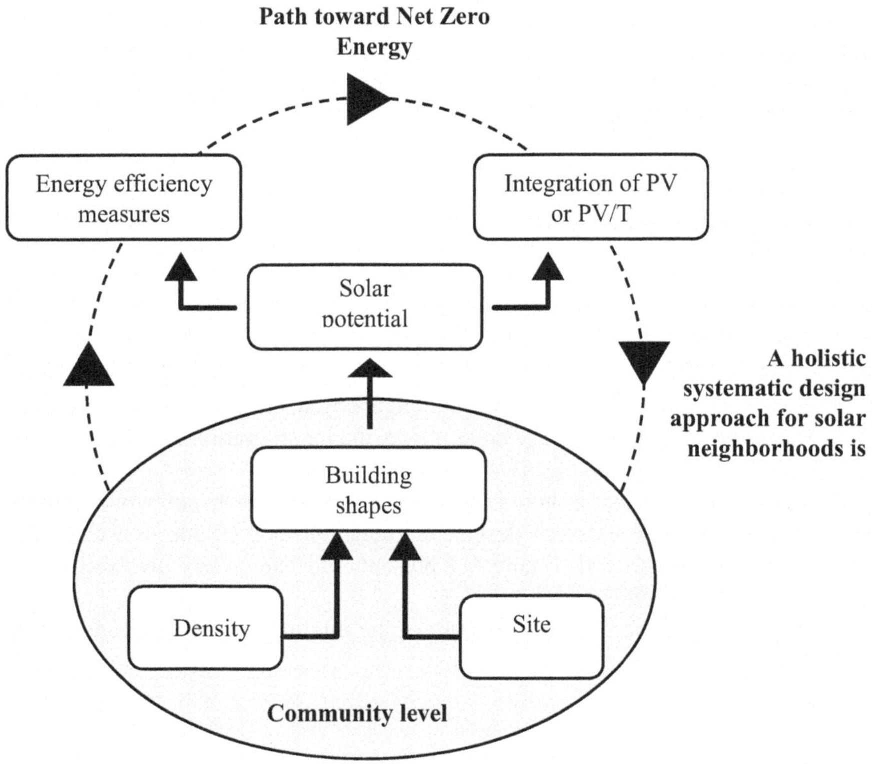

Fig. 6.4 Various neighborhood parameters affecting the design of solar neighborhoods

but reduce also solar access and hence the availability of sunlight and daylight on building facades. Size and shape of a site, as well as the layout of streets within this site can influence orientation of buildings and therefore their accessibility to solar radiation [48].

In addition, air pollution and noise which negatively impact the built environment are strongly dependent on urban form. Urban environment with poor air quality and/or high noise level affects the potential of implementing natural ventilation in buildings [9], increasing the dependence on mechanical systems.

Below is a summary of the main neighborhood parameters and a discussion on their impact on the solar access and overall solar potential.

Building Type and Size

The ratio of a building's envelope area to its overall volume (S/V ratio) has strong effect on heat transfer into or out of a building. Heat transfer through the building envelope can have various impacts on building performance, depending on the building types. For instance, residential buildings with reduced internal heat gain are more affected by heat loss and gain through the envelope (see Chap. 2).

For residential buildings, the energy performance of a detached single-family or a multifamily house is inherently linked to this S/V ratio. The envelope of these types of buildings is exposed to the outdoor environment and thus depending on its thermal characteristics can be major source of heat gain and loss. For a given building volume, a single-family detached unit has a higher S/V ratio than a multifamily unit, resulting in higher susceptibility to heat loss and gain, and associated increase of energy consumption.

Other types of housing such as multifamily units and single-family attached housing units have less outdoor surface areas and have less heat transfer through their outdoor envelope.

In addition to the S/V ratio, the shape of building itself has an impact on its potential to capture and utilize solar energy (as discussed in Chap. 5).

Density

Energy planning of urban areas including energy demand and energy generation potential from renewable sources is significantly affected by building density.

Two different methods are generally employed to specify density, the *population density* or *building unit density* and the *compactness*. The building density is associated usually with the planning side of a neighborhood, while compactness describes the physical architectural side. Population density is associated with both size and type of residential units. For instance, single-family houses in suburban areas are generally larger in terms of floor area than urban multifamily houses, resulting in lower suburban building density [49, 50]. Figure 6.5 presents an example of building density associated with various types of residential buildings, and the relative lot area utilized.

Building density can be measured employing different methods including the plan area density, defined as the ratio of built area to the total lot area [5], and the frontal area density which is the ratio of the windward-facing facade area to the building area [9].

Floor area ratio (FAR) defined as the ratio of the total floor area of a building to the land area employed is another density parameter that influence renewable energy resources, at the urban level. FAR does not, however, reflect the height or shape of buildings, nor the open space between them [9], which are density related parameters of significant impact on energy performance. Other indicators are thus required to be considered together with the FAR in order to adequately plan urban energy system, especially the design and integration of renewable energy technologies.

Determining the exact impact of density on the energy performance of urban areas is not a straightforward task as it depends strongly on the context of these urban areas, as well as on many other interrelated factors. Determining density impact has raised some controversy in the literature. There is, for instance, disagreement regarding the effect of density on energy consumption by buildings, and the magnitude of the effect. Some empirical studies found no significant increase in energy use at higher density [52]. Other literature sources suggest that higher building density leads to higher nighttime urban air temperature, increasing the urban heat island effect. This,

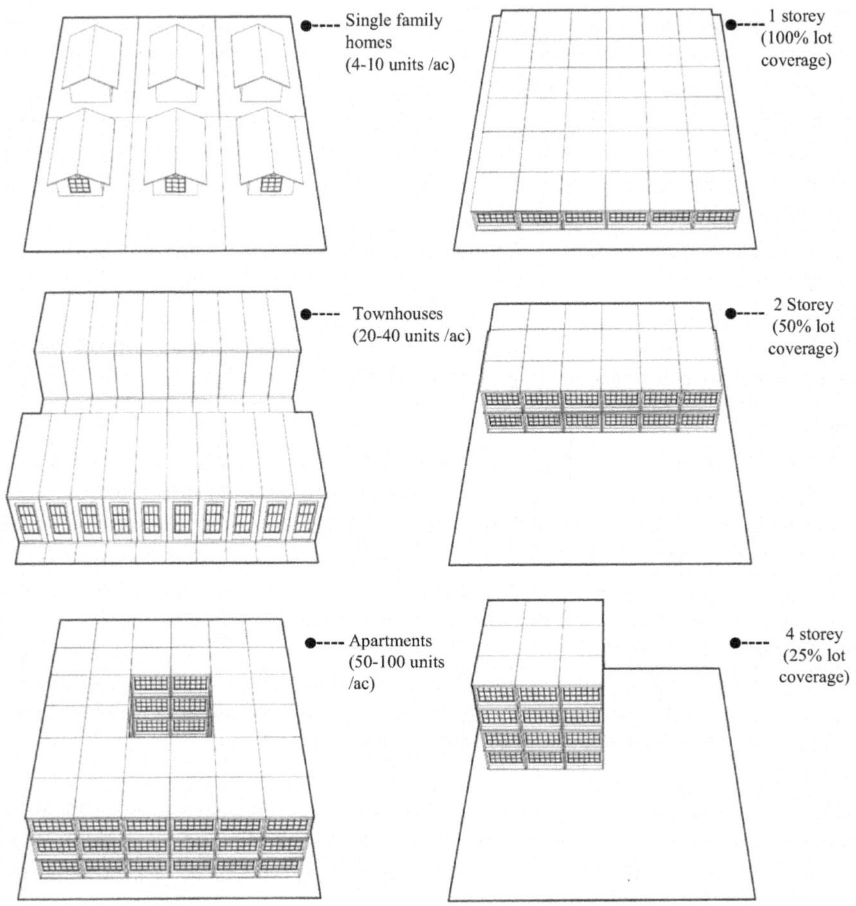

Fig. 6.5 Building density associated with various types of residential buildings with the same lot area

in turn, may increase cooling loads and decrease, often not significantly, the heating load of buildings [53].

In addition to energy demand, density affects the type and size of renewable energy systems that can be integrated in the neighborhood, both in buildings and in the surrounding landscape. Building density affects available surface for integration of solar technologies (e.g., on roofs, facades, or in open public areas).

Neighborhood Layout

Neighborhood layout is determined by a number of interrelated factors such as the position of the buildings on the site, the configurations of these buildings in terms of their height and volume, the layout of the streets, the proportion of public green areas, and others. Neighborhood layout can significantly affect the solar access of buildings and open public areas, within this neighborhood. Figure 6.6 presents an example of

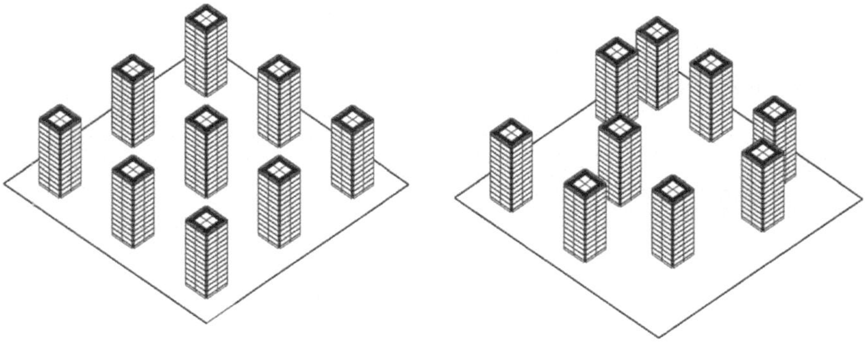

Fig. 6.6 Examples of different building layouts

two neighborhood layouts created by the layout of buildings on the site and their configurations. These layouts can affect the mutual shading between buildings, and thus solar radiation potential on roofs and facades.

Below is a discussion of the main factors that affect the performance of a neighborhood, particularly in terms of capture and utilization of solar energy.

Street Layout

The layout of streets within a neighborhood (or urban area) play a crucial role in determining the orientation of buildings along them, and often the configurations of these buildings. This, as discussed above (Chaps. 1 and 5), has a major impact on the energy performance of these buildings, and their potential to capture solar radiation. The setting of buildings on a site, distance between buildings, and facades orientation, significantly affects the amount of solar radiation received by these buildings [54–56].

Research on solar neighborhoods suggests that streets should be oriented in general along the east–west axis. This allows the opportunity of designing buildings with major south-facing facades, since in an urban setting, buildings are typically orientated toward the street. Buildings can be thus designed to present a compromise between the optimal orientation and a required orientation, such as that imposed by the street direction. This is discussed in more detail in Chap. 7.

Impact of street layout on solar radiation and energy consumption of buildings in different types of neighborhoods is further discussed in Chaps. 7 and 8.

Planting and Surface Coverage

Diverse research methods, including experimentation and simulation, demonstrate the impact of trees on reducing cooling loads, and in some cases heating loads, in urban areas. Open space planning, tree planting, and surface coverage also have a significant influence on urban microclimate, particularly on solar access and the urban heat island effect, which is becoming a major issue in urban areas.

Carefully selected tree types can control solar access when it is not required, enhance natural ventilation in buildings, and mitigate the urban heat island effect. The selection of trees with reduced shading in the heating periods, such

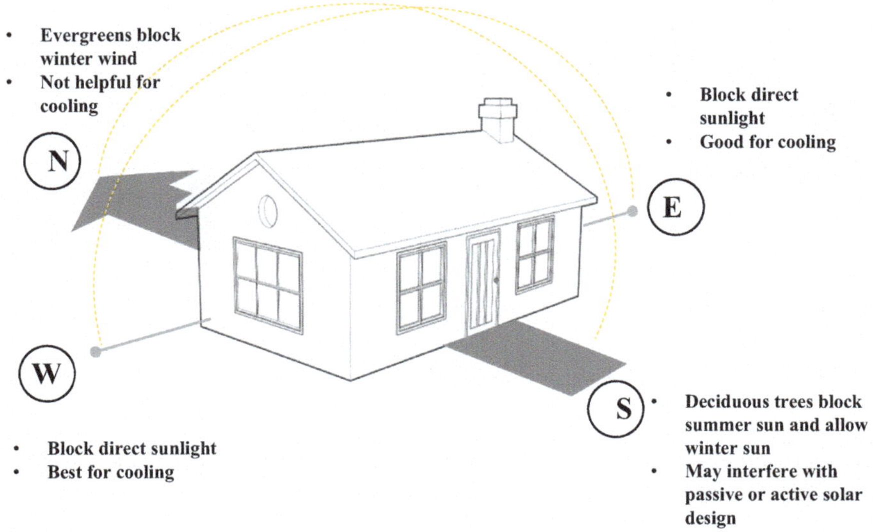

Fig. 6.7 Desired tree types on each orientation of a building

as deciduous trees, can mitigate potential increase in heating demand of urban areas. Figure 6.7 presents a diagram summarizing the desired tree types, on each orientation of the building.

Additionally, reducing exposed surface area through various measures such as reduced lot sizes, and increasing tree canopy cover is recommended to reduce urban heat island effects.

6.2.2 Advanced Neighborhoods

This section presents advanced neighborhood concepts, such as net-zero energy and carbon-neutral neighborhoods. Such neighborhood concepts strongly rely on maximizing the solar energy potential and solar access of buildings and sites to increase the overall energy efficiency and renewable energy generation.

Net-Zero Energy neighborhoods

A net-zero energy neighborhood can be defined as a neighborhood that has the capacity to generate on-site renewable energy to balance its total annual energy requirement. To achieve net-zero energy neighborhood, a two-folded approach should be adopted: reducing the energy consumption to the minimum through various efficiency measures, and generating sufficient renewable energy to offset the energy demand of the neighborhood.

A small-scale, low-density residential neighborhood of single-family houses can reach net-zero energy status, assuming a careful design of building shapes and orientation. In contrast, a mixed-use neighborhood, which combines residential and commercial buildings, in a higher density setting, is significantly less amenable to achieving such status.

Applying the net-zero energy concept at urban scale can provide opportunities for large-scale deployment of renewable energy technologies, such as seasonal storage, implementation of smart grids for power sharing between buildings, controlling peak electricity production timing, and reducing utility peak demand.

High density neighborhoods, and those encompassing diverse types of buildings, with different energy demand require considering various types of renewable energy and potentially alternative energy sources to achieve net-zero energy status. Thermal collectors coupled with seasonal thermal storage become an important aspect of the energy scheme of such neighborhoods. This can be relevant in extreme cold climate (in high latitudes). Cold thermal storage can be an additional benefit, especially for neighborhoods of high density, encompassing office, and institutional buildings. Various technologies including heat pumps, combined heat and power and smart controls, can support the development of net-zero energy communities. Some of these technologies are discussed in Chap. 8.

Planning Net-Zero Energy Neighborhoods

Planning and delivering net-zero energy neighborhoods require a holistic approach to the design of the community and its energy systems. Such approach is multifaceted, integrating building and urban design, energy efficiency measures, implementation of renewable and alternative energy resources, sharing of energy resources, transportation modes, energy storage and management. It should also aim at providing methods and instruments to master planners, decision-makers, and stakeholders.

The design of these types of advanced neighborhoods should meet diverse requirements. Such requirements are not restricted to the technical and economic aspects but encompass architectural, financial, legal, and others. While most of these requirements are closely interrelated, they may compete for the same resources (including space, budget, etc.). This can lead to assuming contradictory measures, unless as mentioned above; a holistic approach, which underlines a coordinated, interdisciplinary effort, is adopted since the early stages. Such approach should be based on a futuristic vision of the community, defining a realistic long-term objective and a roadmap to attain it.

Initiatives of Solar Net-Zero Energy Neighborhood

Existing examples of net-zero energy communities have presented both challenges and opportunities. Established opportunities include increasing energy savings, enhancing human comfort and quality of life, reduced greenhouse gas emissions and negative environmental impacts, and enhancing the potential of local production that can boost local economies.

Initiatives of solar net-zero energy neighborhood are undertaken, worldwide [36]. In addition to adopting the goal of achieving net-zero energy status, many communities are adopting various models of generating renewable energy and reducing energy demand. Several communities are developing business models based on generating solar electricity that can be sold directly to local grids [57]. A number of other initiatives focus on reducing the energy consumption of existing neighborhoods through the implementation of various retrofit measures, and on introducing PV systems to the existing building surfaces. Although such approach results in significantly increased performance of these neighborhoods, the solutions presented are not always optimal, since they have to deal with existing building settings and designs. Designing new net-zero energy neighborhood can be less challenging, especially if efficiency measures and renewable generation are planned at early stages.

Some of the main observations generally found in existing initiatives are discussed in the following.

- **Urban planning for PV integration**. In many of existing initiatives, the decision of integration of PV systems is taken at a late stage in the urban planning process, after the site selection and sometimes after construction of the buildings. This restricts significantly the availability of surfaces for the installation of PV systems and their electrical output.
- **Optimization of the design process**. PV systems are generally installed on surfaces that present good solar exposure. However, building geometry and roof designs were not specifically designed to maximize the solar potential. This results in less than optimal solutions in terms of orientation and tilt angles required for enhanced energy generation. In addition, in some situations, installing PV systems on existing buildings requires additional structural components or modifications in the building envelope, which implies increased cost.
- **Role of government and utilities**. Governments together with utility companies play an important role in various net-zero initiatives. In some cases, the utility companies own the PV systems, and therefore the cost of the PV systems is provided by these companies, while the electricity generation is fed directly to the grid. In other cases, the government contributes in subsidies to install the PV systems. The excess of the electricity is in most cases sold to the grid with a price at least equal to the tariff of use.

Urban Energy Systems

Urban structures are not always suitable to accommodate the required size of solar collectors to achieve the net-zero target. Other energy renewable energy (RES) or alternative energy sources (AES) can be beneficial in complementing solar energy generation. Types of RES and AES are briefly discussed above (Sect. 6.1.2, Energy supply).

Developing adequate urban energy system assists in managing energy on an urban scale, allowing thus to design net-zero energy neighborhoods. Figure 6.8 presents an example of urban energy system that integrates various technologies including energy generation methods and storage.

Carbon-Neutral Neighborhoods

The universal efforts toward radical reduction of GHG emission are motivated primarily by the hazards of climate change, which are becoming more tangible. A

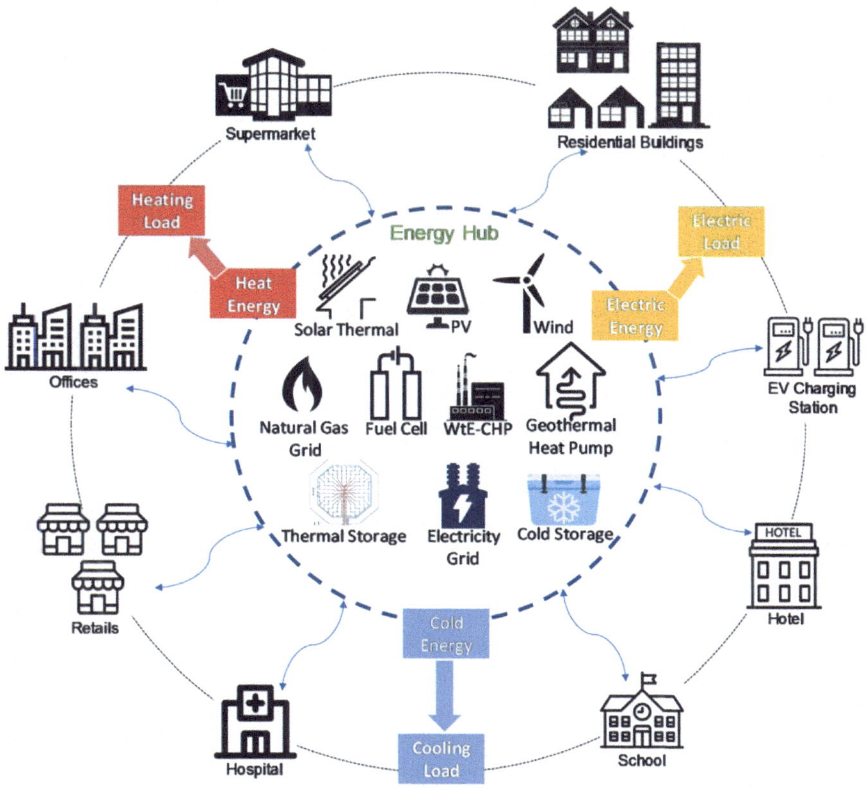

Fig. 6.8 Integrated urban energy system

carbon-neutral neighborhood or net-zero emissions neighborhood is defined in this document as a neighborhood that reduces or eliminates GHG emissions throughout buildings' life cycle [58]. It should be borne in mind, however, that although such neighborhoods require reducing GHG emissions of the building sector, other sectors such as transportation and its impact on emissions should be taken into accounts. This is discussed in more detail in Chap. 8.

To reduce the impact on climate change, there is an urgent need to stabilize the CO_2 atmospheric concentrations. Such an objective necessitates the application of strict carbon management principles in sustainable urban planning, aiming at drastic cuts in emissions.

In order to transform the ambitious targets of low or zero carbon emissions into meaningful and feasible concepts in the context of urban planning, a carbon accounting framework needs to be rigorously defined and adapted to the urban scale.

Measuring GHG Emissions by Buildings

Most of the existing GHG protocols adopt a process that estimates GHG emissions during buildings' occupation phase, including operation, maintenance, and retrofits. Below are three scopes usually employed by these protocols [59].

- Scope 1: Direct on-building-site GHG emissions, consisting of GHG emissions within the boundaries of a building or building cluster. This includes emissions associated with fuel combustion for building operations (such as for heating, cooling, DHW), as well as other emissions released intentionally or unintentionally.
- Scope 2: Indirect on-building-site GHG emissions, also termed indirect energy emissions. Indirect emissions are released outside the specific building site boundary but are linked to energy consumed on this site. These include emissions produced by generation of energy employed for the operation of buildings, such as electricity or steam for cooling, ventilation, or heating.
- Scope 3: Other indirect GHG emissions: not covered in scope 2- activities that are relevant to building performance and not included in the common carbon metric. Examples of these include upstream and downstream emission, which mainly depict the before-use phase, such as extraction of raw materials, GHG emissions for the transportation of materials, and the disposal of waste.

Carbon emission during various stages of the building life cycle is associated with different processes. These include the preconstruction phase, where emissions are mostly related to processing raw materials into building materials. The construction phase entails emissions related to the transportation of building materials into site, site excavation for construction, and the operation of equipment. The building operation phase includes emissions associated with energy consumed to operate mechanical systems, lighting, and appliances. In addition, GHG is emitted during the end of building life, throughout the demolition phase. Energy used in building operation accounts for up to 70% of the total carbon emissions associated with a building [60].

Some definitions of carbon-neutral architecture only consider emissions related to energy supply and consumed by buildings. For example, Architecture 2030 defines a carbon-neutral building as one in which no fossil fuel GHGs are emitted to operate the building (e.g., heating, cooling, lighting, and appliances). Carbon neutrality may be accomplished by implementing innovative sustainable design strategies, improving efficiency by implementing energy conservation measures including passive heating and cooling strategies, and finally installing on-site renewable energy systems.

6.3 Simulation of Neighborhood Performance

Studies of energy performance and solar potential of neighborhoods are usually carried out using various simulation and modeling tools. Modeling urban areas allows taking into account various levels of complexity of the built environment, including climatic conditions, location, and buildings and urban characteristics. Modeling local solar resources and their potential is particularly beneficial prior to planning and installing solar technologies, allowing to understand prospective issues such as shading cast by neighboring structures, as well as estimation of expected electrical output of various buildings and neighborhood surfaces.

In addition to modeling solar potential and how it is affected by the design of a neighborhood, energy modeling should be capable of analyzing various energy opportunities that can be implemented within urban areas. Such opportunities include district energy systems for heating and cooling, thermal storage, microgrid applications, and others. Urban energy modeling should be able to accommodate some (or all) of these urban design variables. In contrast, the majority of available modeling techniques assess energy demand of buildings in an isolated mode, without consideration for urban design, urban energy systems, storage, and energy exchange potential, and their impact on the performance of individual buildings or the neighborhood as a whole [5].

A number of studies focus on developing specific modeling procedures to analyze energy performance of urban areas. Two major methods are adopted in general to model urban-scale energy consumption: top-down and bottom-up [61].

- Top-down methods consider building clusters as an energy sink. This approach assumes the whole studied building cluster or neighborhood as an entity. As such, the neighborhood provides the general energy demand profile, without consideration of energy characteristics of specific buildings contained within it. In such approach, buildings are treated as black boxes and therefore cannot provide information on the environmental impact of building design options, including adopting various passive design strategies and technologies at the individual building level. The underlying models of this method are based on statistical data and economic schemes.
- Bottom-up models utilize sets of buildings representative of actual practice, which can be modeled using building performance simulations. These simulations allow understanding the impact of building design, including retrofitting measures (for existing buildings) on energy consumption at the urban level. Other research joins various models and modeling tools to obtain the desired information. For instance, some studies couple models of individual buildings simulated in whole building energy simulation programs, with models of district energy plant, developed employing other specialized tools [62, 63].

Simulation tools for analyzing actual energy performance of buildings and neighborhoods are inherently based on multiple assumptions and statistical data, and therefore their predictive reliability is sometimes debatable. Occupant behavior is a prominent factor, affecting the energy performance of buildings and can result in significant discrepancy between the actual performance of the building and the modeled performance. Simulation models should be thus considered as analysis tools that allow to obtain comparative relative results, rather than indicative of actual performance.

Tools aiming at modeling energy performance, including solar access, daylighting, and energy consumption, are in continuous development. Efforts are aimed at increasing models accuracy and flexibility while reducing time of computation and complexity of data input and output. This will encourage more professionals to use simulation tools to evaluate the impact of urban planning and building designs, on energy specific energy (and performance) indicators, especially at the early design stages.

6.3.1 Solar Potential in Urban Areas

Modeling solar radiation in urban areas can be beneficial to plan the integration of solar technologies within building and urban surfaces, and optimizing their electrical output. To achieve such benefits, the model should be able to perform the following tasks:

(1) Conducting accurate computation of solar radiation within the built environment, taking into considerations various climatic conditions and geographic data that can affect solar irradiance intensity (see Chap. 1);
(2) Identifying various building and urban surfaces suitable for the integration of solar technologies and determining their orientation and tilt angle;
(3) Computation of solar radiation incident on specific surfaces, taking into account their tilt and orientation angles [5].

This section summarizes methods of establishing solar access in urban areas, including techniques to minimize mutual shading by buildings and to determine insolation or shading in a given urban area. Such methods employ a number of tools to address some of the tasks presented above.

Solar Irradiance Computation

Models to compute solar irradiation are continuously progressing. Early techniques concentrated on 2-D models of specific surfaces (mostly roof tops), in individual buildings. Such models did not have the capacity to model full 3-dimensional urban scenarios [e.g., 64–66]. The enhancement in computation capacity and continuous development of modeling techniques are allowing for more comprehensive models that take into account urban settings and complexities. Various tools are employed in urban modeling, ranging from those that can handle small-scale urban models (e.g., computer-aided design (CAD) software [67, 68]) to macroscale models based on geographical information system (GIS) tools [69–71]), capable of processing large amounts of data. GIS data have been used to map insolated and shaded areas of a development to inform design decisions [72]. Despite the significant progress in computation tools, determining solar radiation within a 3D setting is still presenting some challenges.

Additional challenges facing solar irradiation modeling and their accuracy are caused by the characteristics of the modeled surfaces, their tilt angles and orientations (of an individual building or block of buildings). Basic models that do not capture such characteristics can produce results significantly different from actual values.

Selected Solar Access Models

In the last couple of decades, numerous research efforts have attempted to develop relatively simple models to determine the solar access of urban areas, without having to perform rather complex simulations (as discussed above). Such efforts include developing models to define the maximum allowed height of a building to avoid overshadowing its surroundings [73]. Another well-known approach, termed *solar envelope*, was developed to assure solar access to each building in a community [74]. *Solar envelope* is defined as a theoretical geometric surface, which outlines the maximum heights of new or proposed structures in a development such that they do not substantially shade existing buildings [74]. The solar envelope model was adapted to define maximum heights of new/proposed structures such that they allow for a predetermined mean annual horizontal irradiance (W/m^2) for existing buildings [75].

The concept of *solar volume* [76] has been proposed to determine solar access and solar rights volume within the built environment. This volume is comprised of a *solar rights envelope*, which is the upper bounds to building heights/positions where they do not violate solar rights of surrounding buildings, and the *solar collection envelope, which* is lowest locations for solar collectors/windows that will still receive sunlight in the winter. These envelopes, based on predetermined solar access values (commonly 4 h of sunlight during the Winter solstice), do not account for light intensity or angle of incidence, however, and the result on building energy consumption is dependent on many variables including building construction and climate.

Some of the simulation programs employed for the investigation of solar access of buildings within urban context include the ray tracing program RADIANCE, which simulates the irradiation on facades [77, 78]. Digital elevation models (DEMs) are also employed in some cases to find the effect of urban texture on building energy consumption. These DEM models are mainly based on image processing and were employed in lieu of detailed numerical simulation of radiation exchange [9].

6.3.2 Energy Performance

Two approaches are adopted to analyze energy performance of urban areas. The first approach uses models, representative of real urban design morphologies. These cases are usually limited in their ability to generalize the findings, unless the studied morphologies can represent ubiquitous prototypes of urban developments. The second approach is based on simplified urban morphological archetypes which can be easily parameterized, both for performing sensitivity analyses and for urban morphology optimization. The major problem of this approach is the risk of not representing realistic urban design forms. The most frequently assessed archetypes are pavilions, including shape variations for high-rise buildings, courtyard configurations, row houses, and urban street canyons.

The urban morphology parameters most widely investigated can be divided into three categories, depending on whether they describe (1) building form only; (2) the morphological surrounding of a given building; (3) the morphological patterns of an entire neighborhood. These are described below:

(1) Parameters relating to individual buildings, including: wall surface area, ratio of envelope area to floor area, building orientation, and ratio of passive to non-passive floor area. Passive solar floor area is defined as the area of the floor adjacent to the equatorial facade, having a total width of about double the interior height (measured from floor to ceiling). This method is used to estimate the solar radiation penetration [79].

(2) Parameters relating to the morphological surroundings of an individual building include: obstruction angle defined as "the smallest angle with the horizontal under which the sky can be seen from the lower edge of a vantage point, usually an opening in a building" [75]; urban horizon angle which combines: orientation of the building; elevation of the obstruction and elevation of the sun (depends on the latitude of the urban area) [79]; sky view factor, defined as the ratio of the radiation received (or emitted) by a planar surface to the radiation emitted (or received) by the entire hemispheric environment; and the ratio of the building height to its width (H/W ratio).

(3) Parameters characterizing the morphological patterns of an entire neighborhood including: the site coverage, defined as the portion of a site occupied by buildings or structures for human occupancy, and the typology (including heights) of clusters of buildings.

A number of tools, employing simplified algorithms, are established to compute heating demand of buildings in specific neighborhoods [e.g., 64]. More inclusive tools based on advanced energy simulations engines (such as EnergyPlus and Radiance/Daysim) are being developed as well to estimate various energy related components such as operational energy and daylighting [e.g., 13]. In addition, urban building energy models based on GIS (Geographic Information System) are proposed to estimate city wide hourly energy demands at the building level [80].

Several existing tools provide explicit models for technologies such as solar thermal collectors and geothermal heat pumps. For instance, EnergyPlus and TRNSYS were employed to perform a feasibility analysis of zero energy houses with renewable electricity, solar hot water system, and energy-efficient heating systems [81]. However, the capability of modeling innovative technologies or interactions between multiple pieces of equipment is still restricted.

Despite the availability of powerful simulation programs, such as mentioned above, some of these tools lack the ability to model some passive and active solar potential in conjunction with some specific building system (e.g., HVAC systems, heat pump systems, etc.). Two major categories of potential improvements to solar energy-efficient neighborhood tools can be identified: improving the interface between various tools to complement each other's capability, and enhancing the potential of models to represent various technologies and their integration.

References

1. Britter R, Hanna S (2003) Flow and dispersion in urban areas. Annu Rev Fluid Mech 35:469–496
2. Srebric J, Heidarinejad M, Liu J (2015) Building neighborhood emerging properties and their impacts on multi-scale modeling of building energy and airflows. Build Environ 91:246–262
3. Hang J, Li Y (2012) Macroscopic simulations of turbulent flows through high-rise building arrays using a porous turbulence model. Build Environ 49, 41–54
4. Monteiro CS, Pina A, Cerezo C, Reinhart C, Ferrão P (2017) The use of multi-detail building archetypes in urban energy modelling. Energy Procedia 111:817–825

5. Zhang X, Lovati M, Vigna I, Widén J, Han M, Gal C, Feng T (2018) A review of urban energy systems at building cluster level incorporating renewable-energy-source (RES) envelope solutions. Appl Energy 230:1034–1056
6. Hachem C (2016) Impact of neighborhood design on energy performance and GHG emissions. J Appl Energy 177:422–434
7. Monteiro CS, Pina A, Cerezo C, Reinhart C, Ferrão P (2017) The use of multi-detail building archetypes in urban energy modelling. Energy Procedia
8. Ko Y (2013) Urban form and residential energy use: A review of design principles and research findings. J Plan Lit 28(4):327–351
9. Ratti C, Baker N, Steemers K (2005) Energy consumption and urban texture. Energy Build 37(7):762–776
10. Rode P, Keim C, Robazza G, Viejo P, Schofield J (2014) Cities and energy: urban morphology and residential heat-energy demand. Environ Plan 41(1):138–162
11. Christensen C, Horowitz S (2008) Orienting the neighborhood: a subdivision energy analysis tool, ACEEE Summer Study on Energy Efficiency in Buildings Pacific Grove, California, 17–22, August
12. Reinhart C, Dogan T, Jakubiec JA, Rakha T, Sang A (2013, August). Umi-an urban simulation environment for building energy use, daylighting and walkability. In: 13th conference of international building performance simulation association, Chambery, France
13. Aydinalp M, Ugursal VI, Fung AS (2004) Modeling of the space and domestic hot-water heating energy-consumption in the residential sector using neural networks. Appl Energy 79(2):159–178
14. Iqbal MT (2004) A feasibility study of a zero energy home in Newfoundland. Renew Energy 29(2):277–289
15. Vakiloroaya V, Samali B, Fakhar A, Pishghadam K (2014) A review of different strategies for HVAC energy saving. Energy Convers Manag 77:738–754
16. De Boeck L, Verbeke S, Audenaert A, De Mesmaeker L (2015) Improving the energy performance of residential buildings: a literature review. Renew Sustain Energy Rev 52:960–975
17. Kamel RS, Fung AS, Dash PR (2015) Solar systems and their integration with heat pumps: a review. Energy Build 87:395–412
18. Natural Resources Canada (NRCan), Office of Energy Efficiency (2008) Heating and cooling with a heat pump. http://oee.nrcan.gc.ca/sites/oee.nrcan.gc.ca/files/pdf/publications/infosource/pub/home/heating-heat-pump/booklet.pdf
19. Natural Resources Canada (NRCan), Office of Energy Efficiency (2009) How a heat recovery ventilator works. http://oee.nrcan.gc.ca/residential/personal/new-homes/r-2000/standard/how-hrv-works.cfm?attr=4
20. Xue X, Wang S, Sun Y, Xiao F (2014) An interactive building power demand management strategy for facilitating smart grid optimization. Appl Energy 116:297–310
21. Braun JE (2003) Load control using building thermal mass. J Sol Energy Eng 125(3):292–301
22. Kemp WH (2006) The renewable energy handbook: a guide to rural energy independence, off-grid and sustainable living. Aztech Press, Tamworth, ON
23. Pelland S, Poissant Y (2006) An evaluation of the potential of building integrated photovoltaics in Canada. In: 31st annual conference of the solar energy society of Canada (SESCI). Aug 20–24th, Montréal, Canada
24. Loonen R, Trčka M, Cóstola D, Hensen J (2013) Climate adaptive building shells: state-of the-art and future challenges. Renew Sustain Energy Rev 25:483–493
25. Rogers JC, Simmons EA, Convery I, Weatherall A (2008) Public perceptions of opportunities for community-based renewable energy projects. Energy Policy 36(11):4217–4226
26. Cuthbert R, Pan F, Nieminen K, Friedrich K, Wilkinson D, Cotton JS (2013) A solar PV-thermal energy design optimization study of a building footprint limited net-zero energy facility. ASHRAE Trans 119(1):188–202
27. Gavioli S, Seynhaeve J, Bartosiewicz Y (2012) Dynamic simulation of an ejector-based cooling system for residential solar air-conditioning. In: ASME 2012 international mechanical engineering congress and exposition (IMECE2012), 2012. https://doi.org/10.1115/imece2012-88867

28. Hemidi A, Seynhaeve J, Bartosiewicz Y (2008) Designing and rating a Tritherm solar ejector system for residential cooling. An energetic and exergetic evaluation. In: 1st international conference on solar heating, cooling and buildings-EUROSUN 2008, Lisbon, Portugal, du 07/10/2008 au 10/10/2008
29. Rae C, Bradley F (2012) Energy autonomy in sustainable communities—a review of key issues. Renew Sustain Energy Rev 16(9):6497–6506
30. Walker G, Simcock N, Smith SJ (2012) Community energy systems. In: Smith SJ (ed) International encyclopedia of housing and home, 1st edn. pp 194–198
31. Debbarma M, Sudhakar K, Baredar P (2017) Thermal modeling, exergy analysis, performance of BIPV and BIPVT: a review. Renew Sustain Energy Rev 73:1276–1288
32. Jelle PB (2016) Building integrated photovoltaics: a concise description of the current state of the art and possible research pathways. Energies 9:21–51
33. Gu Y, Zhang X, Myhren JA, Han M, Chen X, Yuan Y (2018) Techno-economic analysis of a solar photovoltaic/thermal (PV/T) concentrator for building application in Sweden using Monte Carlo method. Energy Convers Manage 165:8–24
34. Shukla AK, Sudhakar K, Baredar P (2017) Recent advancement in BIPV product technologies: a review. Energy Build 140:188–195
35. Yan Y, Qian Y, Sharif H, Tipper D (2012) A survey on smart grid communication infrastructures: motivations, requirements and challenges. IEEE Commun Surv Tutor 15(1):5–20
36. Koirala BP, Koliou E, Friege J, Hakvoort RA, Herder PM (2016) Energetic communities for community energy: a review of key issues and trends shaping integrated community energy systems. Renew Sustain Energy Rev 56:722–744
37. O'Dwyer E, Pan I, Acha S, Shah N (2019) Smart energy systems for sustainable smart cities: current developments, trends and future directions. Appl Energy 237:581–597
38. Koirala B, Chaves Ávila J, Gómez T, Hakvoort R, Herder P (2016) Local alternative for energy supply: performance assessment of integrated community energy systems. Energies 9(12):981
39. Huang P, Sun Y (2019) A clustering based grouping method of nearly zero energy buildings for performance improvements. Appl Energy 235:43–55
40. Harvey LD (2010) Energy and the new reality 2: carbon-free energy supply. Routledge
41. Zahedi A (2011) Maximizing solar PV energy penetration using energy storage technology. Renew Sustain Energy Rev 15(1):866–870
42. van der Schoor T, Scholtens B (2016, February) Community energy: a critical review of the literature. In: Dynamics of energy, mobility and demand (DEMAND) conference 2016: what energy is for-the making and dynamics of demand
43. Chmutina K, Goodier CI (2014) Alternative future energy pathways: assessment of the potential of innovative decentralised energy systems in the UK. Energy Policy 66:62–72
44. Geelen D, Reinders A, Keyson D (2013) Empowering the end-user in smart grids: recommendations for the design of products and services. Energy Policy 61:151–161
45. Vigna I, Pernetti R, Pasut W, Lollini R (2018) New domain for promoting energy efficiency: energy flexible building cluster. Sustain Cities Soc 38:526–533
46. Chatzipoulka C, Compagnon R, Nikolopoulou M (2016) Urban geometry and solar availability on façades and ground of real urban forms: using London as a case study. Sol Energy 138:53–66. https://doi.org/10.1016/j.solener.2016.09.005
47. Steemers K (2003) Energy and the city: density, buildings and transport. Energy Build 35(1):3–14
48. Knowles RL (1981) Sun rhythm form. The MIT Press, Cambridge, Massachusetts
49. Ewing R, Rong F (2008) The impact of urban form on US residential energy use. Hous Policy Debate 19(1):1–30
50. Kaza N (2010) Understanding the spectrum of residential energy consumption: A quantile regression approach. Energy policy 38(11):6574–6585
51. Macdonald RW, Griffiths RF, Hall DJ (1998) An improved method for the estimation of surface roughness of obstacle arrays. Atmos Environ 32(11):1857–1864
52. Ko Y, Radke JD (2014) The effect of urban form and residential cooling energy use in Sacramento, California. Environ Plan 41(4):573–593

53. Li C, Song Y, Kaza N (2018) Urban form and household electricity consumption: a multilevel study. Energy Build 158:181–193
54. Hachem C, Fazio P, Athienitis A (2013) Solar optimized residential neighborhoods: evaluation and design methodology. Sol Energy 95:42–64
55. Littlefair PJ (2000) Environmental site layout planning: solar access, microclimate and passive cooling in urban areas. BRE publications
56. Cheng V, Steemers K, Montavon M, Compagnon R (2006) Urban form, density and solar potential (No. CONF)
57. Oteman M, Wiering M, Helderman JK (2014) The institutional space of community initiatives for renewable energy: a comparative case study of the Netherlands, Germany and Denmark. Energy Sustain Soc 4(1):11
58. La Roche PM (2017) Carbon-neutral architectural design. CRC Press
59. Gupta R, Garrigan C (2013, July) Developing and testing a global common carbon metric approach for measuring energy use and greenhouse gas emissions from building operations. In: Integrated approaches to sustainable building: developing theory and practice through international collaboration and learning. Proceedings of sustainable building and construction conference, pp 3–5
60. Nässén J, Holmberg J, Wadeskog A, Nyman M (2007) Direct and indirect energy use and carbon emissions in the production phase of buildings: an input–output analysis. Energy 32(9):1593–1602
61. Swan LG, Ugursal VI (2009) Modeling of end-use energy consumption in the residential sector: a review of modeling techniques. Renew Sustain Energy Rev 13(8):1819–1835
62. Hachem C, Athienitis A, Fazio P (2011) Parametric investigation of geometric form effects on solar potential of housing units. Sol Energy 85(9):1864–1877
63. Huber J, Nytsch-Geusen C (2011, November) Development of modeling and simulation strategies for large-scale urban districts. In: Proceedings of building simulation, vol 2011, pp 1753–1760
64. Goretzki P (2013) GOSOL–Solarbüro für energieeffiziente Stadtplanung. http://www.gosol.de/index.html
65. Peckham RJ (1990) Shadowpack–P.C. version 2-0 user's guide
66. Rich P, Hetrick W, Saving S (1995) Modeling topographic influences on solar radiation: a manual for the SOLARFLUX Model. Los Alamos, NM
67. Skelion (2013) Skelion 5.0.7 user's guide
68. Ecotect (2010) Autodesk ECOTECT analysis
69. Carneiro C, Morello E, Desthieux G, Golay F (2010) Urban environment quality indicators: application to solar radiation and morphological analysis on built area. In: Advances in visualization, imaging and simulation, pp 141–148
70. Hofierka J, Zlocha M (2012) A new 3-D solar radiation model for 3-D city models. Trans GIS 16(5), 681–690
71. Jakubiec JA, Reinhart CF (2013) A method for predicting city-wide electricity gains from photovoltaic panels based on LiDAR and GIS data combined with hourly Daysim simulations. Sol Energy 93, 127–143
72. Chow A, Fung A, Li S (2014) GIS modeling of solar neighborhood potential at a fine spatiotemporal resolution. Buildings 4(2):195–206
73. Freitas S, Catita C, Redweik P, Brito MC (2015) Modelling solar potential in the urban environment: state-of-the-art review. Renew Sustain Energy Rev 41:915–931
74. Knowles RL (2003) The solar envelope: its meaning for energy and buildings. Energy Build 35(1):15–25
75. Morello E, Ratti C (2009) Sunscapes: "Solar envelopes" and the analysis of urban DEMs. Comput Environ Urban Syst 33(1):26–34
76. Capeluto IG, Shaviv E (1997) Modeling the design of urban fabric with solar rights considerations. In: Proceedings of the ISES 1997 solar world congress. Taejon, Korea, pp 148–160
77. Compagnon R (2004) Solar and daylight availability in the urban fabric. Energy Build 36(4):321–328

78. Kämpf JH, Montavon M, Bunyesc J, Bolliger R, Robinson D (2010) Optimisation of buildings' solar irradiation availability. Sol Energy 84(4):596–603
79. Baker N, Steemers K (2003) Energy and environment in architecture: a technical design guide. Taylor & Francis
80. Remmen P, Lauster M, Mans M, Fuchs M, Osterhage T, Müller D (2018) TEASER: an open tool for urban energy modelling of building stocks. J Build Perform Simul 11(1):84–98
81. Basarkar M, Pang X, Wang L, Haves P, Hong T (2011) Modeling and simulation of HVAC faults in EnergyPlus (No. LBNL-5564E). Lawrence Berkeley National Lab. (LBNL), Berkeley, CA (United States)

Chapter 7
Residential, Low-Density Neighborhoods

This chapter presents the main principles in the design of solar residential, low-density neighborhoods. The main parameters influencing the capture of solar energy are discussed. These parameters are mostly associated with density and street layout, as well as building shapes. The chapter employs a number of hypothetical case studies of small-scale residential neighborhoods to demonstrate various design principles and their impact. The selected examples aim at presenting flexibility of design, while promoting energy conservation, and maximizing solar capture and utilization. They highlight the importance of three interconnected design considerations in planning of residential neighborhoods mentioned above–building shape, density, and site layout.

Design considerations discussed for new neighborhoods are relevant to existing neighborhoods to understand their solar potential and solar access.

7.1 Parameters of Solar Energy Access

Neighborhood designs are characterized by the layouts of roads along which these neighborhoods are located, by the shape of dwelling units and their density [1–4]. Three site layouts are considered in this chapter to present examples of the most generic street layouts: straight road, south-facing semi-circular road, and north-facing semi-circular road. The basic straight road site runs east–west, but deviations from this direction are also studied. Housing density is considered through detached configurations, representing lower density, and by attached configurations, representing intermediate density, within the specific housing types considered in this chapter (2-storey, single-family housing units). Effect of rows of housing units is also considered for the straight road site.

This section summarizes the three main design considerations (building shape, neighborhood layout, and density), and their impact on the solar potential of the whole neighborhood. This discussion is based on sample small-scale residential neighborhoods, designed to reflect various potential site layouts, densities, and building shapes.

C. Hachem-Vermette, *Solar Buildings and Neighborhoods*, Green Energy and Technology, https://doi.org/10.1007/978-3-030-47016-6_7

Analysis performed on these scenarios is used to formulate recommendations for enhanced solar capture and energy performance of small-scale neighborhoods, designed for cold northern climate, as presented in Sect. 7.3.

7.1.1 Urban Characteristics of the Selected Examples

Existing guidelines for the design of new communities are scarce, in particular, those that aim at increasing solar access. A number of guidelines of urban design and by-law zoning are combined to generate general characteristics of the neighborhood examples discussed below [5, 6]. Table 7.1 summarizes the urban planning characteristics of the discussed neighborhoods, incorporating the most relevant parameters in the design of small-scale residential neighborhoods. A general approach is employed in these neighborhood designs, consisting of first determining the site layout and density, then designing the housing unit shapes to conform to this layout.

A combination of various building shapes is assumed to demonstrate the interaction, in some situations, between building shape and site layout.

Table 7.1 Urban planning characteristics of the studied neighborhoods

Land use designed	Building total floor area			120 m^2 (designed)
	Lot Coverage Ratio $\left(LCR = \frac{\text{Ground floor area}}{\text{Lot area}}\right)$			37% (calculated)
	Distance from sides	Front		4 m
		Back		6 m
		Sides	Units positioned with respect to a straight road	2 m
			Units positioned with respect to curved road	2–3.5 m
Roads based on [5, 6]	Road width	Neighborhood streets[a]		12–15 m
		Gravel alley[b]		4 m
	Cul de sac	Diameter see Table 7.2		$D1 = 42$ m $D2 = 52$ m
Density based on [7]	Low/medium–Low– outer suburban area (detached units)			5–9 u/a
	Medium–high- outer suburban area (attached units)			16 u/a
	Medium–inner suburban (row townhouse)			up to 35 u/a

[a]Access to residential, includes street width with parking on one or two sides, planting strip, sidewalks on both side

[b]Access to sanitation and utilities, garages, backyard, and secondary units

7.1.2 Housing Units' Shapes

Dwelling units considered in this study are two-storey single-family houses with a total floor area of 120 m^2. Dwelling units' shapes include basic shapes and variations on some of the basic shapes. Basic shapes are rectangle, which serves as a reference, and L-shape [2]. Variations of L-shape consisting of varying values of the relative dimensions of shading and shaded facades and variations to the angle enclosed by the wings of the L-shape (see Fig. 7.1).

These building shapes are selected as representing convex and non-convex shapes for passive solar design. Other basic shapes can be derived from combination or variation of these shapes. Discussion on impact of building shapes is presented in Chap. 5 (Sect. 5.3.1).

Non-convex dwelling units employed in the neighborhood study are shown in Table 7.2. L-shaped units consist of a main wing, which has the shaded facade and a branch containing the shading facade.

The building is positioned with the main wing in E-W orientation and the branch perpendicular to the street, to satisfy architectural and planning decisions, as well as to maximize solar exposure. The direction of the branch (and terminology–see below) depends on which side of the road the unit is situated– south if the unit is located at the north of the road (Fig. 7.2a), and north if the unit is on the south of the road (Fig. 7.2b).

The L-shape and its variations are characterized by a shading depth ratio of ½. This ratio is selected, based on practical, functional considerations, and reduced shade cast on the shaded facade, within the main wing (see Chap. 5). L-variants are characterized, in addition to the depth ratio, by the angle β–the deviation from 90° of the angle enclosed by the branches of the L (Table 7.2). Four values of β are considered in this study–15°, 30°, 45 °, and 60°. L-variants are identified by the letter V followed by a series of characters specifying the position and angle of the branch (Table 7.2). For instance, V-WS30 is a variant with a branch attached to the west end of the main wing, facing south and having an angle $\beta = 30°$.

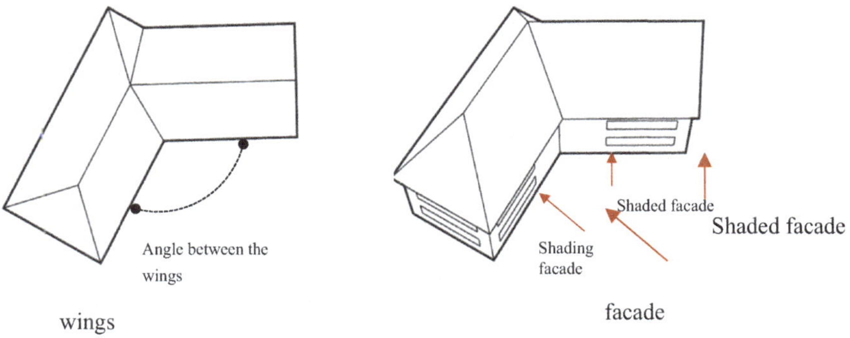

Fig. 7.1 Illustration of L-shape variation

Table 7.2 Summary of shapes studied in the neighborhood, and their characteristics

Direction of L Branch	Shape					
	L-shape	Variations of L-shape				
		L-variant (V)				Obtuse angle
South	(L-WS[a]) 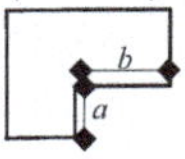	$\beta =$ 15°–West (V-WS15) 	$\beta =$ 30°–West (V-WS30) 	$\beta =$ 45°–West (V-WS45) 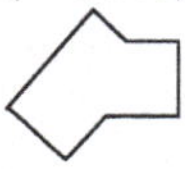	$\beta =$ 60°–West (V-WS60) 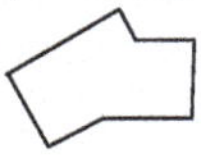	(O-S) 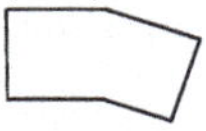
		$\beta =$ 15°– West (V-WS15) 	$\beta = 30°$ – East (V-ES30) 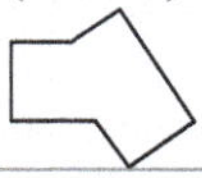	$\beta = 45°$ – East (V-ES45) 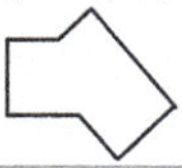	$\beta = 60°$ – East (V-ES60) 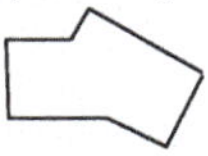	
North	(L-WN) 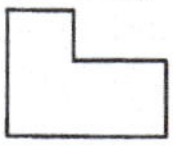	$\beta = 15°$ – West (V-WN15) 	$\beta = 30°$ – West (V-WN30) 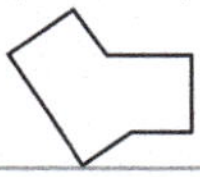	$\beta = 45°$ – West (V-WN45) 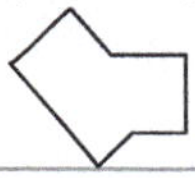	$\beta =$ 60°– West (V-WN60) 	(O-N) 
		$\beta = 15°$ – West (V-EN15) 	$\beta = 30°$ – East (V-EN30) 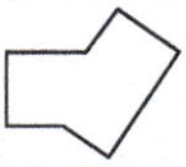	$\beta = 45°$ – East (V-EN45) 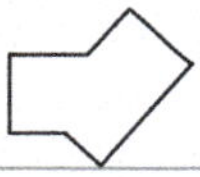	$\beta = 60°$ – East (V- EN60) 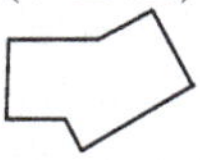	

[a]Solar potential and energy demands of L-E and L-W are not significantly different (with L-W performing slightly better than L-E)

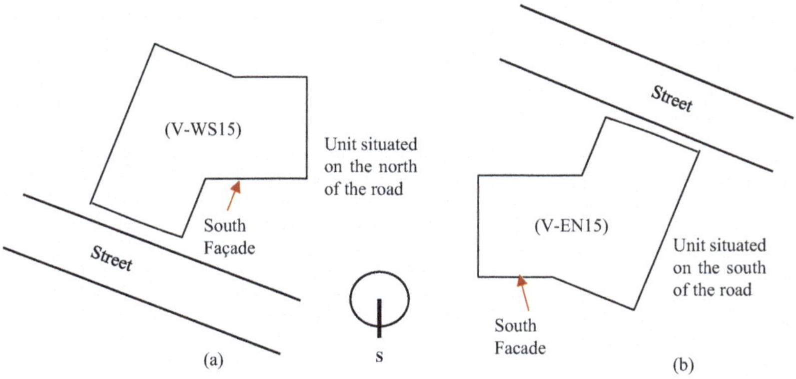

Fig. 7.2 Illustration of L-shape design with respect to road orientation

An additional shape *Obtuse angle* (O) represents a special variant of L-shape, with larger values of the angle β ($\beta = 70^\circ$ is adopted). The obtuse shape may be with the main wing facing south direction–O-S or north direction–O-N.

7.2 Main Parameters and Their Impact

This section presents the main parameters in the design of small-scale residential neighborhoods. As discussed above, three main design considerations affect the solar potential of a neighborhood, namely, building shape, density, and layout of the site. Among these, two design considerations, building density and the site layout are related to the spatial design of the neighborhood. Each of these parameters is associated and interacts with several secondary parameters, such as units' shapes, their orientation, and their relative position.

These design parameters affect the amount of solar radiation reaching the housing units, not only through an adequate exposure, but also because of shading impact. A summary of shading effect on solar radiation is followed by a more detailed analysis of the impact of site layout and density.

The impact of site layout is determined by analyzing various representative sites and road designs, and the arrangement of housing units with respect to these roads. Impact of increased density is studied through reducing the spacing between the housing units, increasing thus the number of houses within specific area, as well as adding row of houses (on the south–north axis).

The discussion presented below is based on simulations of hundreds of neighborhood patterns employing various combinations of parameters (Tables 7.2 and 7.3).

7.2.1 Shading Effects

A major effect on solar potential of residential neighborhoods is mutual shading by adjacent dwelling units. Two parameters define the relative position of the shaded and shading units: the angle of obstruction and the distance between the units. The *planar obstruction angle* (POA) is the angle between the line connecting center of the south facade of the shaded unit and the closest corner of the shading unit and the south direction. The second parameter is the distance (d) between the shaded and shading units along the same line (Fig. 7.3).

The impact of mutual shading on the solar potential of identical housing units, as well as on energy demand is summarized below.

Solar Potential

The solar potential is studied through two main metrics: Solar irradiation and potential generation of electricity from PV integrated on southern roof surfaces.

Table 7.3 Main parameters employed in the site layout study

Parameters and values			
Shape	Site layout	Density	
		Spacing effect (s)	Row effect (r)
R* – Rectangle/Trapezoid L – L V – L-variant O – Obtuse angle	I* – Straight East–west (refernce case); Inclined (±30°, ±45°, ±60°) II – Curved south, with diameter: $D1 = 42$ m (associated with s_1) $D2 =$ 52 m(associated with s_2) III – Curved north ($D1$, $D2$)	$s_0 = 0$ (attached) $s_1* =$ { 4m – site I, detached rectangles in sites II, III; 4m-7m detached L – variants in sites II, III $s_2 = 2.s_1$	r_0* – no 2nd row r_1–5m r_2 – 10 m r_3– 15 m r_4 –20 m

Parameters marked * serve as reference for assessing effects

Solar Irradiation
The analysis of the effect of obstructing a rectangular dwelling unit by an identical unit, at different angles of obstruction and different distances, shows that the largest yearly incident radiation reduction occurs when the shading unit is aligned with the shaded unit. The effect of obstructing a dwelling unit by two identical units placed symmetrically with respect to the shaded unit is almost double the effect of a single shading unit.
The shading effect is inversely related to POA and *d*. For 5 m distance, at 15° POA, the annual reduction in incident and transmitted radiation is about 45%, while for a 15 m distance the reduction is about 10%.
The most significant impact of mutual shading occurs during the winter period, due to the position of the sun affecting the length of the shadow (the longest shadow in the year occurs on the 21st of December).
Energy Generation
The analysis of the impact of the mentioned parameters on PV electricity generation shows that for a distance larger than 5 m, no shadowing effect on electricity generation is observed. In addition, electricity generation of non-aligned units is not significantly affected by the POA and distance between the units (≤3%).

Energy Demand
Mutual shading between housing units affects the energy demand for heating and cooling. Heating load decreases while cooling load increases with increasing distance (d) between the units. For instance, the increase in heating load as compared to

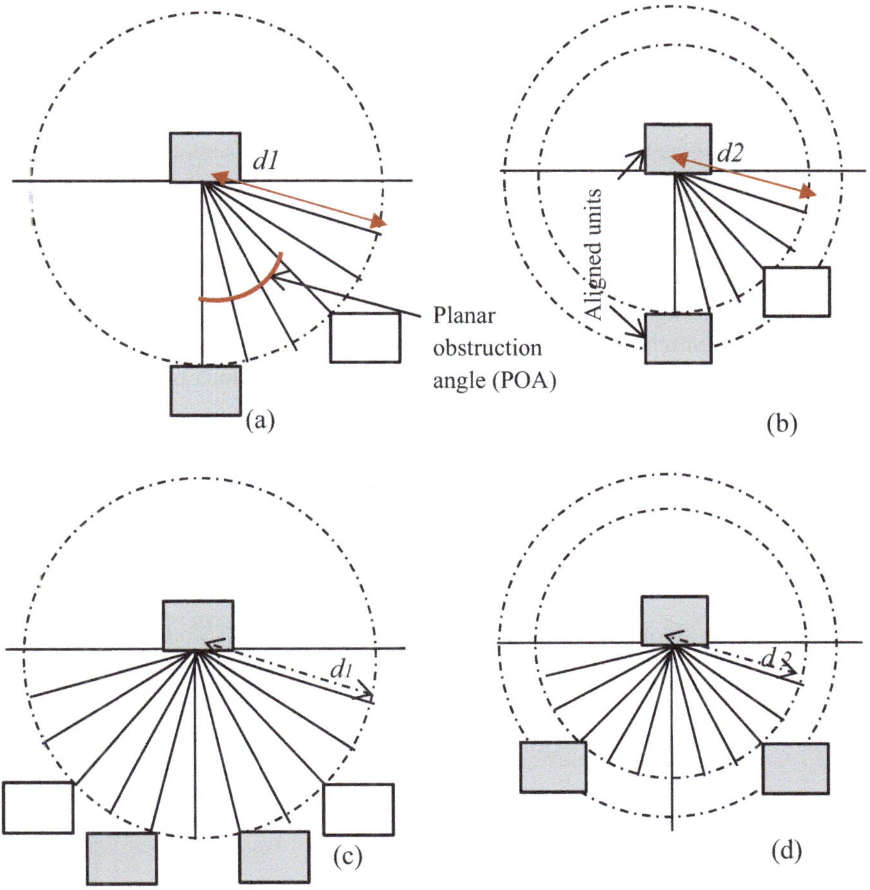

Fig. 7.3 POA concept, shading and shaded identical units are represented by solid color; shaded unit is in the center of the circle; **a** and **b** single shading unit and different distance, **c** and **d** two shading units, and different distance

unobstructed unit can reach 35% for a POA of 15° or less, at a distance of 5 m. In scenarios where the dwelling unit is shaded by two obstructing units, the increase in heating load, relative to the unobstructed unit is significantly higher (almost double).

7.2.2 Site Layout

Three basic site layouts are selected, to present as general approach as possible, capturing representative designs of small-scale residential neighborhoods. Site I is characterized by a straight road, while sites II and III feature curved roads facing south and north, respectively. The curved roads are selected to represent scenarios of *cul-de-sac*. Figure 7.4 presents the sites and configurations studied to assess the impact of the site layout.

In addition to these basic sites, variations on site I incorporating varying straight road orientations, are also discussed (Fig. 7.5).

Table 7.3 presents the main parameters and their values associated with building shape, site layout, and density (spacing and row effect, detailed below), and assumed in the examples presented below. Spacing between houses, on east–west axis, is referred to by s, while the spacing on the north–south axis is identified by r. The effect of density (i.e., r and s, in Table 7.3) is discussed below in Sect. 7.2.3.

The effects of site layout on the solar potential and energy performance of neighborhoods are highlighted below. The impact of various sites layout is determined by comparing specific designs of layouts of site I (with various road orientation),

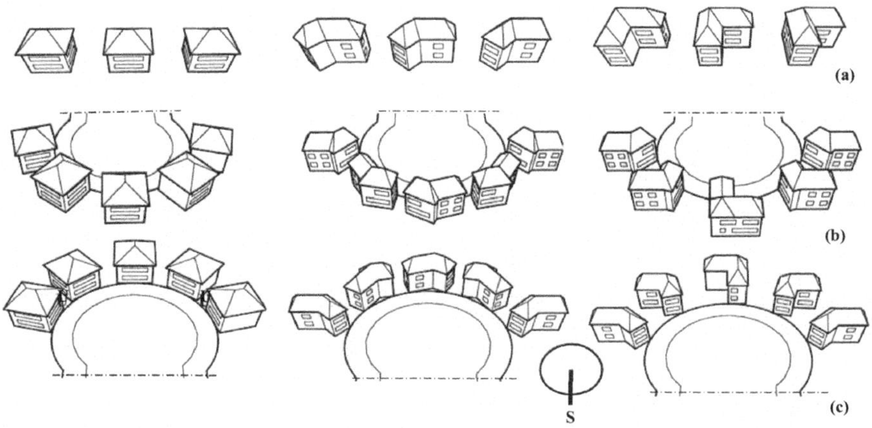

Fig. 7.4 Configurations of shapes in different site layouts: **a** Site I; **b** Site II; **c** Site III

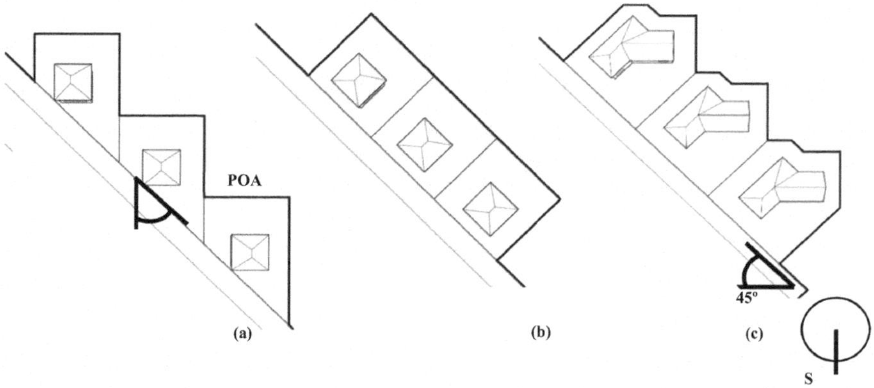

Fig. 7.5 Variation of site I, **a** south-facing rectangle, **b** rectangles-oriented to the street (R-O), **c** L-variants (V)

site II, and site III, to a site with straight east–west (E–W) oriented road. The site layouts are compared for the three housing units' configurations that are common to the three sites (see Fig. 7.4). These configurations are detached rectangles, detached L-variants, and attached L-variants, with spacing s_1 (Table 7.3).

For each site layout, the effects of varying house unit' shapes are also summarized.

Straight Road–Site I

Six directions of the straight road are studied, in addition to the east–west running road (Fig. 7.5), with the E-W road rotated by 30°, 45°, and 60° in each of clockwise (+) and anticlockwise (−) directions.

For each of the inclined layouts of site I, 3 housing units' configurations are analyzed: south-facing rectangle (R), rectangle-oriented toward the street (R-O), and L-variants (V), with the angle between the wings to conform to road direction, with the shaded wing facing south. For this site analysis, only units north of the road are considered (see Sect. 7.1.2, and Fig. 7.2).

Layout with R-O-shape (Fig. 7.5b) is employed for assessing the effect of road orientation without interaction of shape effect

Effect of Road Orientation

The effect of road orientation on solar potential and energy demand is summarized below for detached units (Fig. 7.6).

Solar Potential

Solar Irradiation. The effect of road orientation on solar radiation on facades of units along the road is influenced by the orientation of the units and their facades, as well as by mutual shading by adjacent units. For example, for the configuration shown in Fig. 7.5a, presenting a road inclined by 45°, the solar radiation incident on the south-facing rectangular shape is reduced by a maximum of 9%, relative to the E-W orientation, due to shade cast by adjacent units. For the configurations

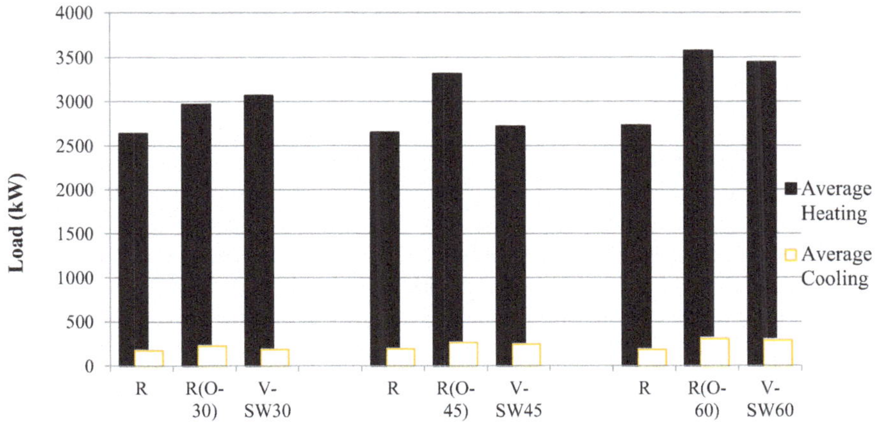

Fig. 7.6 Comparisons of heating and cooling loads of all configurations (R-south-facing rectangles; (R-O) rotated rectangles and V-shapes), of the inclined road sites

where the whole rectangular unit is oriented toward the street (Fig. 7.5b), there is no shading effect and the radiation on the facade changes according to the orientation of the housing unit (as explained in Chap. 5).
Solar radiation incident on L-variants (Fig. 7.5c) is only affected by the shape of unit itself–self-shading. This is mostly based on the angle between the wings and the depth ratio (see Chap. 5).
Energy Generation. No significant effect is found in the case of the south-facing rectangles or L-variants. In configurations where rectangular units are oriented to face the road, electricity generation is only affected by the orientation. Deviation from south orientation leads to some reduction in the total yearly energy generation; however, it allows some spread in the time of energy generation during the day. This can be beneficial in grid connected PV systems, as it allows supplying excess of electricity to the grid at different time of the day, thus reducing potential stress on the local grid (as discussed in Chap. 5).
Energy Demand
The effect of road orientation on the energy demand for heating and cooling of selected units is summarized in the following:

- The energy demand for heating of the rectangular south-facing configurations (R) is not significantly affected by the road inclination. Change in demand (depicted in Fig. 7.6) is mainly due to the shade cast by adjacent units on windows, reducing thus potential of solar heat gain.
- Rectangular configurations that are oriented parallel to the road (R (O)) require significantly larger heating energy (by up to 35%) as compared to the south-facing rectangular E-W road configuration.
- L-variant (V-WS60) requires up to 30% more energy than the same unit simulated in isolation that does not take into account adjacent buildings. This increased energy demand is primarily due to the relative position of the units, where a large portion of the south-facing facade is shaded. This effect can be reduced by increasing the distance between units.

Figure 7.6 presents the effect of variations of site and residential units' position on average heating and cooling loads of all studied units shapes.

Curved road–Sites II and III

Site layout II and III incorporate semi-circular roads, facing south and north, respectively. The housing units are positioned with respect to the shape of the roads, in both curved sites. Configurations of these sites include rectangular shape (R), combination of L-shape and its variants, and a configuration of obtuse angle shapes.

The effects of density and site layout are strongly coupled for curved layouts. The main comparison for overall site effect is between sites with curved road (II and III) and site I (with E-W road). This comparison can only be applied to configurations of similar unit shapes and density. The effect of curvature on solar performance of individual (detached) units of a given shape in the neighborhood can be assessed by reference to the isolated, south-oriented unit of the same shape.

The analysis in this section is restricted to the detached (lower density scenarios). Impact of layout associated with higher density is presented in the section "Density", below.

Solar Potential

Solar Irradiation

In each detached assemblage, the incident radiation on the near south facades of individual units and transmitted by their windows is compared to isolated south-facing units to assess the effect of shade from adjacent units.

The rectangular units of site I are positioned in straight layout facing south and therefore the south-facing facades of detached units are not affected by adjacent units. The rectangular units of site II and III are positioned around the curved roads. The impact of site curvature on the rectangular building configurations is obtained by comparing sites II and III to site I. The main impact on solar radiation is due to the orientation of the units from due south. Reduction of incident radiation on the south-facing facades of the rectangular building shape, on a winter day, ranges from about 4% on the central units to up to 30% for some adjacent units. This depends on the distance between the units and on the POA value which determines the extent of shading.

For detached L and L-variants in site II and III, similar to the rectangular configurations, the main effect, as compared with site I, is due to the rotation of units relative to the south, along the curved layout. In addition, some units in these configurations cast shadows on facades of adjacent units.

Energy Generation

The response variable for the comparison with the reference (site I) is electricity generation per unit area averaged over all units in a neighborhood. No significant effect of the site layout is indicated on overall electricity generation per unit area. A maximum reduction of about 3% is observed in the generation of the detached rectangle configuration in sites II and III as compared with the similar configuration in site I.

An important result of the interaction of site layout and configurations is the shift of peak electricity generation. A significant shift of the profile of the electricity generation is obtained by the BIPV of different units. As discussed above, in site I the difference in timing of peak electricity is due to the rotation of the south wing of L variation shapes. A maximum shift of 3 h is obtained in site I (with E-W road). In sites II and III, an additional source is the rotation of whole units. A difference of peak time of up to 6 h is observed in the configurations of site II and site III.

The graph of Fig. 7.7 shows the electricity generation profiles of the hip roof of wings of units of detached L-variant configurations of site II for the winter design day (WDD). The hip roof production profile (and not the whole electricity production) is shown because only the hip portions of the roof change orientation in the L-variants. It should be noted, however, that the hip constitutes a small portion of the electricity generating roof surface and the overall effect on total generation would be reduced.

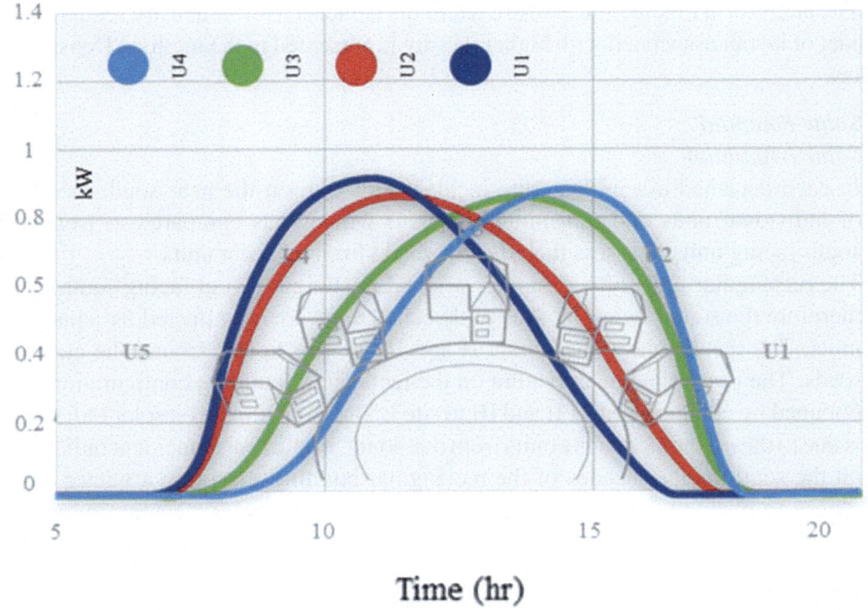

Fig. 7.7 Hourly electricity generation (from 4–6 AM to 6–8 PM) (kW) for site II, on a WDD on the hip of L-variants of detached L-variants

Energy Demand

Only the cooling load increases in site II and III, relative to site I. For instance, the average cooling load of the rectangular configurations is increased by approximately 45% and 48% for site II and site III, respectively. However, the energy demand for cooling is low relative to heating (<10%, for the rectangular configuration in site II). Figure 7.8 presents the comparison of heating and cooling load between different configurations of site II and site III, and those in site I.

One important reason for the increase of cooling load in sites II and III, particularly for rectangular units, is the rotation of units around the curve toward east or west. Since these units are originally designed to be south facing (with 35% south-facing windows), their orientation toward the east or west results in increased transmitted solar radiation in the morning and the evening, when the sun is at low altitude during the summer period. This can be resolved by modifying the window area for the rotated units.

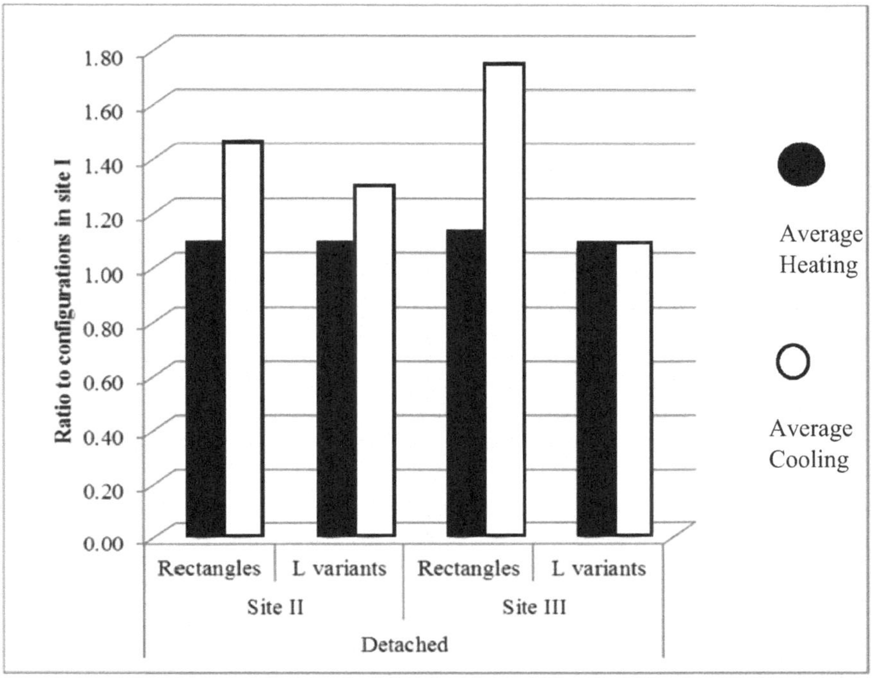

Fig. 7.8 Comparison of thermal load of configurations of sites II and III, to those of site I

7.2.3 Density

Density is considered by the spacing between units in a row (s) and by the spacing between rows of units (r). The impact of various levels of density is presented below.

Spacing of Units

Three values of spacing are adopted for each site: s_1, the basic spacing of detached units is assumed as 4 m in site I. The spacing between detached units in a curved site varies (4–7 m), depending on the curvature of the road and the shape of the units. In order to assess the influence of increased spacing on energy demand, a second spacing $s_2 = 2s_1$ is adopted (see Table 7.3).

At the other extreme, the highest density is obtained by attaching units in triplex, quadruplex, or quintuplex configurations, with $s_0 = 0$.

Three shapes are employed in attached configurations of each site. In site I, these shapes are rectangle, L and L-variant shapes (Fig. 7.9a–c). In sites II and III, the rectangle is replaced with trapezoid to allow the attachment of units along the curve. The south-facing curve of site II implies that the narrower side of the trapezoid is south or near south facing (Fig. 7.9d), whereas for site III the wider side faces south (Fig. 7.9f). The other two shapes are L-variants and obtuse angle (Fig. 7.9e–i).

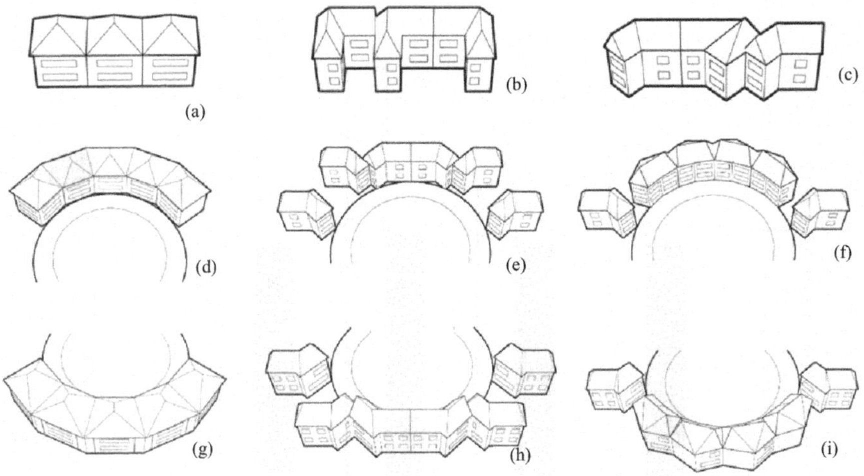

Fig. 7.9 Attached units in sites I, II, and III. Site I: **a** rectangular, **b** L-shape, **c** L-variants; Site II: **d** trapezoid; **e** L-variants; **f** obtuse angle. Site III: **g** trapezoid; **h** L-variants; **i** obtuse angle

Layouts of attached non-convex shapes in sites II and III include, in addition to the four central attached units, two detached units at the extremes of the curve for improved site design (Fig. 7.9e–i). These detached units are not included in the analysis for density effect.

Rows

The effect of obstructing the south facades of selected configurations by a row of similar housing configurations–the *row effect*–is investigated for site I. The minimum distance between the two rows to avoid shading is estimated, based on the shadow length for the 21st of December, associated with the lowest sun altitude at solar noon. The minimum spacing to avoid row shading for the units studied is *ca.* 25 m.

To assess the effect of shading, four values of row spacing (r) are studied: 5 m, 10 m, 15 m, and 20 m, corresponding to 20%, 40%, 60%, and 80%, respectively, of the minimum spacing between rows to avoid shading. The studied configurations are the detached and attached rectangular units, the detached and attached L units, and the detached and attached configurations of L-variant (V-WS30) (Fig. 7.10). Two configurations of L units are studied, with the branch facing south L-S and the branch facing north L-N.

Effects of Spacing

Solar Potential

Density effect is strongly coupled with the site effect and some aspects related to this factor are considered under site effects, above. A summary of the main impact of spacing between the units on solar irradiation and energy demand is presented below.

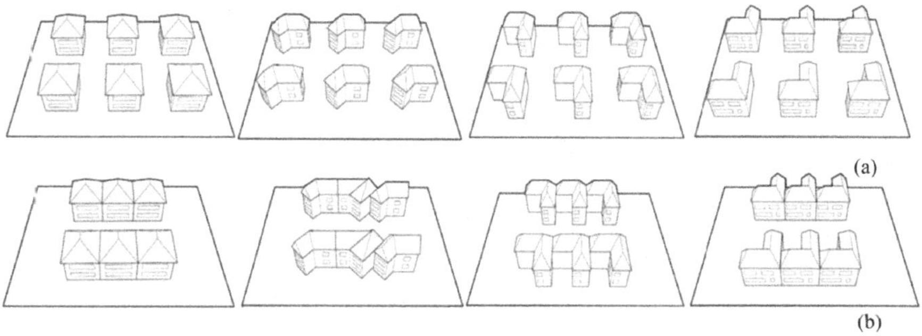

Fig. 7.10 Row configurations of all studied shapes; **a** detached configurations; **b** attached configurations

Solar Irradiation
The effect of spacing on solar irradiaiton is mainly observed in sites II and III, and variations of site I (inclined road variations). The effect is mostly related to increased mutual shading between units. In sites where L-variants are included (Figs. 7.9c, f, i), self-shading of units can further reduce solar irradiation on south facades.
Energy Generation
There is no significant difference in average electricity generation per unit surface area between attached and detached configurations of a given shape. A maximum reduction of 3% or less of the average annual generation is observed in the attached units of L-shape due to self-shading of the protruding wing roofs.
Due to the replacement of rectangular with trapezoid shapes in sites II and III south-facing roof surfaces in site II are reduced and in site III increased. As a result, average annual electricity is reduced by up to 10% for site II and increased by similar amount for site III.
It should be noted that in many cases attaching units enable larger roof area. Roofs can be redesigned to obtain a continuous roof (as compared to individual hip roofs in the present neighborhoods) which enables the design of larger PV system.
Similar to the detached configurations presented above, the shift of peak of electricity generation is noticed in these attached configurations. Examples are presented in Fig. 7.11, for the attached rectangular configuration and attached L-variants. A maximum shift of peak of about 6 h can be achieved.
Energy Demand
Comparison between Detached and Isolated Units. In general, heating load increases for detached units in a neighborhood while cooling load decreases, as compared to isolated units (with no neighboring buildings) (Fig. 7.12). One reason for this effect is the shade cast on the east and west facades, in all configurations, and partially on south-facing facades in sites II and III. The shade on south-facing facades depends on the POA effect described above.
Comparison between Attached and Detached Units. Energy demand for heating and cooling of attached units is lower than for the corresponding detached

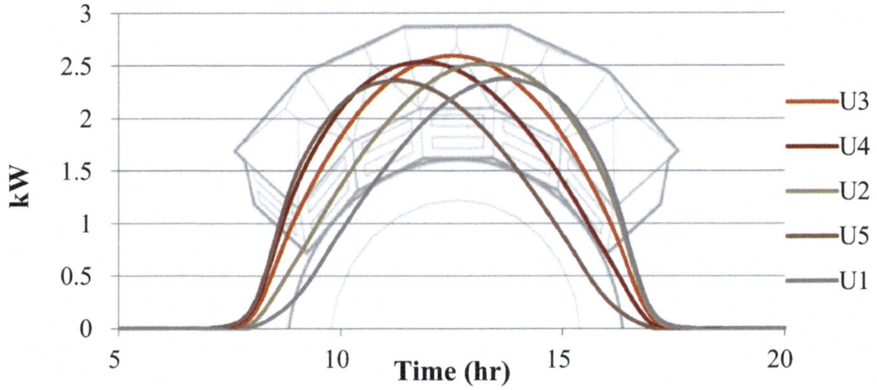

(a)

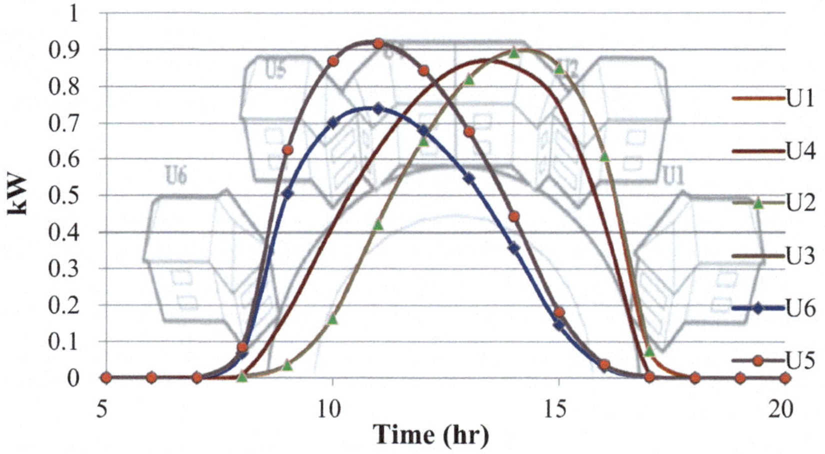

(b)

Fig. 7.11 Hourly electricity generation (from 4–6 AM to 6–8 PM) (kW) for site II, on a WDD: **a** on the total south roof of attached rectangular (trapezoid); **b** on the hip of L-variants of attached L-variants

configurations. For instance, heating demand of the attached studied rectangles and attached obtuse angle configurations in site II is reduced by 35% and 20%, respectively, relative to the detached units.

On the other hand, doubling the spacing of units on the east–west axis (from s_1 to $s_2 = 2s_1$) does not affect significantly the heating energy consumption. Figure 7.13 presents the impact of spacing between units on heating and cooling loads.

For the attached configurations, the method of assemblage of L-variant results in self-shading geometries (non-convex configurations). The decrease of solar radiation, relative to the corresponding detached unit, can reach up to 30%, in

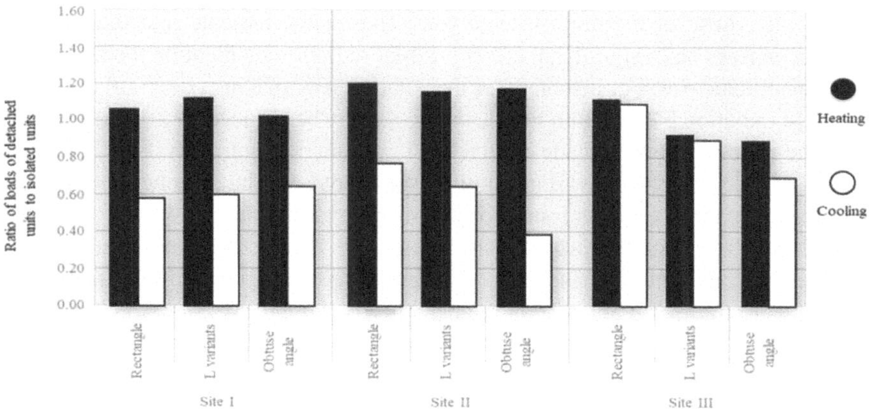

Fig. 7.12 Comparison of heating and cooling demand between isolated units and detached units in a neighborhood

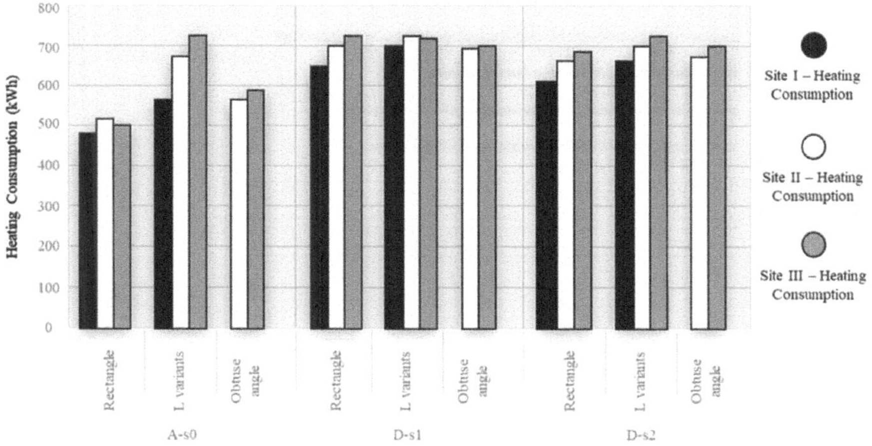

Fig. 7.13 Heating consumption at different spacing between units

some of the studied cases, depending on the ratio of the dimensions of mutually shading facades (within the assembly of building units).

Row Effect

The row effect on solar potential and energy demand for heating and cooling of various configurations is summarized below.

Solar potential

Solar Irradiation

The row effect on incident and transmitted radiation in site I is assessed by comparison of the corresponding radiation levels on south facades of the obstructed row with those of the unobstructed row. The results indicate that the distance between rows has a significant effect on solar radiation incident on the facades of the obstructed row. For instance, for the smallest row spacing ($r = 5$ m) the transmitted radiation by the south facade windows may be reduced by as much as 95% on the exposed branch facade of attached L-variant configuration (Fig. 7.14a), by 60% for the detached L-variants and by some 85% for the detached rectangular shapes. A similar effect is found for the incident radiation.

At 20 m, corresponding to a density of about 8-10 u/a, the maximum effect is about 13% reduction of incident and transmitted radiation on the attached L-variant configuration and 9% on the detached rectangle. At 20 m, the row effect on the shaded facades of L-variant configurations is small, reaching a maximum reduction of about 6% for WDD.

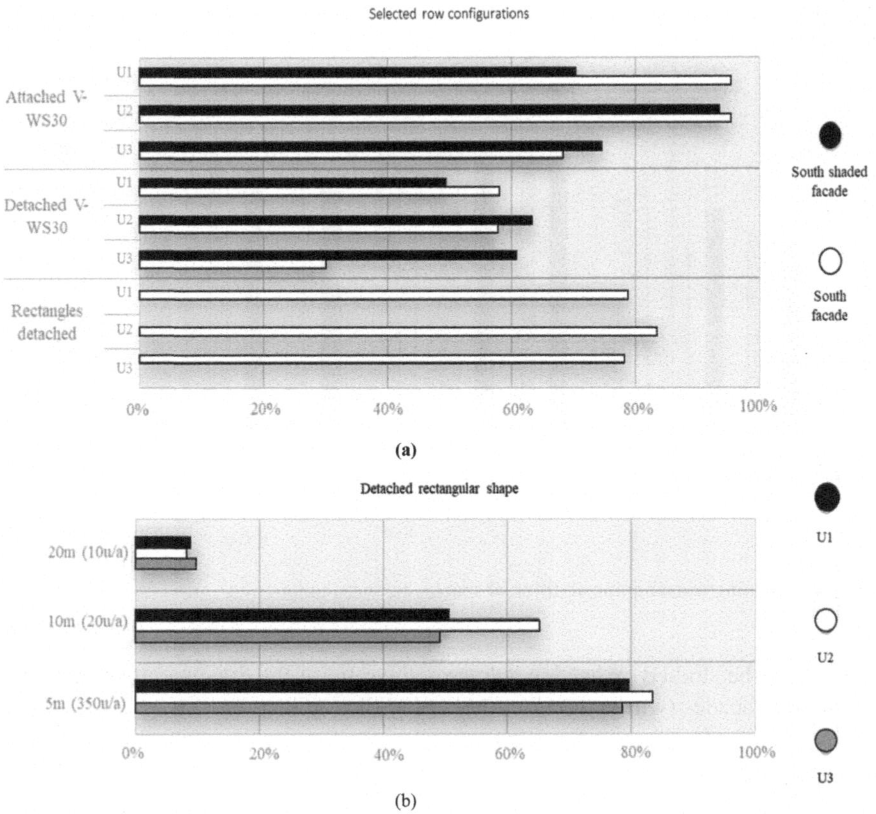

Fig. 7.14 Reduction in transmitted radiation due to row effect for WDD. U1, U2, and U3 are the units of the shaded row (U2 is the middle unit): **a** Effect on selected configurations at 5 m row separation; **b** Effect on detached rectangular units, at separations of 5, 10, and 20 m

Figure 7.14a and b represent the results in terms of the reduction of solar radiation transmitted by the windows of the obstructed row as compared to the corresponding unobstructed row, for some selected configurations. For non-convex shapes (VWS30), the effect is given for the shaded and for the exposed facades.

Energy Generation

The row effect on electricity generation is measured by comparing the electricity generation of the roofs of the obstructed rows to that of the exposed row. The examples show that there is no significant impact.

Energy Demand

The row effect on heating and cooling loads is assessed for site I by comparing the loads of obstructed and obstructing rows to the corresponding isolated row. Results indicate that, generally, the average heating load increases for the units of the obstructed row (R2), while the cooling load decreases.

The heating load of the obstructed row of detached rectangular units (Fig. 7.15) increases by ca. 50% at 5 m row spacing and by 25% at 10 m spacing. For attached rectangular units (Fig. 7.16), the increase in the heating load of the obstructed row is about 70% at 5 m row spacing and 30% at 10 m spacing. At 20 m there is no significant effect.

Similar effect is observed for all other types of housing units. The impact on L-variant is somehow smaller, where the obstructed row of detached units requires 25% more heating at 5 m spacing, and 10% at 10 m. The attached units of L-variant in the obstructed row require 35% more heating at 5 m, and 15% at 10 m spacing.

The impact of spacing between rows of housing units is presented in Figs. 7.15 and 7.16, for attached and detached configurations.

Figure 7.17 presents the ratio of heating load of the obstructed to isolated row, for detached rectangular shape, expressed as function of the distance between rows. This distance is presented as percentage of the minimum distance between rows to avoid shading ($x = 25$ m). The figure presents also the regression equation of the exponential curve and its value at $x = 25$ m.

7.2.4 Evaluation of Energy Balance of Neighborhoods

In this section, energy consumption and supply are compared for the different configurations analyzed in this chapter.

Isolated Units

The total consumption of electricity for lighting, DHW, and appliances, in addition to the computed heating and cooling energy consumption, for isolated south-facing units of each building shape employed in the neighborhood example (Table 7.2) is presented in Fig. 7.18, alongside the energy production by the BIPV systems of the corresponding units. The electricity production of a rectangular layout with hip roof

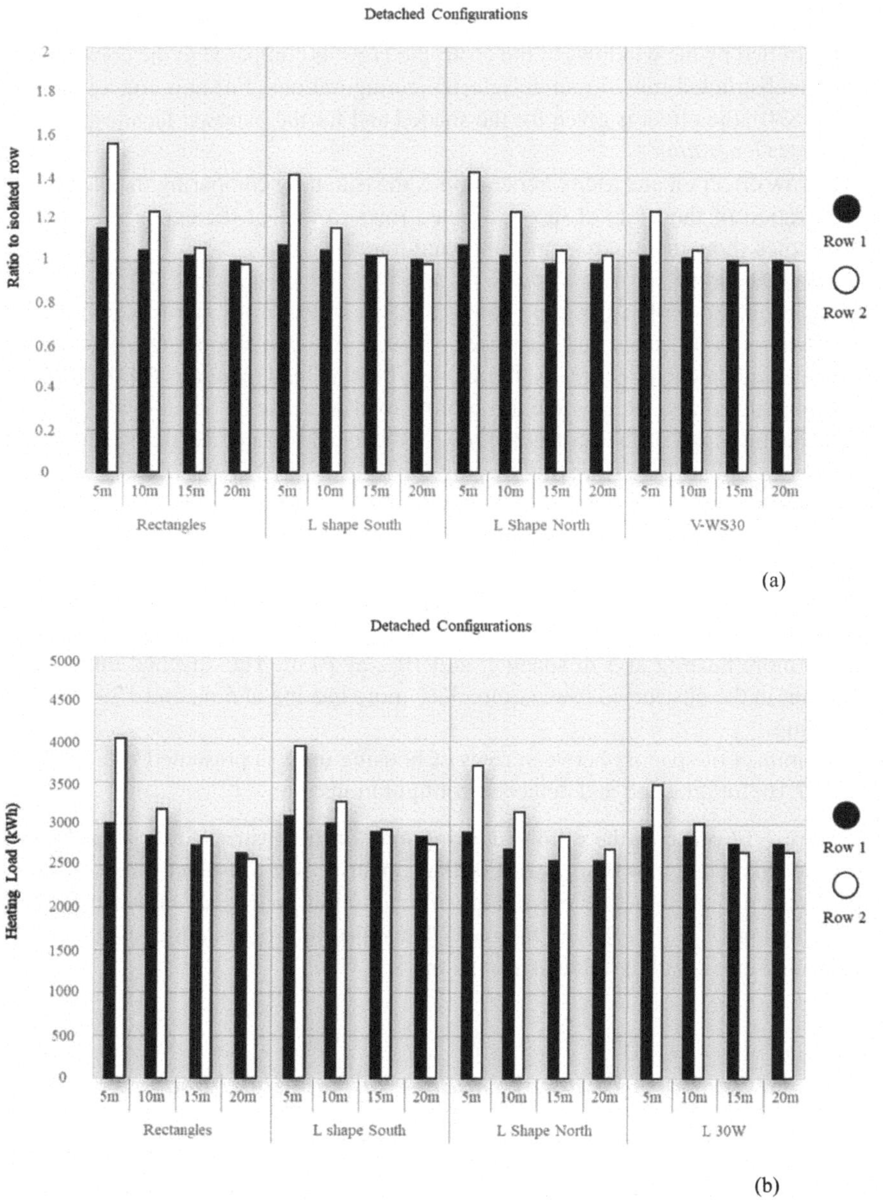

Fig. 7.15 Detached units **a** Heating load of two rows relative to isolated rows, **b** Heating load of the two rows of detached units

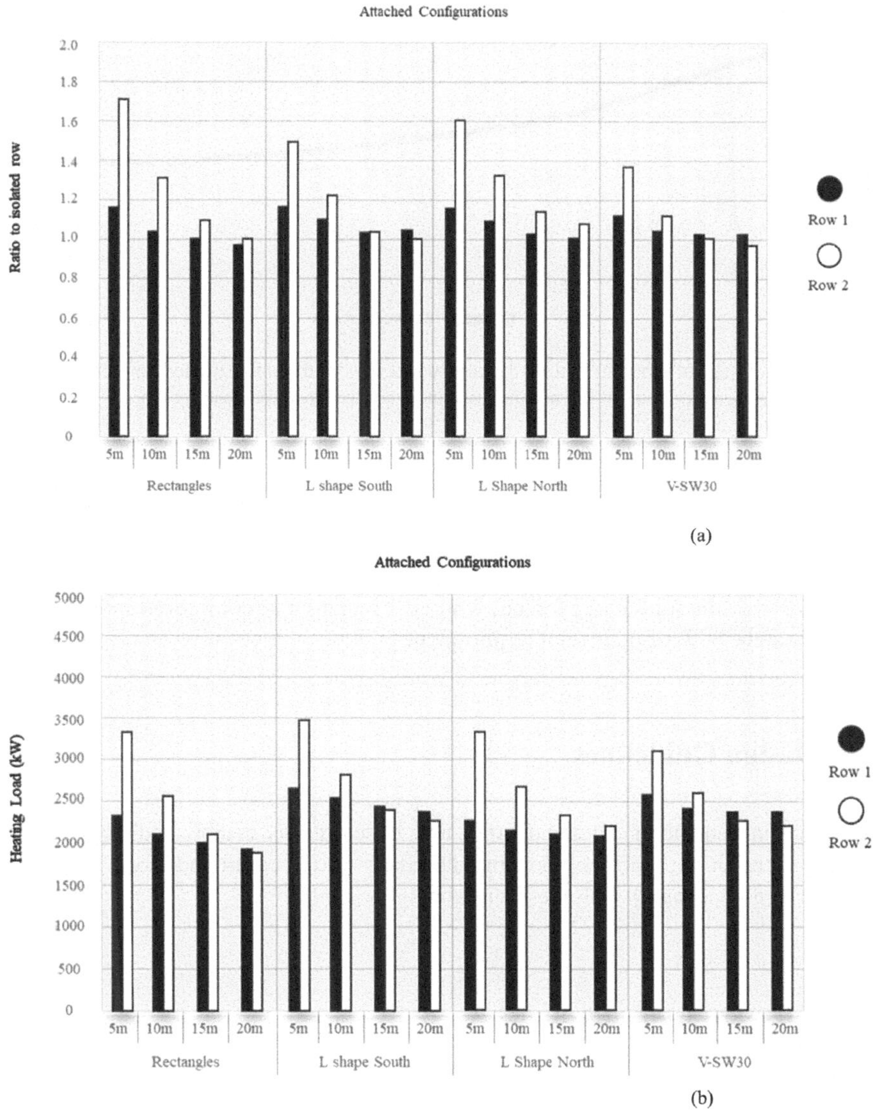

Fig. 7.16 Comparison of the row effect in site I – R1 exposed row, R2 obstructed row: **a** Comparison to isolated row, **b** Heating loads of the two rows

is some 65% of consumption. Some L-variants, such as V-EN30W, produce up to 96% of total consumption.

Neighborhoods

The analysis of the total energy supply and consumption balance for assemblages in all sites shows that these configurations can supply significant portion of their energy

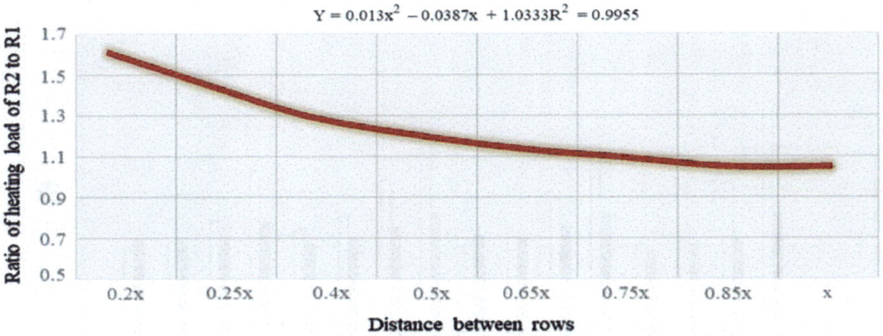

Fig. 7.17 Ratio (y) of heating load of the obstructed row to the unobstructed row of detached rectangular units, as function of the minimal distance required to avoid shading (x)

demand. For instance, L-variants in both site I and site II generate around 80% of their total energy consumption. In site III, L-variant shape is optimal for detached configuration while the obtuse angle is optimal for the attached configuration. These configurations generate 85% of the total energy consumption. In site I, L-variants can supply 79% of the total energy need, while the rectangular configuration generates approximately 65% of their total requirement.

7.3 Design Guidelines

This section assembles all the results and observations established above, and present them in the form of design guidelines and recommendations for solar high-performance small-scale, residential neighborhood.

7.3.1 Design Considerations

Table 7.4 summarizes the main design considerations for solar optimized dwellings and neighborhoods. These design considerations are suggested based on various objectives, for instance, minimizing heating and cooling loads, maximizing energy generation, and enabling shift of peak energy generation.

Design considerations, relating to building shapes (presented in Chap. 5, and summarized below) and to the impact of site layout and density, as they interact with each other and with the building form, are also summarized below.

Building shape
Several parameters should be considered in the optimization of building shapes for passive solar design. Although rectangular layout is considered the optimal shape

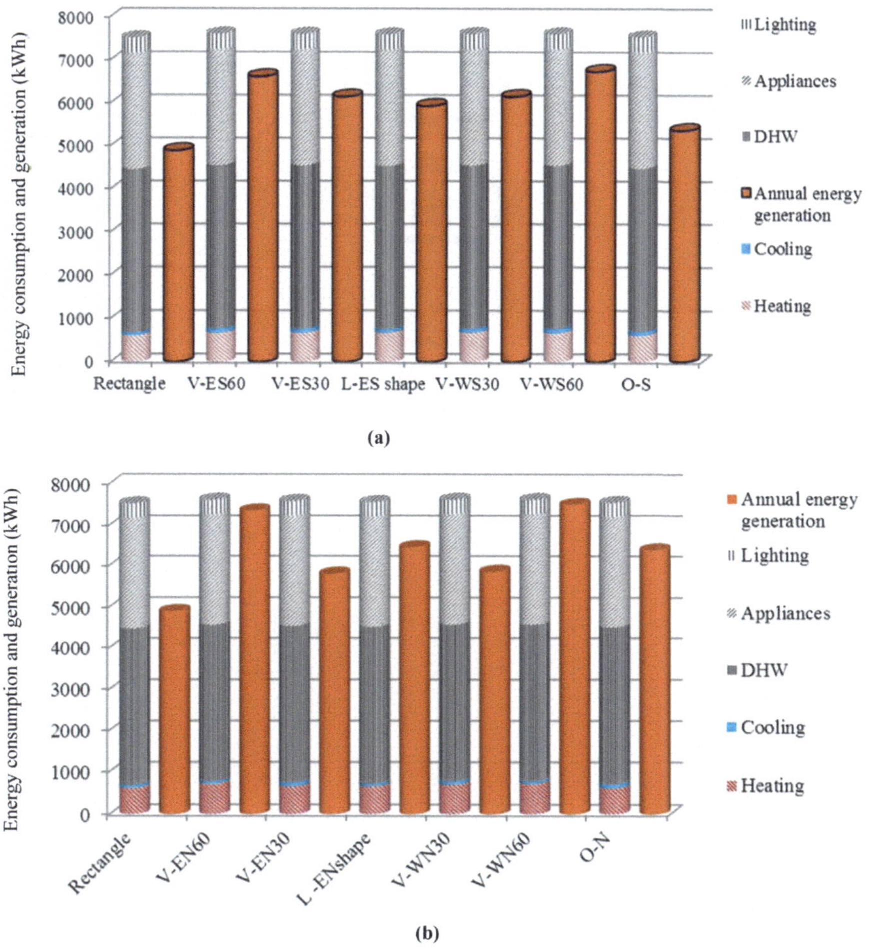

Fig. 7.18 Energy consumption and production for isolated units of different shapes: **a** Shapes of sites I and II; **b** Shapes of site III

for energy efficiency, non-rectangular and particularly non-convex shapes offer a wider flexibility in architectural as well as solar design. Non-convex building shapes involve usually some design complexity, and thus efficient performance of these shapes is governed by a number of design parameters.

The design parameters affecting the performance of rectangular shapes include the aspect ratio, the orientation of the building and its main facades, and the window area. Parameters affecting non-convex shapes are mostly the shading depth ratio, number of shading facades, and angle between these facades. These parameters are presented in Chap. 5. Considerations of building shapes in the context of neighborhoods are summarized in Table 7.4.

Table 7.4 Considerations for solar neighborhood design

Site	Objective	Density		
		Low	Medium/High	Row configuration
Straight E-W road	Minimize heating and cooling energy consumption	**Convex shape** Aspect ratio should be (1.3–1.6) • South-facing window: 35% is a good compromise to reduce heating load while not significantly increasing cooling load • Orientation up to 30°, otherwise L-variant should be used **Non-convex shapes**: • SDR ≤1/2 • Wing-oriented north more advantageous Use of L-variants with: • Rotation of the wings ≥30° from south • Windows on the rotated facade should be readjusted (from 35%) to reduce cooling load.	**Attached units** • Rectangular shapes are optimal, for reduction of heating and cooling load; energy generation is enhanced when continuous roof surface is applied (gable roofs–Chap. 5) • Attached L-variants (see recommended design of L-variants) are good alternatives, enabling larger roof area, for increased potential electricity/heat generation • Attached L-shapes are advantageous when a small shading depth ratio is used. Otherwise, L-shape with south branch should be avoided, to eliminate shading on south facade from the wings of the adjacent units 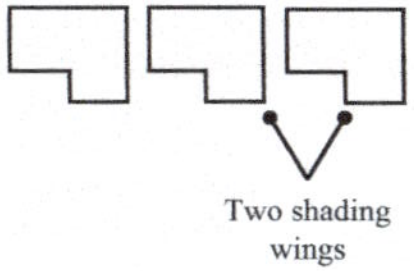	**Detached units** • Distance between rows should be at least twice the height of the shading house, to avoid significant increase of heating load. Staggering detached units in the exposed and shaded rows can present also advantages 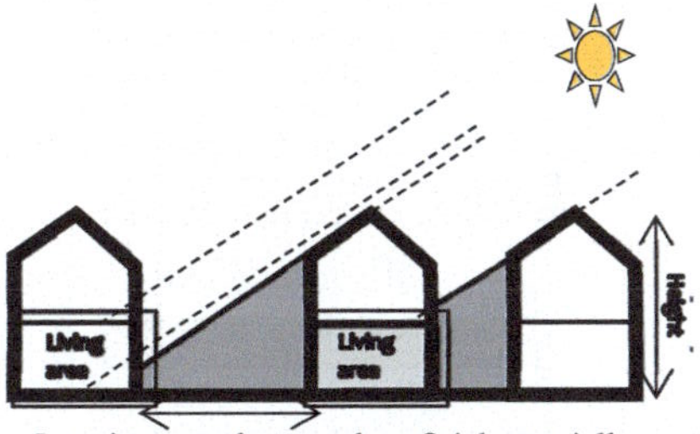 • L-variants can be more beneficial especially when the land available is small. Due to the fact that the facades are not coplanar, a smaller distance between the rows is required  **Attached units**: Design recommendations for attached units should be followed, in addition to respecting the distance between rows.

(continued)

Table 7.4 (continued)

Site	Objective	Density		
		Low	Medium/High	Row configuration
Straight E-W road	Maximize electricity generation	Designing large south roof areas for the integration of PV systems, with appropriate tilt angle (which depends on the geographical location of the building): • Hip roofs: large side angle or/and large aspect ratio should be used, allowing larger south-facing roof area. Gable roofs (with side angle of 90°) can be recommended as simple solution to maximize the surface area • In case of limited roof area (associated with a small floor plan) multifaceted roofs that join different tilts, with different orientations may be considered. These can increase significantly the total roof area, in addition to optimize the tilt angle for summer and winter solar radiation • L-shape and L-variants: Design the shapes to reduce shade on south roof (as detailed above) 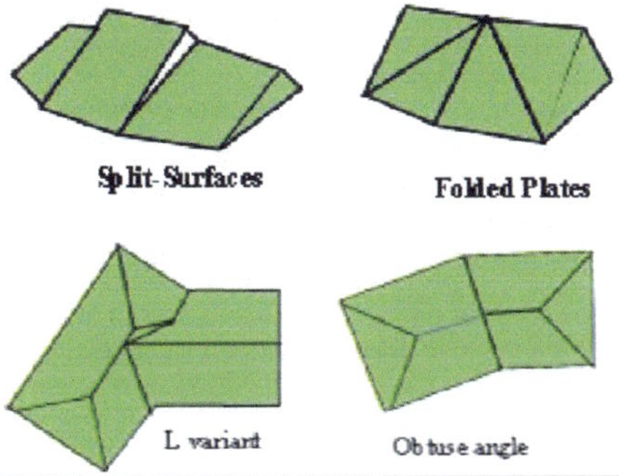 	• For rectangular shapes, it is recommended to use more advanced roof design for increased productions (as shown in Chap. 5) • Gable roofs should be used instead of hip roofs, to increase roof area • For L and L-variants, recommendations for detached units apply to attached units. Careful design of the number of shading facades and the depth ratio (DR) is suggested (see above)	• For buildings with the same height, distance between rows does not affect electricity generation • For attached units, careful design of the number of shading facades and the DR (same as for attached units)

(continued)

Table 7.4 (continued)

Site	Objective	Density		
		Low	Medium/High	Row configuration
	Peak shift	• For rectangular units, split-surface or folded plates are recommended to realize shift of peak • L-variants with various angles of orientation are recommended to obtain various peaks • If the cooling load is large, orientation as much as 45–60° west can be recommended, to produce energy at the time when the cooling load is large (i.e., in the afternoon hours) • Multiple orientations of the BIPV systems are advised especially in a neighborhood, since this enables counterbalance between winter and summer production and shift of peak	Same as for detached	Same as for detached
Inclined road	Design recommendations for layout with E-W road can be applied to layouts with inclined roads, however, a tradeoff should be made between the total orientation of rectangular units (if they are oriented to face the road) and partly oriented units (such as L-variants). For instance, it is recommended to use L-variants with angle between the wings $\geq 45°$ to replace a rectangular unit of orientation $\geq 45°$. The POA and distances between units should be also taken into consideration			

(continued)

Table 7.4 (continued)

Site	Objective	The same design recommendations apply for sites with curved road facing south or north	
		Density	
		Low	High
Curved south and north	Minimize heating and cooling energy consumption	• In the case of convex shapes or non-convex shapes with orthogonal wings, the whole unit can be oriented along the curve provided that the orientation from south is still within the optimal range • Beyond certain angle of orientation (30°), it is recommended to design partly oriented units, such as L-variant (where only one wing is oriented toward the curve, while the main wing is south facing). For instance, a rectangular shape with 45° orientation requires up to 20% more heating energy than an L-variant where only the wing is oriented at 45° 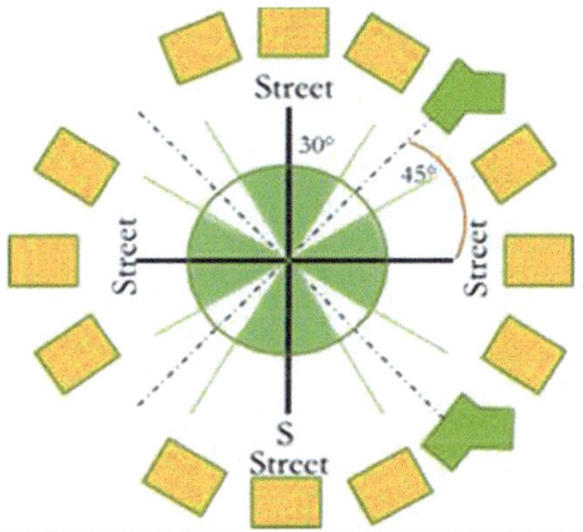	• Rectangular shapes are modified to fit the curve into trapezoid. In south-facing curve the trapezoid has the smallest faced south facing and therefore it is not an optimal design. Vice versa for north-facing curve • L-variants configuration are optimal for minimizing heating load in south-facing curve • Obtuse angle configuration is optimal for minimizing heat load in north-facing curve

(continued)

Table 7.4 (continued)

Site	Objective	The same design recommendations apply for sites with curved road facing south or north	
		Density	
		Low	High
	Maximize electricity generation	• Avoid rectangular shape with high angle of orientation (e.g., >45°) • Recommendations for straight road layout are valid	• Avoid trapezoid, with smaller south-facing surfaces, in south-facing curve (see above) • Recommendations for attached units in straight road layout apply to curved layouts
	Peak shift	• Recommendations for straight road layout are valid • Combinations of orientation of the units and of the roofs enable large shift of peak generation	• Recommendations for attached units in straight road layout apply to curved layouts

The roof design of a building plays an important role in the design of solar neighborhoods. Solar optimized roof design requires optimal choice of orientation and tilt angle of roof surfaces, as well as the available surface area. Considerations of roofs' design to optimize solar capture are presented in Chap. 5. Table 7.4 summarizes some of these considerations in a neighborhood context.

Site Layout

The main design considerations of site layout for solar neighborhoods are presented below and summarized in Table 7.4.

- The positioning of housing units around curved road involves buildings' rotation, in addition to mutual shading of some detached units or some wings of attached units. This effect can lead to a significant reduction of irradiation on some surfaces. Two effects should be accounted for in the design–the relative position (angle and distance) between adjacent units in detached configurations and the relative dimensions of adjacent, mutually shading facades of different units, in attached configurations.
- The main effect of site layout on electricity generation, other than the shape effect outlined above, is the shift in peak generation time on surfaces of different orientations of roof surfaces. In the straight road (east–west) site, where the different orientations are due exclusively to roof surfaces' orientations of non-convex shapes, the time difference in peak generation (between main wing and branch) can reach 3 h, in the studied example. In curved road sites, where rotation of whole units together with wing rotation produces a wide range of orientations, the difference in peak electricity generation time can reach 6 h.
- Units in curved layouts have generally larger heating and cooling loads than in a straight road configuration. One reason for the increase of loads in curved roads is the mutual shading of units, for instance, in a north-facing curve, where L-variants may shade significantly each other. This shade can be reduced by more careful design of the relative ratio of self-shading surfaces. Cooling load is increased since the units are originally designed to be south facing, implying large window size on the south facades. In the curved layouts, some of these units are oriented toward west or east, resulting in increasing transmitted radiation in the morning and evening, when the sun is at low altitude during the summer period.

Density

The impact of higher density is summarized below and presented in Table 7.4.

- A higher residential density can be achieved by attaching units in multiplexes (up to 16 u/a maximum density) or in row townhouse configurations (up to 35 u/a density). Assembling units in multiplexes may increase shading on some surfaces of non-convex shapes.
- Another effect on irradiation is shading by parallel rows of units (designed on the south–north axis). This effect is strongly dependent on the distance between rows of units and is most significant in winter.

- In some specific applications, attaching units in multiplex configurations can give some benefits for electricity generation, as in the studied example. Attaching some L configurations may produce some mutual shading.
- Row assemblage does not have significant effect on electricity generation for a row distance larger than 5 m, due to the uniform height of all units assumed in this study.
- Heating and cooling loads depend strongly on unit density in a site. Attaching units in multiplexes reduces heating and cooling loads (by up to 30% and 50%, respectively, in the studied example), compared to the detached configurations of the same site. Heating and cooling loads of detached units are not highly sensitive to the spacing of units (on the east–west axis).
- Arranging the units in south-facing rows affects significantly energy demand of the obstructed row, due to shading. The heating load is directly influenced by shading and is inversely related to the distance between rows, while the cooling load of both exposed and obstructed rows is significantly lower than for the single row configuration. For instance, with a distance of 10 m between rows of rectangular units (about equal to the height of the buildings in the studied example), the heating load of the obstructed row can increase by up to 25%, relative to single row configuration. At 20 m distance the effect is negligible.

7.3.2 Solar Neighborhood Design Methodology

A heuristic methodology is presented below to assist the design of near optimal small scale, residential solar neighborhoods, whereby initial designs are evaluated for energy performance (energy consumption versus generation) and selected designs are progressively modified for improved performance. The methodology outlines each step of the design process, highlighting alternatives that may offer good solar potential and presenting systematic methods for comparison and evaluation of these alternatives based on predetermined selection criteria (see infogram of Fig. 7.19). The design procedure allows for various site layouts and density levels. Stages of the design procedure are detailed below. While the design stages are generally applicable, the details of implementation are related to the climatic conditions at the basis of this research–mid-latitude northern location. Following are the main steps of the design methodology:

- The design process starts with a brief that includes data related to the site and to the housing units, such as road layout, density, number of units, number of stories of housing units, and functional floor area.
- The next step is, for the given site layout and density to prepare a number of design alternatives that fit the layout and have potential beneficial energy profile, based on the finding of this (or similar) investigation. Design alternatives should include different shapes and orientations.

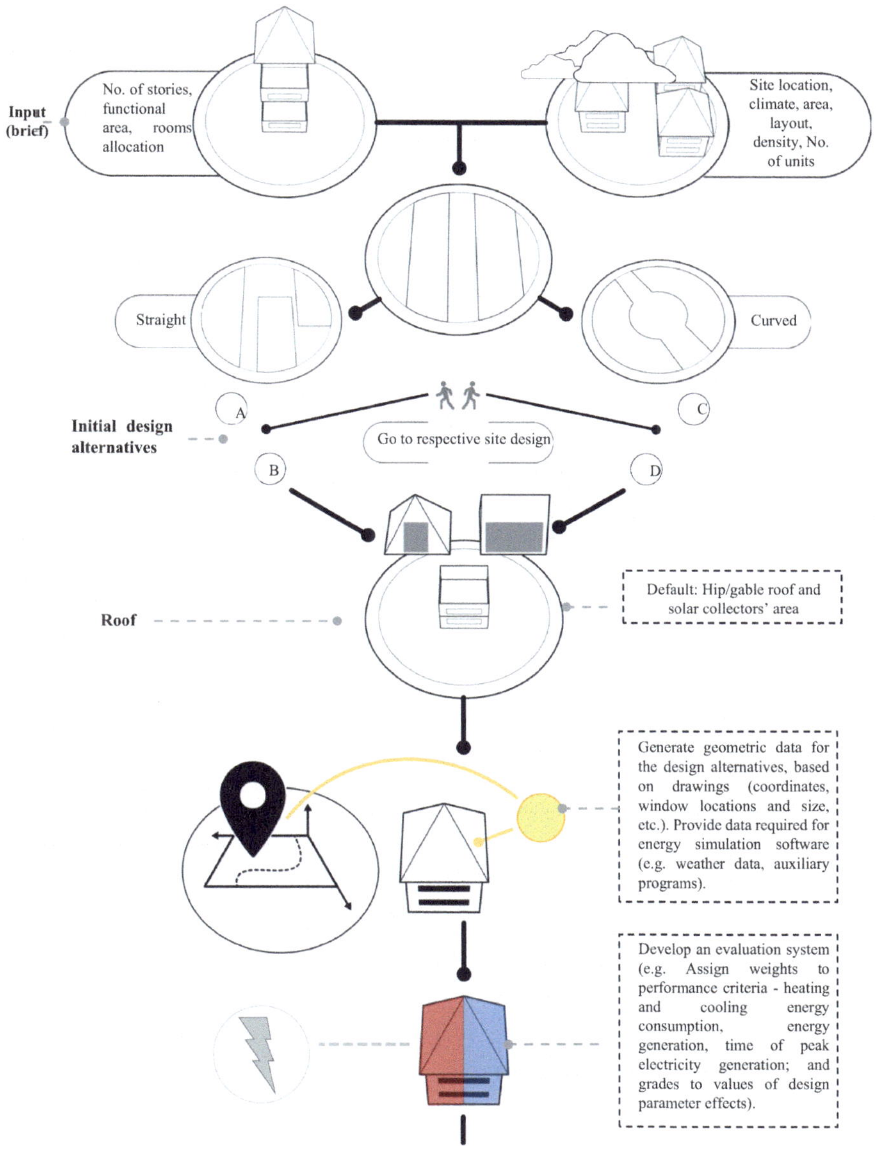

Fig. 7.19 Infogram illustrating the design process of solar energy-efficient residential neighborhoods

- Each of the design alternatives is analyzed for energy performance (consumption vs. generation) by means of a simulation program, (such as EnergyPlus, in this study). Other performance criteria, such as cost, may also be included in the analysis.

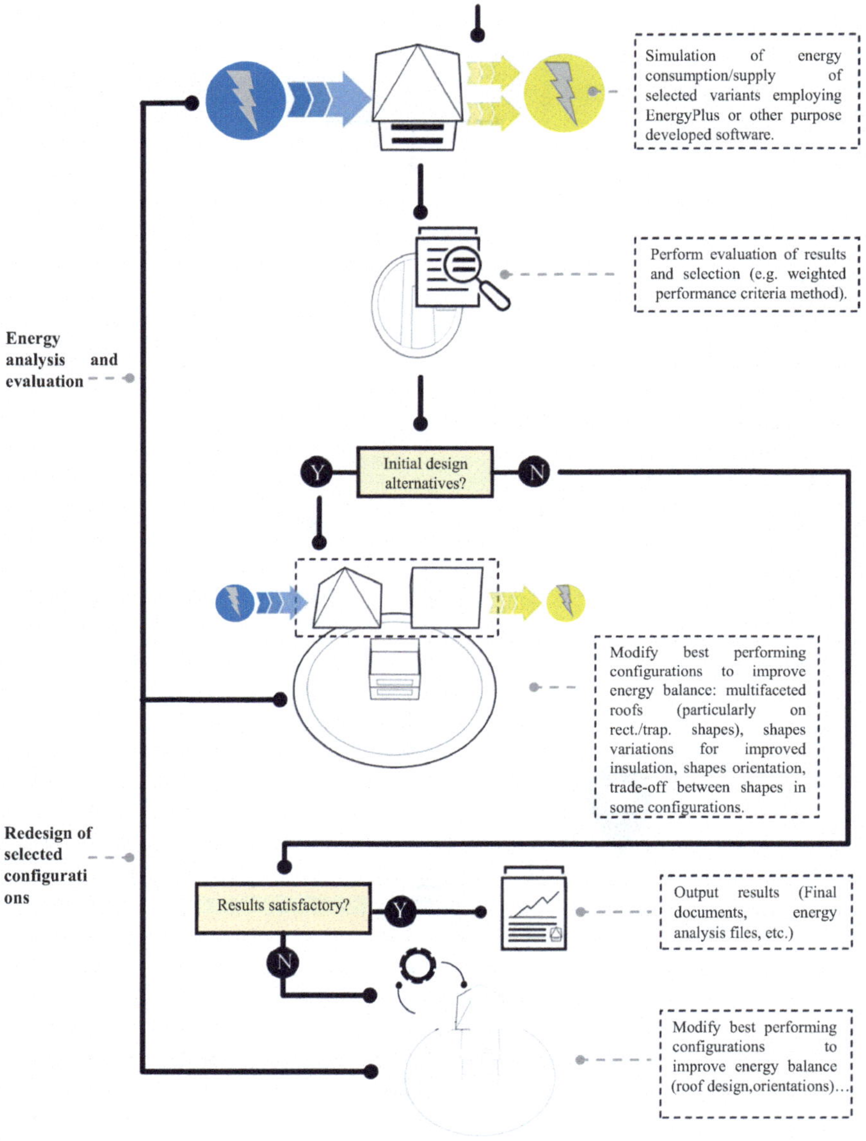

Fig. 7.19 (continued)

- Design alternatives are evaluated and compared based on the objectives of the specific project.
- A number of the most promising alternatives (depending on the size of the project) should be then selected for further analysis. Modifications are made to the original design, as appropriate, in an attempt to improve performance and simulations are performed again.

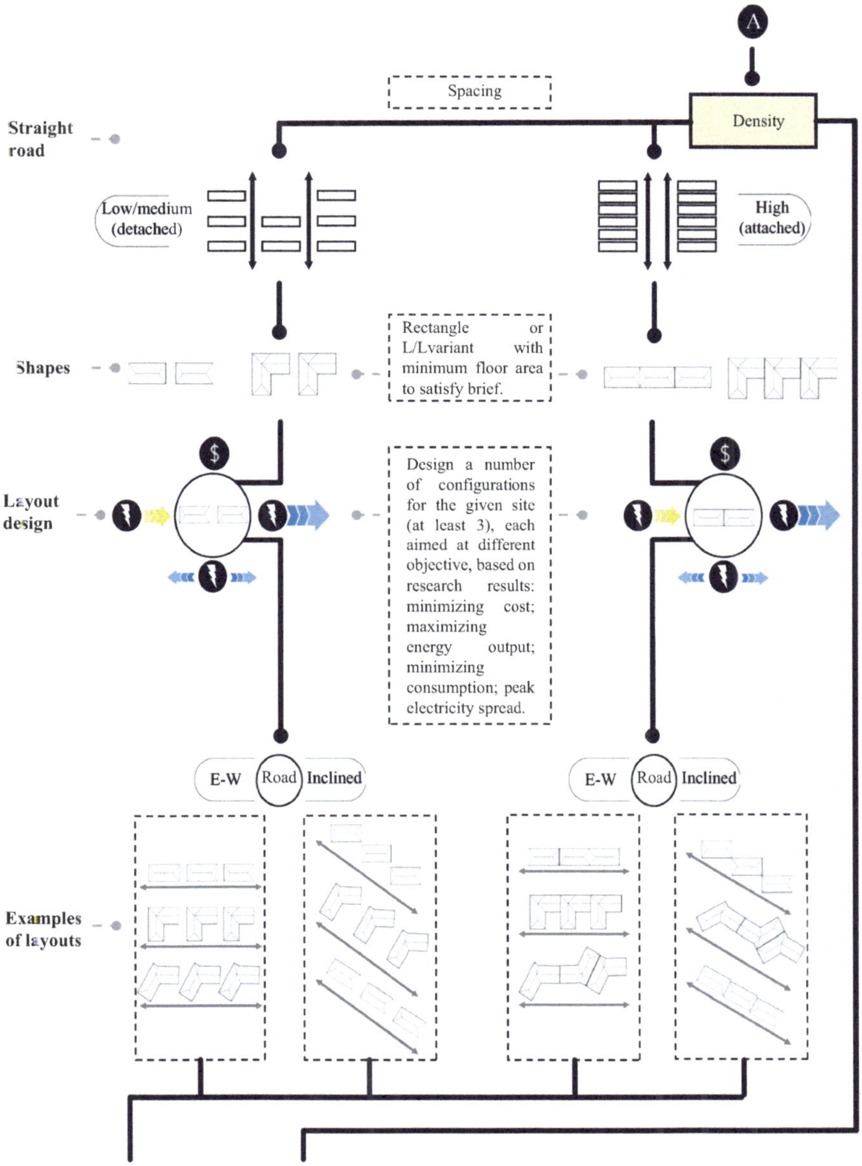

Fig. 7.19 (continued)

- The process of evaluation, selection, modification, and re-analysis is repeated until the optimal design is reached for final preparation of documents.

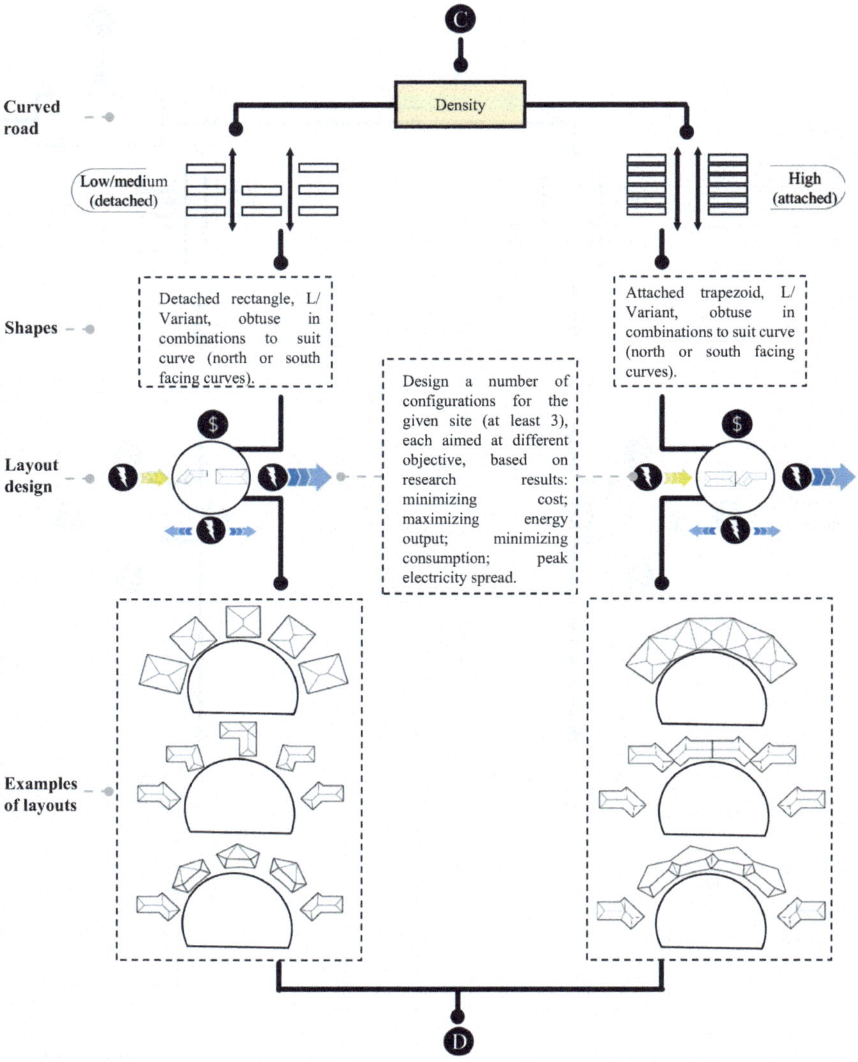

Fig. 7.19 (continued)

Application to Other Climates

The effects studied in this research are specific to the climatic conditions–mid-latitude northern climate. However, the nature of the design parameters and the design methodology are generally applicable. While the nature of performance criteria–heating/cooling load, energy potential etc.–does not change, their relative importance is climate dependent. Similarly, the effects of design parameters, such as dwelling shapes, site layout, density, are similar in different climates, but the objectives change with climate. For instance, while mutual shading is undesirable in cold climate it may be desirable in hot climate and will have a positive effect on cooling load, which is a major performance criterion.

References

1. Hachem C, Fazio P, Athienitis A (2013) Solar optimized residential neighborhoods: evaluation and design methodology. Sol Energy 95:42–64
2. Hachem C (2012) Investigation of design parameters for increased solar potential of dwellings and neighborhoods (Doctoral dissertation, Concordia University)
3. DeKay M, Brown GZ (2013) Sun, wind, and light: architectural design strategies. Wiley
4. Ko Y (2013) Urban form and residential energy use: a review of design principles and research findings. J Plan lit 28(4):327–351
5. Burden D, Wallwork M, Sides K, Bright H (1999) Street design guidelines for healthy neighborhoods. Rue for Local Government Commission, Center for Livable Communities
6. Cohen A (2000) Narrow streets database, congress for the new urbanism. www.sonic.net/abcaia/narrow.htm. Accessed 23 Oct 2010
7. Teed J, Condon P, Muir S, Midgley C (2009) Sustainable urban landscape neighbourhood pattern typology. The University of British Columbia James, Produced for the Sustainable Development Research Institute

Chapter 8
Mixed-Use Solar Neighborhoods

This chapter presents an overview of some of the main design issues and opportunities in planning mixed-use solar communities. The chapter discusses the impacts on solar access of neighborhood's shape, layout of streets, density of buildings, and of the combination of various types of buildings. The application of energy generation and storage technologies, such as thermal storage and PV integration applications, is presented as well. Design considerations for enhancing energy and environmental performance and resilience of neighborhoods are discussed. These include optimal mixture of building types, the impact of neighborhood design on transportation and related energy consumption and GHG emissions, as well as the impact of street layout on the overall resilience of the neighborhood for natural or other disasters.

8.1 Introduction to Mixed-Use Neighborhoods

This section presents an overview of mixed-use neighborhoods, including the various design trends and their main characteristics. It also highlights benefits and challenges associated with these types of neighborhoods.

8.1.1 Overview of Mixed-Use

Urban forms have a significant impact on the balance of energy consumed by buildings and transport [1–3]. Mixing land use within urban developments is becoming a key planning principle that can assist in reducing the negative environmental impact of the built environment [4].

A mixed-use development is generally defined as a development that contains a diversity of land use, including residential, commercial and public facilities, and sustaining a mixture of population, income, and transportation modes.

C. Hachem-Vermette, *Solar Buildings and Neighborhoods*, Green Energy and Technology, https://doi.org/10.1007/978-3-030-47016-6_8

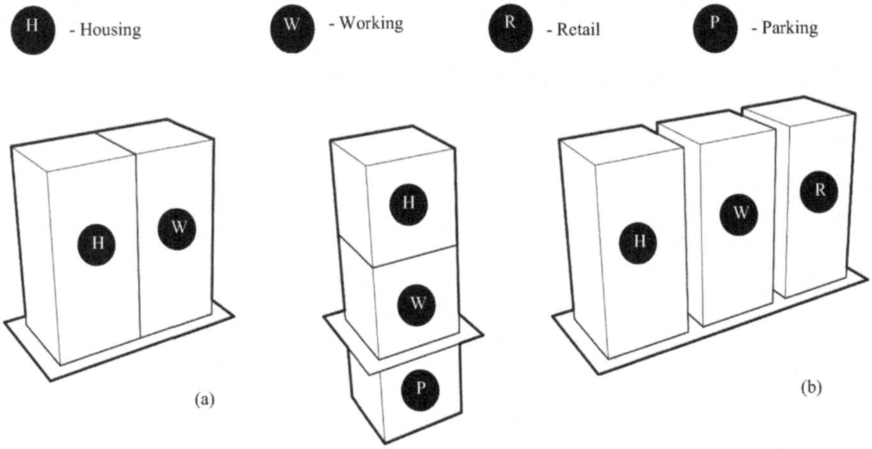

Fig. 8.1 **a** Mixed-use of the same building, **b** in adjacent buildings

Mix of land use aims ultimately at containing diverse residential building types, and various amenities (including work, commerce, education, and leisure) within close proximity. Conceptually, mixed-use of urban land, which serves complementary functions, can enhance the utility of each of these functions [5, 6]. For example, planning commercial areas in proximity to a residential neighborhood provides various services including social interaction and activity [6, 7]. Likewise planning residential areas around commercial developments can motivate businesses, potentially contributing to the survival and success of these businesses.

Mixed-use may be developed at a range of scales: mixed-use buildings, mixed-use parcels or sites, and mixed-use walkable or transit areas. Figure 8.1 illustrates various types of mixed-use, within a building or in adjacent buildings of a cluster. This chapter concentrates on mixed-use sites and neighborhoods.

Development of Mixed-Use Strategies

Mixed-use was the standard form of urban development before modern zoning and land use practices. Examples of mixed-use commercial and residential areas were often seen at intersections and transit stations.

Modern zoning practices assigned land uses according to function. Houses were segregated from commerce, work, and school. From the 1910s through the 1950s mixed land uses were rare in new developments. In the last decades, mixed-use has emerged as a means of solving urban problems associated with urban sprawl [8–10]. These problems are related to diverse social, economic, and environmental issues, such as social segregation, decline of city centers, reduction of open spaces, and others.

The concept of mixed-use is a key component of various trends of sustainable urban developments such as transit-oriented development (TOD), traditional neighborhood development (TND), and many other developments. These mixed-use communities stimulate a number of shared objectives, including increased intensity and diversity of land use, and integrating segregated uses [10].

Below is a summary of two trends, commonly discussed in the pertinent literature, of mixed-use developments–traditional neighborhood development and transient-oriented development.

Traditional Neighborhood Development (TND)

A traditional neighborhood development, or TND, is a trend of mixed-use community, which promotes diverse types of residential and commercial buildings accommodating various amenities [10, 11]. TND is characterized by a compact area, an active urban center, and sustainable transportation methods including walkability and public transportation. Due to these characteristics, TND is defined often as a village-style development. This type of neighborhood can be designed as infill in an existing urban area or as a new development.

Figure 8.2 Illustrates the main differences between segregated development and a mixed-use development.

Transit-Oriented Development (TOD)

Transit-oriented development (TOD) aims at integrating sustainable public transport with mixed-use of land [12]. It concentrates high density, mixed-use urban development in nodes associated with public transportation (transit) stations, such

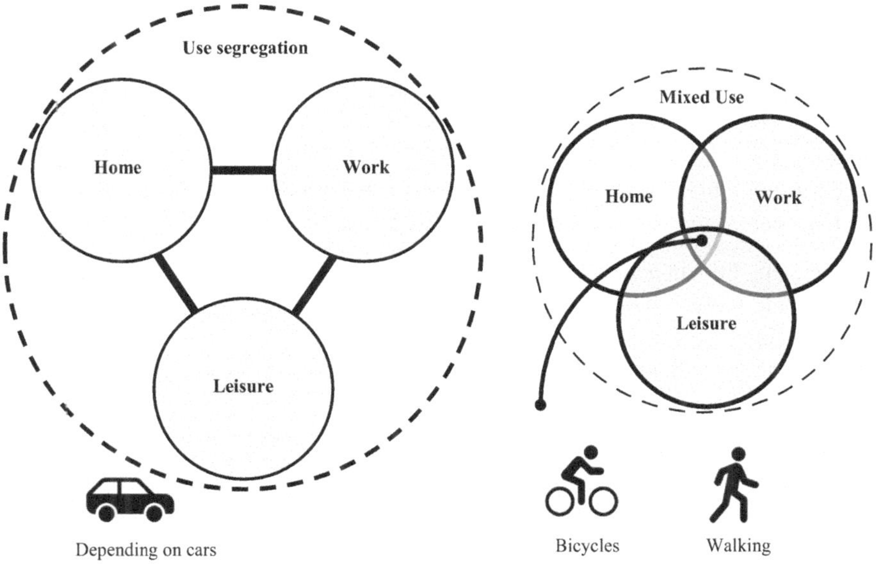

Fig. 8.2 Illustration of mixed-use as compared to segregated neighborhood

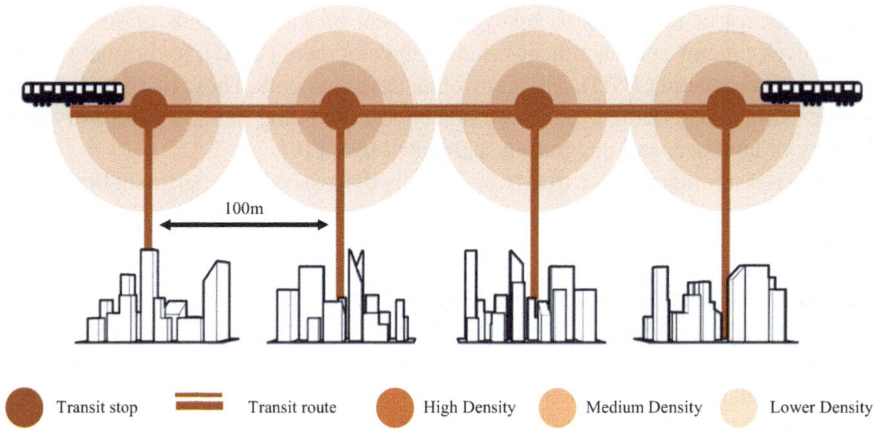

Fig. 8.3 Illustration of TOD

as train or bus stations. Such transit nodes are designed to be within a walking distance (500–800 m) from various residential and commercial areas. TOD encourages diverse types of amenities around the transit station such as offices, retails, and entertainment. Lower density residential areas are designed around the edge of the node. Figure 8.3 shows the gradient density of the community from the center of the node, around the transit station, outwards [13–15].

8.1.2 Benefits and Challenges

Benefits

Mixed-use neighborhoods present a number of benefits related mainly to reduced car dependency and car use and other potential benefits including the following:

- Increasing human activities in urban areas during extended hours of the day;
- Providing increased housing options for diverse household types;
- Enhancing travel and public transportation options;
- Encouraging walkability and reducing car dependence, by providing housing near commercial and civic activities;
- Increasing affordability and equity due to mixing housing types.

Additional benefits of mixed-use developments are related to energy efficiency and increased potential of integration of renewable and alternative energy sources. Alternative and renewable energy possibilities at neighborhood scale include solar electricity generation, solar heating, and storage, combined heat and power generation, and use of waste for heat and electricity. Energy options should, however, be considered in the context of the local energy landscape, to determine their environmental and economic impact. This is discussed in more detail below.

Challenges

A number of challenges face the design and implementation of the mixed-use communities concept. Some of these challenges are associated with the advantages and benefits that a mixed-use development offer. For example, from a social point of view, mixed housing, especially those associated with various income levels, often does not provide the anticipated benefits to low-income groups. Research examining mixed housing discloses that social issues related to employment, income, educational outcomes, and others are not significantly improved, especially for marginalized groups [16, 17].

From spatial and functional standpoints, a successful mixed-use neighborhood design should ensure a proper connection with the surrounding and with neighboring communities. It should also provide adequate access to transit and nodes of transportation. In addition, the advocacy for reduced parking-dedicated space can cause significant issue as inhabitants of the neighborhoods or visitors struggle to find lots to park their vehicles, increasing the cost of mixed-use development [18].

Other challenges are related to solar access and energy consumption patterns. Challenges in the design of solar mixed-use community include the existence of various types and forms of buildings, with different energy needs, as well as the restriction of solar access due to mutual shading of buildings. These challenges are discussed below.

8.2 Solar Access and Energy Performance

Solar access can be significantly affected by the shape and height of buildings and their spacing, in urban context. Mixed-use communities, incorporating various types of buildings with various heights and layouts, require thorough design effort to ensure an optimal level of solar radiation.

Solar access is widely studied in the literature. Various methods are proposed to quantify solar radiation in urban areas, as well as methods of controlling solar access. Some of these methods are aimed at defining heights and spacing of buildings in a neighborhood, in order to ensure adequate solar access [19]. Some of these methods are discussed in Chap. 6.

Several approaches have been applied to identify and measure the inter-building effect on energy consumption and solar PV production, including the use of geographic information system (GIS) (see Chap. 6). Genetic algorithms have been developed to investigate optimal solar irradiation arrangements of multistorey buildings [20].

The impact of major design parameters on solar access and energy performance of buildings in mixed-use neighborhoods is reviewed below.

8.2.1 Density

Urban density refers to the magnitude of the ratio of the total built area to the area of the given site. The negative impact of increased building density on solar potential, including access to daylight is highlighted extensively in the literature, as well as the correlation between density and energy performance of buildings (e.g., [21–24]). In addition to the impact on building performance, increased density can affect the urban microclimate, as well as the outdoor human thermal comfort [25].

Nonetheless, many studies associate an increased built density with urban environmental sustainability, especially at the city scale [26]. In temperate and cold climates, where enhancing solar availability is a high priority, the counterbalance of the negative impact of increased density is sought through the deliberate manipulation of urban layout [27]. For instance, for a given density, the level of solar radiation can be manipulated through combinations of site coverage and building heights [28]. Increasing spacing between buildings allows better solar access to buildings, and thus increases their potential to utilize solar radiation for passive hearting and daylighting, while also increasing solar availability at ground level.

Effects of density on solar access and energy performance and means of mitigating them are illustrated below.

Mutual Shading Effects

Distance between buildings plays a major role in creating mutual shading between buildings, and should be determined as a function of the height of these buildings, within imposed constraints (such as density and functionality–building types and use). This effect seems not fully appreciated, even by designers of new developments. An example of the impact of various building heights on solar access is presented below. The example is based on a theoretical study that investigates systematically the effect of buildings' heights and the distance between them, on solar irradiation [29, 30].

Figures 8.4 and 8.5 present the shading effects on solar radiation and on heating load. Figure 8.4 depicts the impact of shading by a 9-storey building on solar radiation incident on the south facade of a building of similar height, positioned to the north, at varying distances between the two buildings. Solar radiation is significantly reduced with smaller distance between the buildings. The effect is more critical to the lower floors of the shaded buildings. The incident radiation effect gives an insight into the anticipated passive heat gain and energy generation potential from the integration of PV systems in the facades of the shaded building.

Figure 8.5 illustrates the effect on heating load of shading of a 9-storey building by a building of varying height positioned at varying distance to the south. Heating

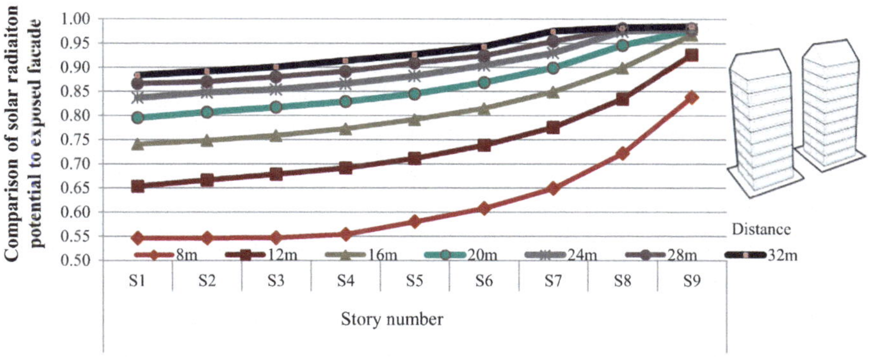

Fig. 8.4 Mutual shading of two identical buildings 9 stories high; shading building is on the south of the shaded buildings

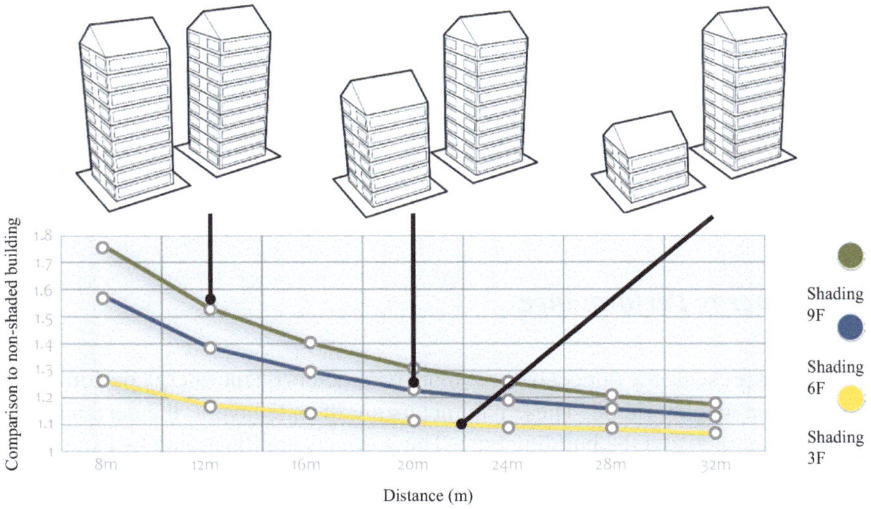

Fig. 8.5 Comparison of average heating load in 9-storey buildings, with different shading scenarios and at various distances

load is increased with smaller distance between the buildings, and with increased height of the shading building.

The effects discussed above are expected to be amplified in case where other adjacent buildings cast shade from other directions, in addition to this studied effect.

8.2.2 Neighborhood Shape Effect

Urban layout refers to the way in which the built volume is distributed spatially within the site, horizontally and vertically. Chapter 7 shows that the layout design and position of buildings with respect to roads affect the performance of small-scale neighborhoods, especially as the orientation of the main facade of these buildings deviates significantly from south. The study summarized below shows that these observations apply as well for mixed-use higher density community.

Certain design considerations should be taken into account to reduce the impact of neighborhood layout on average annual solar radiation of the neighborhood. For instance, a comparative study of annual solar radiation potential of square, radial, and hexagonal neighborhood layouts shows that the square neighborhood outperforms the radial and the hexagonal neighborhood layouts by about 3%. This reduced impact of the neighborhood layout on solar access is achieved by enhancing the orientation of the dominant facades and roofs of all buildings (within the optimal range of 30° East to 30° West from south) [31]. The hourly solar radiation during specific periods of the year indicates some advantage to the circular pattern due to the larger variation in building surface orientations. This can be beneficial to obtain larger spread in the timing of peak electricity generation, when solar PV technologies are integrated within these surfaces. Figure 8.6 shows solar radiation levels on buildings within various layouts of neighborhoods, at 4 days of the year.

8.2.3 Energy Performance

This section presents the impact of three main parameters on the energy performance of mixed-use neighborhood. These parameters are the site layout, the design of the building envelope, and the building type. These effects are illustrated by examples.

Site Layout

An example of design options of a mixed-use district is employed in this section to illustrate layout effects of a site of fixed dimensions and density.

The configuration presented in Fig. 8.7 is a schematic representation of a new large-scale mixed neighborhood that was proposed to be built in a northern mid-latitude region (of Canada). Figure 8.7a represents the original proposed design, while Fig. 8.7b presents a modified option of the design which provides the same overall density, in terms of overall required housing units (see below for more details).

The height of the buildings designed in the first design option (Fig. 8.7a, about 18 floors) and their position on the south side of the site reduces significantly the potential of the townhouses to generate electricity, and to benefit from passive heat gain. For instance, assuming that BIPV/T systems are integrated within the total south or near south-facing roof surfaces, the potential generation can be reduced by about 50% for the houses that are directly next to the tall buildings. In addition, the

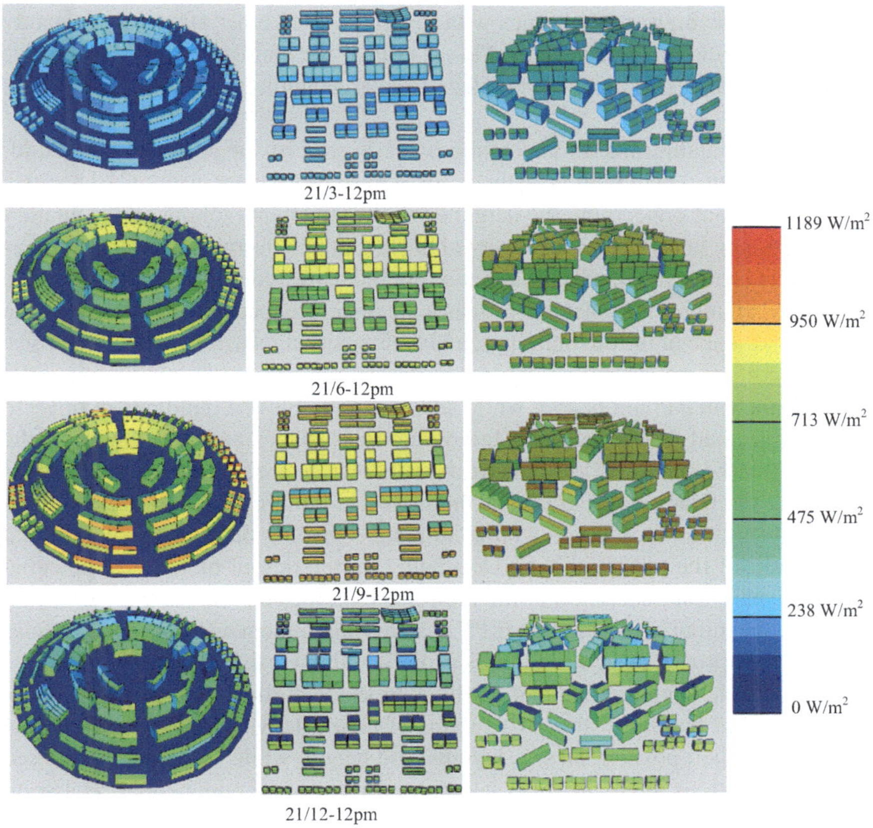

Fig. 8.6 Solar radiation on all neighborhood surfaces, during 4 representative days of the year

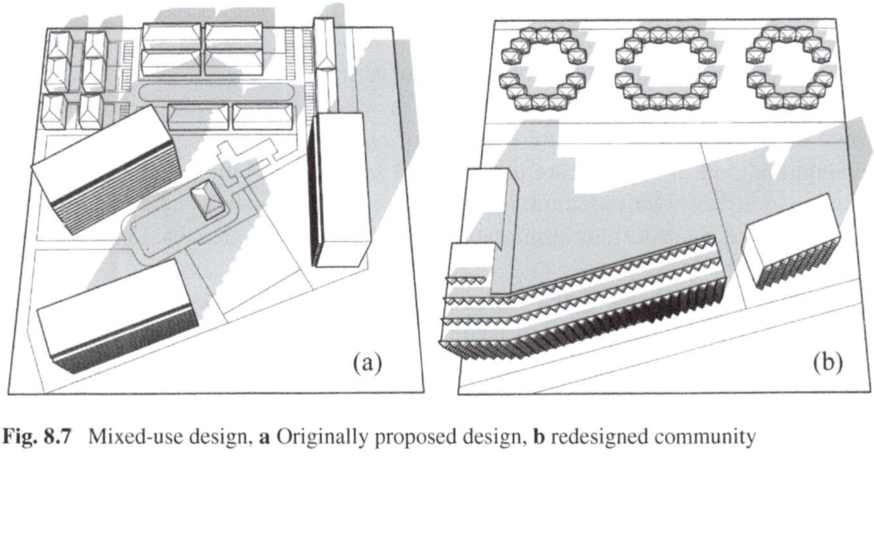

Fig. 8.7 Mixed-use design, **a** Originally proposed design, **b** redesigned community

position of these high-rise buildings with respect to each other affects significantly the solar radiation on their facades. This is critical for multistorey buildings, since facades have the largest potential to integrate solar technologies, due to limited roof surface.

In redesigning this mixed-use neighborhood (Fig. 8.7b), the position of tall buildings on the south, and of houses on the north of the development, cannot be changed, for considerations concerning the neighboring development (zoning and regulatory considerations). In addition, the density of the layout, the overall land area as well as major street layouts are fixed and cannot be altered in the redesign. Design modifications aimed at improving solar performance are therefore focused on rearranging the position and shape of the tall buildings, and their position relative to the townhouses.

One modification consists of altering the design of the high-rise buildings in order to reduce their maximum height, while adding some mid-rise buildings in order to maintain the total functional area. In addition, the shapes and positioning of the buildings can be modified to improve overall solar access. This redesign enables optimizing the near south facades of the high-rise buildings and to eliminate the shade cast on these facades, as well as to reduce shading on townhouses (located on the north of the tall buildings). Depending on the design of facades, potential electricity, and heat generation of the high-rise buildings can increase significantly relative to the initial configuration, if PV systems are implemented on south facades (Fig. 8.7b). Improved design of facades is discussed in Chap. 5.

This example demonstrates that minimizing shading effects can be achieved by manipulation of buildings' design, in terms of their relative positions with respect to each other, their position on site, layouts, and height, while maintaining imposed constraints, such as the main street layouts and position of commercial relative to residential zones.

Building Envelope Role

The impact of building envelope design on energy performance of buildings is surveyed in the first chapters of this manuscript (Chaps. 1, 2, 3, 4 and 5). This section illustrates the role of building envelope within an urban setting.

Design of building envelopes should be an integral part of the neighborhood design as a whole. The exposure of building envelopes to solar radiation, within the neighborhood, should be carefully planned to not compromise their potential to capture and utilize solar radiation, in various applications such as heating, cooling, and daylighting. Such design considerations can reduce energy demand for buildings' operations.

As discussed in Chaps. 3 and 4, building envelope can be designed to actively generate electricity and heat that can be directly linked to building systems, including electrical and mechanical systems, which further improve the overall energy performance of buildings. Reciprocally, energy demand of a building can dictate the size of PV system required to achieve specific performance goal (e.g., net-zero energy status, positive energy, etc.), and consequently the shape of building envelope to accommodate such PV system.

As an example, in mixed-rise neighborhoods, where south facade widths of high-rise buildings are constrained by neighboring lower buildings, these constraints can be mitigated through varying the layout (dimensions, orientation) with height as more space becomes available above a certain height. This concept is roughly illustrated in Fig. 8.8. Tall buildings designed with the long east/west facades change proportions or orientation at higher elevations, reaching a full south exposure.

Building Types

Since mixed-use communities encompass diverse types of buildings, the energy demand profile of the whole neighborhood is different than that of a residential neighborhood. Energy requirements change according to building types. Some commercial buildings are energy intensive, for example, supermarkets and food retails.

Advances in energy efficiency measures, combined with passive design strategies allow reducing energy consumption significantly in residential buildings. High-rise

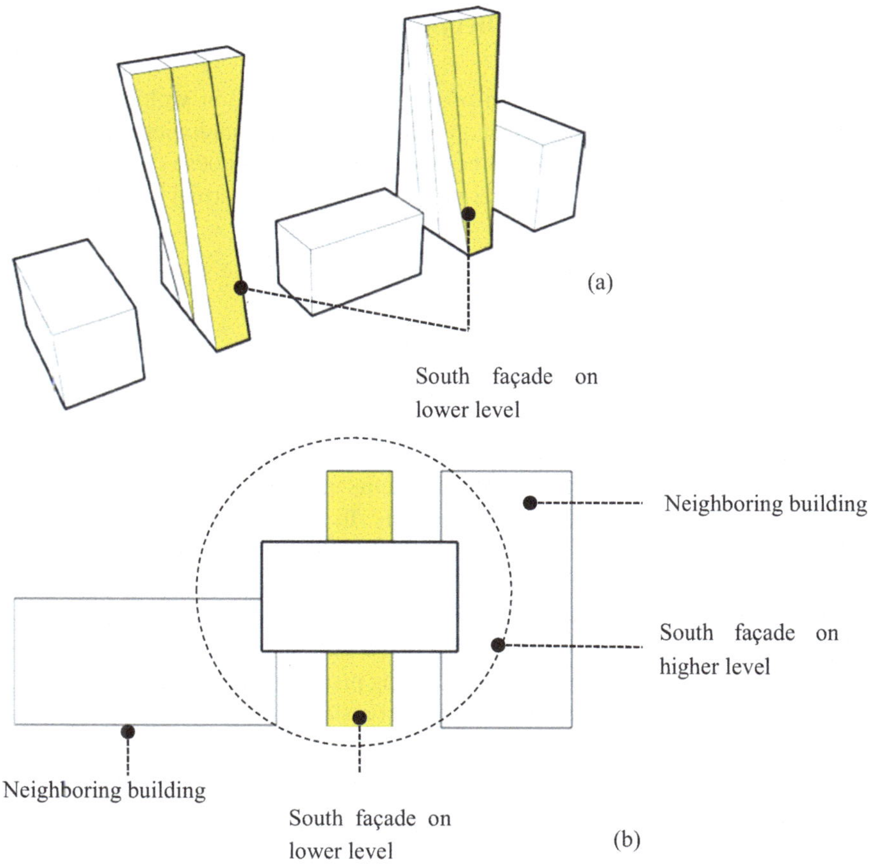

Fig. 8.8 Schematic illustration of a conceptual design of tall building to change the facade exposure with height; **a** 3D view of the development, **b** plan view

apartment buildings have low-energy intensity but also limited capacity to generate PV electricity, due to small ratio of roof to total floor area. This also applies to some types of commercial buildings such as office buildings. Better PV potential for multistorey buildings can be obtained by implementing PV in facades, as discussed in Chap. 5.

Some non-residential low-rise buildings such as schools and supermarkets can be highly energy intensive. However, they usually have large roof area, relative to total floor area, that can be exploited for PV integration and solar thermal collectors. In a mixed-use neighborhood containing net energy positive buildings, excess energy generation can supplement the local grid, or a micro grid, which contributes to the overall capacity of the neighborhood to generate renewable energy.

An example of energy demand and potential energy generation of various building types in a sample mixed-use neighborhood is provided in Sect. 8.3.3.

8.3 Active Solar Energy Collection

This section presents the integration of active solar collection within mixed-use neighborhoods. This includes thermal collection and PV technologies integrated in buildings and landscape. In addition, the section presents methods of enhancing the solar thermal potential by coupling thermal collectors to geothermal heat pumps and thermal storage.

8.3.1 Thermal Energy Potential

Heating demand and domestic hot water (DHW) constitute a major portion of energy requirement and consumption in cold climate. Employing solar energy for space heating and DHW purposes has significant benefits in reducing the reliance on fossil fuel-based energy to satisfy these requirements. In addition, designing a seasonal thermal storage to store thermal energy for utilization during the cold period, especially when adequate solar radiation is not available, can have substantial impact on the overall performance of the neighborhood.

Solar heating and cooling technologies collect thermal energy from the sun and utilize this heat in diverse applications, including providing hot water, space heating and cooling, and miscellaneous applications, for residential, commercial, and industrial applications. A number of methods can be implemented to enhance the solar thermal collection, on a neighborhood scale, such as coupling geothermal energy and seasonal storage to solar thermal collectors.

Solar Collection and Thermal Energy Storage

Solar thermal collection (STC) may be exploited to supplement the heat supply to a neighborhood-scale district heating system. District heating comprises a network of pipes connecting the buildings in a neighborhood, town center or a whole city, so that they can be served from centralized plants or a number of distributed heat-producing units. This approach allows diverse sources of heat to be applied simultaneously, including solar thermal and geothermal energy [32].

Solar thermal systems consist mainly of solar collectors and storage. An efficient solar thermal system requires the fulfillment of a number of criteria, by both collection and storage systems [33].

Solar collectors need to have a high absorbance capacity of solar thermal energy. As described in Chap. 4, solar thermal collectors absorb solar irradiation as heat, which is then transferred to the solar collector working fluid (air, water, or oil). This heat is utilized to either provide domestic hot water/heating, or to charge a thermal storage tank. Chapter 4 presents various types of collectors and their characteristics and applications.

Thermal storage can be used to store excess thermal energy that is not directly utilized by buildings, at the time of energy collection, and utilized when needed. Various aspects should be considered in the design of a thermal storage system, including technical properties of all parts of the system, cost effectiveness, and environmental impact. Thermal storage requires high storage density and high heat transfer rate that allows absorbing and releasing heat at the required rate, in addition to long-term durability [34, 35].

Due to a seasonal mismatch between solar availability and the demand of heat in buildings for space heating, seasonal storage may be established to raise the fraction of solar heat delivered to a neighborhood district heating system. Large long-term storage may also be useful in matching other heat sources to the network, including geothermal, as outlined below.

Geothermal Energy

Geothermal energy is one of the most environmentally friendly and cost-effective energy sources, allowing to reduce reliance on fossil fuels and assisting in mitigating GHG emissions. Geothermal energy can be used on a small scale to provide heat for a residential unit by using a geothermal heat pump or on a large scale for energy production through a geothermal power plant. Geothermal heat pumps (groundwater and ground-coupled) have become popular in North America and Europe and are generally utilized for both heating and cooling [32].

Geothermal energy resource utilization technologies can be grouped, according to the energy generated, as electrical power generation, direct use of heat, or combined heat and power in co-generation applications. Geothermal heat pump (GHP) technologies are a subset of direct use technologies [36].

For buildings with heating-dominated energy consumption, the combination of a ground-coupled heat pump (GCHP) system with a solar thermal system offers a high potential for energy conservation. Such combined systems have been tested with different designs during the last two decades in several countries [37].

An example of seasonal thermal storage: Drake Landing
The Drake Landing neighborhood-scale thermal storage (Okotoks, Canada) is the first large-scale seasonal energy storage solar heating project in North America. The overall intent of this project was to demonstrate the feasibility of reducing GHG emissions through reduction of fossil-fueled energy consumption for space and water heating, using solar thermal energy collection in conjunction with borehole thermal energy storage (BTES). The Drake Landing BTES project contributes to GHG emissions reduction by some 5 tons per home/year. Figure 8.9 presents an overview of the design of the houses, the solar thermal collectors, and the thermal storage system.

System Specifications and Performance
The STC/BTES system consists of five different components: (1) 798 flat-plate solar thermal (ST) panels; (2) the energy center (EC); (3) the BTES; (4) the district heating supply system; and (5) the 52 homes. The energy center is a major component of the district heating system, accommodating the short-term heat storage tanks and most of the mechanical equipment including pumps, heat exchangers, and controls. The main collection, storage, and distribution loops pass through the energy center. Figure 8.10 presents a schematic of the energy center.
During the hot summer months, solar-heated glycol water is pumped into the center of the BTES field. Heat from the water transfers to the surrounding earth, and as the water moves through the system, it cools and works its way to the outer edge of the field. When homes require heat (in winter), cooler water is pumped to the edges of the field, and the water picks up heat as it flows through the center. This heated water is stored temporarily in tanks (acting as a solar thermal temporary storage (STTS)) in the energy center (EC) and subsequently makes its way to each home using the district heating loop. End of summer temperatures in the center of the field are approximately 80 °C. Heat is supplied to the community through the distributed fan coils connected to the hot water distribution loop.
In cases of prolonged extreme cold resulting in the depletion of stored heat, a gas-fired generator in the EC provides supplemental power. After being fully charged, the system operates at up to 97% solar fraction, which is the ratio of solar energy to total heating energy demand [38].

8.3.2 PV Potential

Designing PV application on a whole urban scale can present many benefits. For instance, integrating PV in buildings with different orientations allows time-spread of electricity generation, reducing the stress on the local electrical grid (as discussed in Chaps. 4 and 5). In addition, new mixed-use neighborhoods provide the opportunity to plan PV installations to allow energy sharing between various types of buildings,

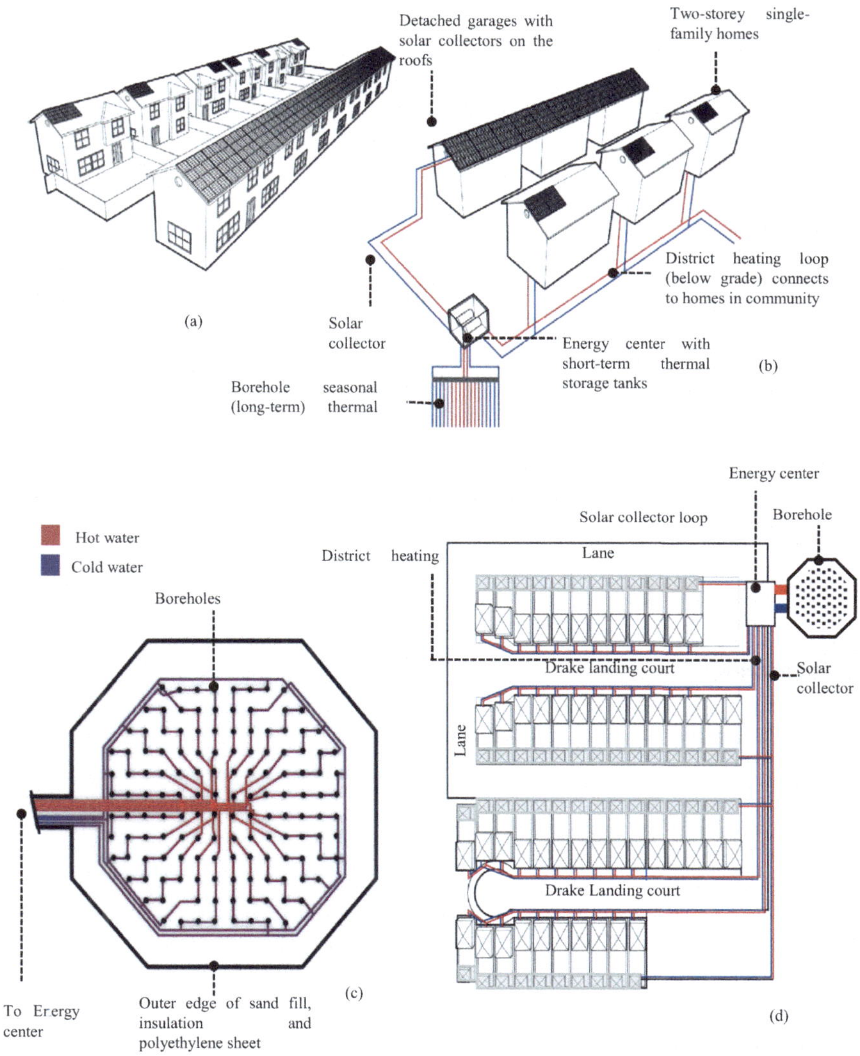

Fig. 8.9 **a** Houses of Drake landing **b** schematic of the BTES system; **c** aerial schematic of the borehole field; **d** site plan of collectors and borehole field

allowing buildings with reduced generation capacity to benefit from excess energy in other buildings.

Meeting the net-zero energy community objective (defined in Chap. 6), particularly in mixed-use neighborhoods, the building's physical boundary is generally not sufficient to integrate adequate amount of PV panels. Exploiting public open space for potential installation of additional solar PV systems can be a key design consideration to generate sufficient energy for the neighborhood.

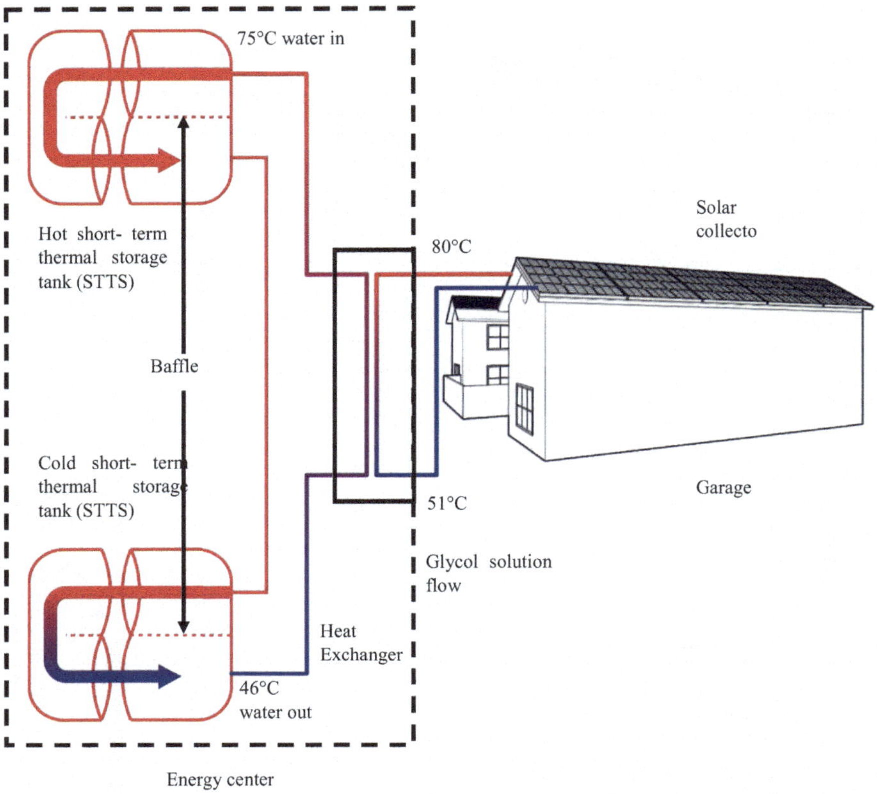

Fig. 8.10 Schematic of the energy center

Figure 8.11 Illustrates the concept of including solar technologies in open public space. PV integration in landscape is presented in more detail below.

PV in Landscape
Landscape can offer solutions for the integration of PV systems. Such systems can be envisioned as a matter of design not only at the architectural scale, but also at the landscape scale, as part of the infrastructures that urban planning needs to accommodate [39].

New issues and opportunities arise in designing public open areas and landscape within the built environment for the integration of PV technologies. Significant challenges are posed by the selection of public areas that offer an adequate solar potential, while avoiding shade from the surrounding buildings. On the other hand, integration of PV structures in the public landscape provides the opportunity to improve the outdoor thermal comfort of the built environment. For example, PV structures can be employed as shading structures in urban landscape. They can be designed as charging stations for electrical vehicles, as bus stations, or as integrated part of public parks to provide weather protection (e.g., shading or rain protection). In addition,

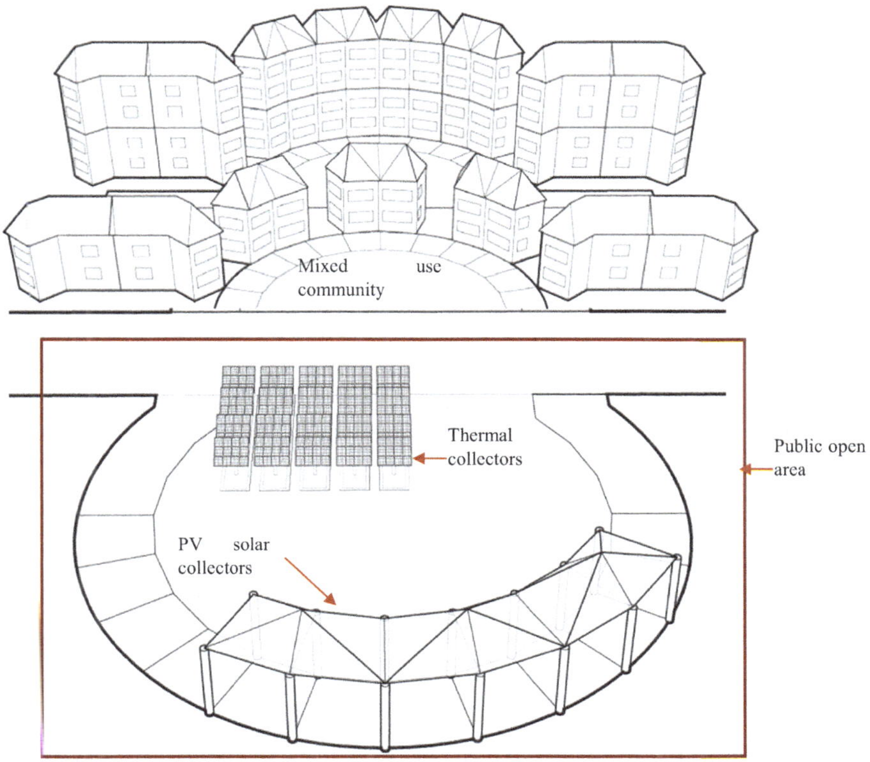

Fig. 8.11 Illustration of the integration of solar technologies in open public areas

PV systems can be integrated on the borders of streets, fulfilling some functions such as noise barriers, while benefiting from high solar exposure. Some integration methods of PV systems as stand-alone structures in public open areas are presented in Fig. 8.12.

8.3.3 Integration of PV and Thermal Collection in a Neighborhood

This section presents a hypothetical prototype neighborhood designed in mid-latitude northern climate, based on the combination of recommendations of available guidelines for sustainable mixed-use communities and on existing standards of community and building design in the selected region. This example has two parts—Part I, which serves as a base-case, presents the passive design and response of individual building types, aiming at increasing the energy performance of the neighborhood; and Part II presents the effect of PV generation and of thermal collection and storage.

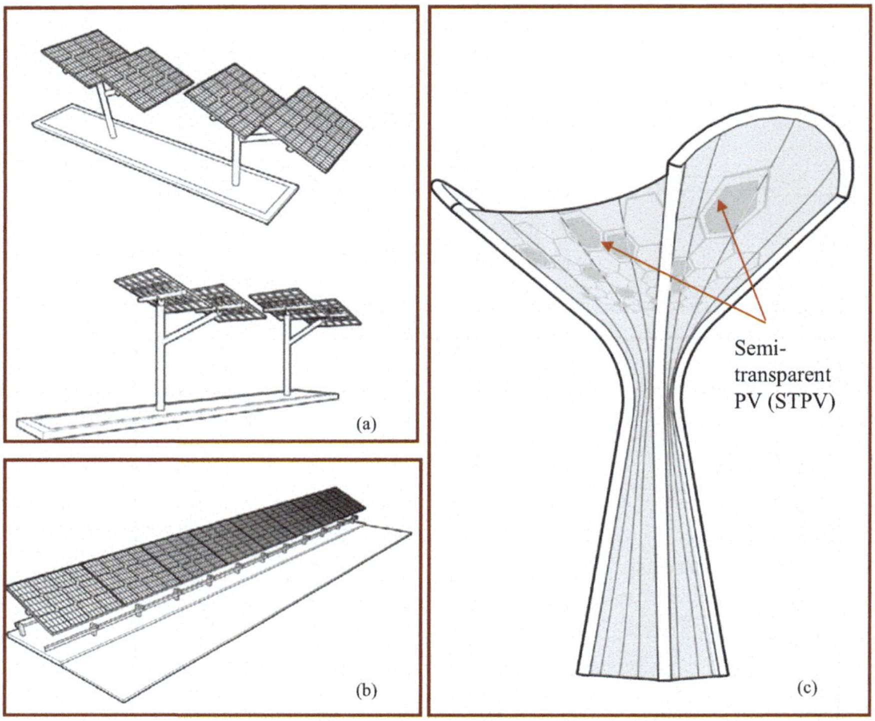

Fig. 8.12 Illustrations of examples of PV in landscape, **a** as parking shading structure, **b** on the border of highways, **c** as stand-alone design (featuring STPV modules)

This mixed-use neighborhood (residential and commercial) includes about 1300 residential units, and 21,000 m^2 of commercial, office, and primary school buildings. The residential units consist of detached houses, attached townhouses, and apartment buildings (of maximum 5-stories). The neighborhood design and its thermal characteristics were developed through a series of studies to achieve a high-performance, near net-zero energy status.

The neighborhood is divided into 16 districts (quadrants), characterized by various types of buildings and densities. While some quadrants are only residential, with detached houses, others are higher density mixed-use (Fig. 8.13).

Part I—Passive design

To illustrate the impact of building types on energy consumption, a comparative study is performed, presuming a northern cold climate (of mid-latitude). Building specifications are assumed according to the North American standards, and national and local codes(of Canada, in particular). The thermal characteristics of all buildings are presented in Table 8.1. This neighborhood is considered as base-case for evaluating the effect of active strategies.

The total yearly energy consumption and consumption per unit area of individual buildings are presented in Fig. 8.14 for the different building types (including

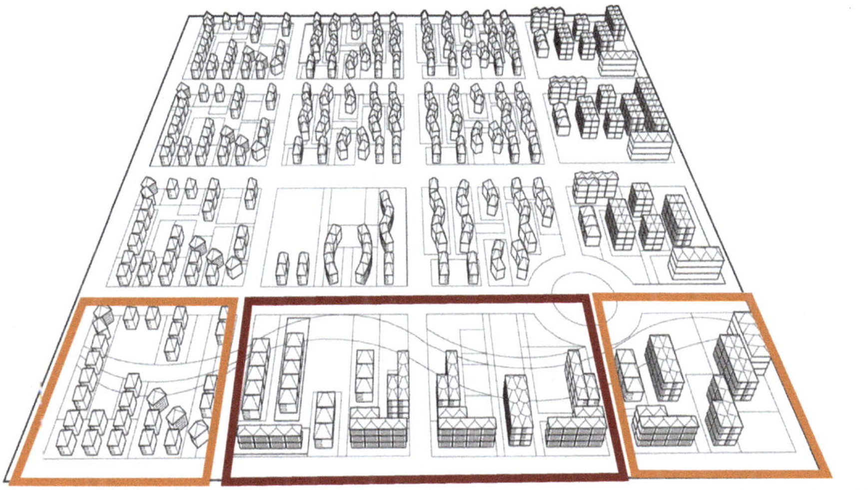

Fig. 8.13 Illustration of the hypothetical neighborhood

houses, apartment buildings, offices, retail, a supermarket, and a school). The total energy consumption for houses is not seen in the graph because it is too small as compared to other buildings types. Energy consumption presented is associated with an all electrical scenario, where electricity is considered as the power source for all appliances and equipment.

The supermarket is one of the most energy intensive buildings, in terms of consumption per unit area, followed by the retail and the school. Research on supermarkets indicates that it is possible to reduce energy consumption of this type of buildings significantly, by adopting various efficiency measures such as LED lighting, heat recovery ventilators, and heat pumps.

Part II—Active Energy Generation

This part focuses on active energy generation of the studied neighborhood example and is divided into PV generation and thermal collection/storage.

PV Integration

PV systems are assumed to be integrated in south and near south-facing roof areas of all buildings. In low-rise residential buildings, such as detached houses and townhouses, PV are integrated within gable roof of 45° tilt angle, representing a near optimal value. For apartment buildings (of 4 stories maximum height) and all commercial building, PV are integrated within the south-facing surfaces of a saw-tooth roof structure. This roof design is adopted to limit the overall height of the roof, while accommodating a tilt angle of 45°.

Table 8.1 Main characteristics and electric loads

Residential units			
Thermal resistance values		Exterior wall: 7 RSI Roof: 10 RSI Slab on grade (for ground floor): 1.2 RSI Slab perimeter: 7 RSI	
Thermal mass		20 cm concrete slab on grade (ground floor) 15 cm concrete slabs (in all apartments except ground floor)	
Window type		Triple glazed, Low-E, argon filled (SHGC = 0.57), 1.08 RSI	
Area of south glazing	Houses	35% of south facades	
	Apartment buildings	30% of south facades	
Shading strategy Shading control		Interior blinds Blinds shut at indoor air temperature of 22 °C	
Occupants		2 adults and 2 children, occupied from 17:00 to 8:00	
Set-point temperatures		Heating set point 21 °C, cooling set point 25 °C	
Infiltration rate		0.8ACH @50 Pa	
Ventilation rate		0.35ACH [40]	
Lighting (houses and front apartments)		3.8 W/m^2 [41]	
Equipment		5.38 W/m^2 [41]	
Commercial and civic buildings			
Thermal resistance values		Similar to residential units	
Window type		Similar to residential units	
Electrical Loads		Lights	Electric equipment
School		Depending on zone activity	Depending on zone activity
Office			
Retail			
Supermarket			

Comparing energy generation potential to the total energy consumption (Fig. 8.15) shows that the detached and attached units can achieve net positive energy status, while the apartment buildings generate about 67% of their total energy consumption.

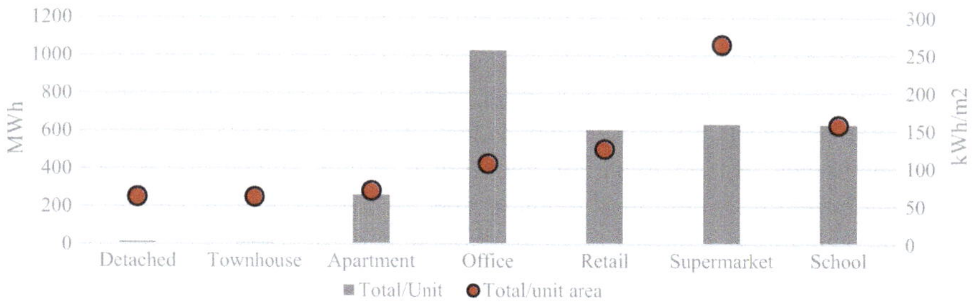

Fig. 8.14 Yearly energy consumption, total and per unit area of building type

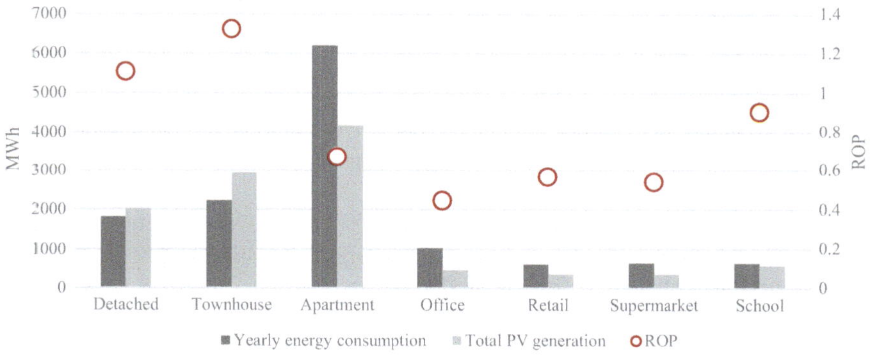

Fig. 8.15 Energy consumption, energy generation, and ROP of all buildings, classified by building type

The commercial buildings generate a portion of their energy consumption ranging between 15% for the office buildings and about 54% for the retail and supermarket (for the studied examples). The school, due to its large roof area, relative to its plan surface, generates about 90% of its energy consumption. Figure 8.15 presents the yearly energy consumption and potential energy generation of all buildings of the neighborhood, classified by building type, together with the ratio of performance (ROP) which is the ratio of energy generation to consumption.

The studied neighborhood as a whole—all buildings combined—has a ROP of 70%, allowing to generate 70% of its energy demand.

Thermal Collection

Thermal collectors are assumed to be installed in public open areas. The design of solar thermal collectors and the sizing of short-term thermal storage is based on the analysis of the thermal loads in each district of the community. The solar collectors' area is designed to generate a slightly higher thermal power than the community loads. The sizing of the volume of the bore hole thermal energy storage (BTES) is determined to guarantee an adequate storage temperature, enabling heat exchange at a temperature of 55 °C. Below is a summary of the potential of STC and BTES, and their impact on the neighborhood energy balance [42].

Since different types of buildings are included in each of the quadrants, the requirement for heating and domestic hot water varies. In addition, the peak energy demand of each of these quadrants occurs at a different time, particularly for residential buildings, where peak demand is in the mornings and evenings. Excess thermal generation is then stored and released as needed.

The utilization of solar thermal collection (STC) combined with the BTES seasonal thermal storage technologies can significantly reduce the energy demand from conventional sources for heating and DHW, and thus the overall energy consumption of the neighborhood. For example, energy requirement of residential buildings can be reduced by up to 50%, and commercial/institutional sectors by up to 30%, for the studied neighborhood.

Implementation of solar thermal collection and thermal storage is particularly advantageous for high density apartment buildings. This is due to their higher energy demand for heating and DHW, which is higher than their potential to generate renewable energy from building-integrated PV systems, due to reduced available building surfaces for PV integration.

Employing STC and BTES contributes over 40% of the overall energy consumption of the whole neighborhood, resulting in energy generation of 20% in excess of consumption (ROP of 1.2, as compared to 0.7 associated with the PV contribution without STC/BTES, see above).

8.4 Building Type Mixture

The composition of building types within a neighborhood can significantly influence the energy consumption by the neighborhood, as well as its potential to generate renewable energy from buildings and neighborhood integrated solar technologies. The proportion of residential to commercial buildings, as well as the design of buildings, affects the potential for alternative energy integration, such as waste to energy, overall energy performance, and the GHG emissions of the neighborhood.

This section presents a summary of the impact of specific design parameters of a mixed-use neighborhood on its energy and environmental performance. Details of this study are reported in [43–45].

The same example of mixed-use neighborhood presented above (Sect. 8.3.3) is employed to analyze the various parameters and their impacts. These parameters include the density and types of residential units, types of commercial buildings, the commercial land fraction (relative to total built area), and the commercial floor area ratio (ratio of a building's total floor area to the lot area).

This study aims at providing an insight into the energy and environmental impact of the composition and land use of a mixed-use neighborhood. It should be borne in mind, however, that energy and environmental impact are not the sole determinant in the design development of such neighborhoods. As discussed in Sect. 8.1, the design of mixed-use neighborhoods is a multifaceted process comprising a multitude of factors, including social and economic aspects. Taking into account these aspects would significantly influence the type of buildings constituting this neighborhood. Conceiving methods to study diverse neighborhood aspects and combining them with a methodology to evaluate the energy and environmental impact of various design decisions, can assist in developing a holistic approach to the design of more sustainability-oriented neighborhoods. Such methodology can be applied to different demographic and climatic conditions, with appropriate modifications.

8.4.1 Summary of the Investigation

Design Parameters

The design parameters are divided into two categories. One category of parameters relates to residential building types, while the other category focuses on commercial buildings. The parameters also include the relative land use proportion of different building types.

Residential Buildings Parameters

Residential buildings parameters include *Housing types, commercial to residential floor area index, and residential unit density index.* Housing units include single detached houses (SD), townhouses (TH), and apartment buildings (AP). Commercial to residential floor area index (C/R) represents the ratio of total commercial to residential floor areas, within the neighborhood. To analyze the residential unit density maximum and minimum densities are assumed. The maximum densities for SD, TH, and AP that can be attained without significantly compromising solar access are assumed as 4, 7.5, and 25 units per acre (u/a), respectively. For the apartment buildings, in addition to solar access considerations, the maximum density is restricted by the maximum height of buildings allowed in the studied region. The assumed minimum densities are 1, 4, and 7.5 u/a for SD, TH, and AP, respectively.

Commercial Buildings Parameters

Commercial building parameters include *Commercial building type*, *Commercial land fraction* (*CLF*), and *Commercial floor area ratio index* (*CFARi*). Three types of commercial buildings are analyzed: offices (*O*), retails (*R*), and supermarkets (*S*), in addition to one institutional building—a school. The area dedicated to the school depends on the residential population of the neighborhood. *Commercial land fraction* (*CLF*) represents the percentage of commercial buildings land relative to the total built land of the neighborhood. *Commercial floor area ratio index* (*CFARi*) is defined as the ratio of the total floor area of a commercial building to its lot area. It indicates the relative area allocated to each building type, taking into account the number of floors and their impact. The restriction of floor number in this study is based on regulations and common practices within the selected pilot location, and it can be modified to respond to various design requirements.

Performance Criteria

A number of energy and environmental performance criteria are employed to evaluate the impact of the design parameters on performance. The performance criteria, listed below, are assessed for all combinations of the design parameters presented above.

- *Net energy consumption* (*NEC*): This indicator is the total energy consumed by the neighborhood for building operations, including heating, cooling, water heating, appliances, and equipment. NEC includes the contribution of alternative and renewable energy (thermal and electrical) to mitigate a portion of (or total) energy consumption. Thermal energy is supplied by thermal collection and storage, as well as by WtE technologies (as described below).
- *PV electricity generation* (*PV*): The potential solar electricity generated by *PV* panels, assumed to be integrated in all available south-facing roof surfaces.
- *Waste to energy generation* (*WtE*): The total heat and electricity derived from the potential total waste disposal of the neighborhood, assuming a community scale combined heat and power (CHP) plant.
- *Ratio of performance (RoP)*: The *RoP* is defined as the ratio of total electricity generated from non-fossil-fuel sources (i.e., *PV* and *WtE)* to total electricity consumption.
- *Net GHG emissions*: The net *GHG* emissions include three sources; (i) building operations, (ii) *WtE* energy production, and (iii) transportation (see below).

Methodology

The analysis of the impact of design parameters on the environmental and energy performance of the mixed-use neighborhood example is performed employing multiple procedures, including energy simulations and statistical methods, to develop various combinations of the studied parameter values, and multi-objective optimization, coupled to decision-making process, to identify optimal solutions.

A number of energy simulation tools are utilized to simulate hourly and yearly energy consumption and energy generation potential from renewable sources, as well as thermal energy collection and storage. These tools include EnergyPlus and TRNSYS. Waste to energy potential is determined based on existing statistical data associated with the assumed region.

The optimization process is conducted to determine sets of most efficient solutions that allow to reduce the environmental impact of the neighborhoods, while increasing their renewable and alternative energy potential. These solutions present combinations of parameter values that optimize specific response variables [43–45].

The decision-making score (*DMS*) method is then applied, to determine optimal and near optimal combinations of neighborhood design variables, by assigning weights (*W*) to each of the optimized, normalized (relative to the base-case scenario) response variables, in accordance with their respective assigned priorities [46, 47].

8.4.2 Main Observations

The main observations of this study are summarized in the following.

Residential Considerations

- Increased residential density, associated with an increased portion of apartment buildings, leads to reduced energy intensity per housing unit, as well as to an overall increase of net energy consumption (*NEC*) and waste generation. Although in such neighborhood energy generation (from PV and WtE technology) is increased, the *RoP* (ratio of energy generation to energy consumption) is reduced. *GHG* emissions of higher density neighborhood are increased.
- Optimal mixtures of residential building types yielding the best *RoP* values consist of approximately 50% of the number of the residential housing units consist of single detached housing units, and 50% of combined townhouses and apartment units (45% TH and 5% AP). Increased detached houses imply, however, a significant increase of land use and reduced overall density.
- The energy consumption of the neighborhood as well as its *PV* and *WtE* generation are increased with the increase of the ratio of commercial to residential building area. The net *GHG* emissions intensify with such design, while the *RoP* is reduced. The optimal commercial to residential floor area that allows to reduce *GHG* emissions and net energy consumptions while maximizing other performance criteria ranges between 0.23 and 0.25.
- The effect of density on the performance of neighborhood is analyzed through two parameters: the housing units per acre and the population size (which can change while the residential units stay unchanged, reflecting an increased number of people per housing unit). Increase of the number of residential units per acre is associated with an increase in the proportion of apartment buildings, resulting in the impact discussed above. Regarding the population density, the type and relative proportion of residential buildings can be decided to accommodate a specific population within a specific land area.

Commercial Composition

- The supermarket building is the highest energy intensive building, per unit area, and responsible to highest *GHG* emission and the largest amount of waste generation.
- The maximum *RoP* of a neighborhood can be achieved by designing a larger portion of office and retail buildings, and reduced supermarket area, within the considered commercial land.
- The optimal combinations of building types—O/R/S—are determined by the relative weight allocated to a specific performance criterion, within the mixed-use

neighborhood. For example, giving high priority to *GHG* emissions reduction would result in higher proportion of office buildings and reduced supermarket.

- Various combinations of optimal proportions of commercial building types can yield similar overall performance. Optimal combinations of commercial buildings include 46–75%, 18–45%, and 6–9%, for offices, retails, and supermarkets, respectively. This allows flexibility in the selection and design of various amenities in a mixed-use neighborhoods [45].
- A higher proportion of commercial buildings area and associated land area within the neighborhood (i.e., *CFARi* and *CLF*) lead to reduced performance ratio (*RoP*) and increased *GHG* emissions.

Optimal Residential and Commercial Mixture

The study suggests that an optimal ratio of commercial land area to overall built land area lies in the range from 23 to 32%. Residential buildings occupy the remaining part of the built area. This optimal range allows reducing the energy consumption and GHG emissions, while increasing the neighborhood potential to generate non-fossil fuel-based energy.

Figure 8.16 presents an illustration of these results. All performance parameters (presented above) are normalized between 0 and 1 and plotted for different commercial floor area ratio index—*CFARi* values. The graphs of Fig. 8.16 show that all optimized performance parameters are positively correlated with the commercial land fraction-*CLF*, except *RoP*. This implies that, while energy generation (PV and WtE) is enhanced with an increased value of *CLF*, energy generation and environmental performance are reduced. The *CLF* values falling within the strips (in the graphs of Fig. 8.16) represent near optimal solutions.

Such commercial to land area proportion is similar to existing recommendations by sustainable mixed-use developments such as the traditional neighborhood development (TND) [37].

8.5 Miscellaneous Impacts of Neighborhood Design

This section presents some other important impacts of neighborhood designs. For example, the position of the business center, and the methods of traveling within a neighborhood can have significant impact on the total energy consumption of the neighborhood and its carbon footprint. Street layout and location of various types of buildings with respect to the streets affect the overall resilience of the community to emergencies.

Fig. 8.16 Optimal commercial land fraction—*CLF* at discrete values of *CFARi* of **a** 0, **b** 0.5, and **c** 1 [43]

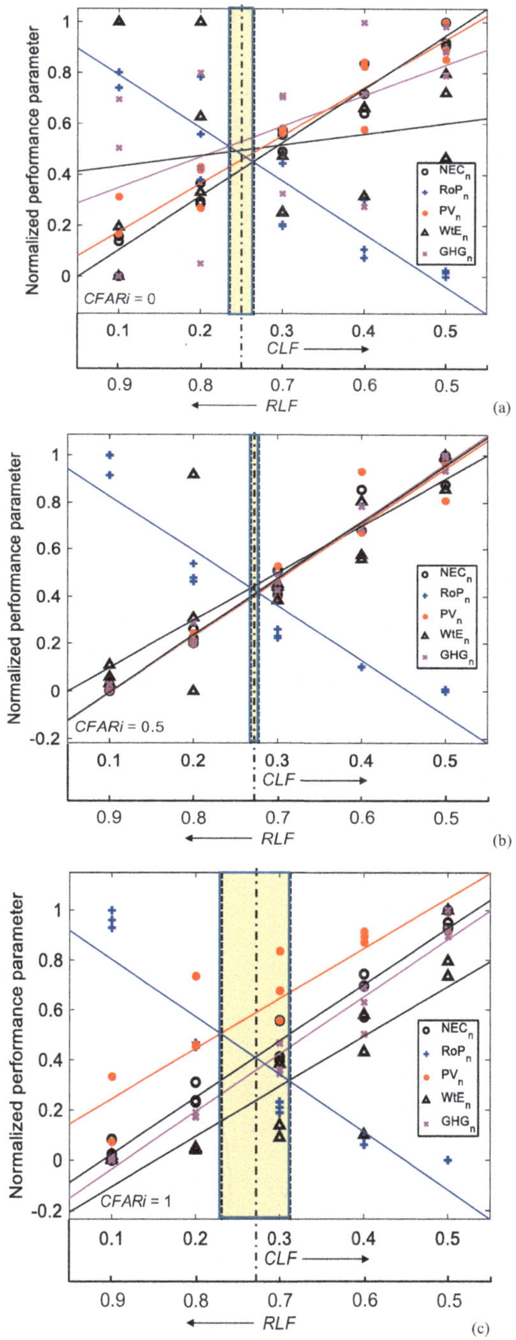

8.5.1 *Impact of Land Use on Energy and Transport*

This section presents the energy and environmental impact of some neighborhood design parameters of the neighborhood case study, discussed throughout the previous sections (Fig. 8.13). The studied parameters are divided into two categories. The first category relates to parameters of the built environment of the neighborhood, including energy efficiency of buildings, density level, and the type of neighborhood (residential as compared to mixed-use). The second category relates to spatial design and includes parameters such as the location of the central business district (see Fig. 8.17) and the design of streets.

The energy performance in this section takes into account the energy consumption by building operations and by the transportation sector, assuming various modes of transport.

Spatial Design

Central Business District (CBD)

The effect of the average distance of the residential areas to the central business district (CBD) is analyzed. According to the type of neighborhood, residential, or mixed-use, two different scenarios are studied. For the mixed-use neighborhood, the CBD is within the community, at the edge of the development (Fig. 8.17a), while the CBD of the residential neighborhood is located outside the limit of the neighborhood. The impact of four average distances of the CBD from the center of the residential neighborhood is studied, 5, 10, 20, and 30 km. The location of the business center with respect to the residential area is expected to significantly affect the number of kilometers traveled per day to various amenities, as well as the transport mode (i.e., private or public transport) and therefore the associated GHG emissions.

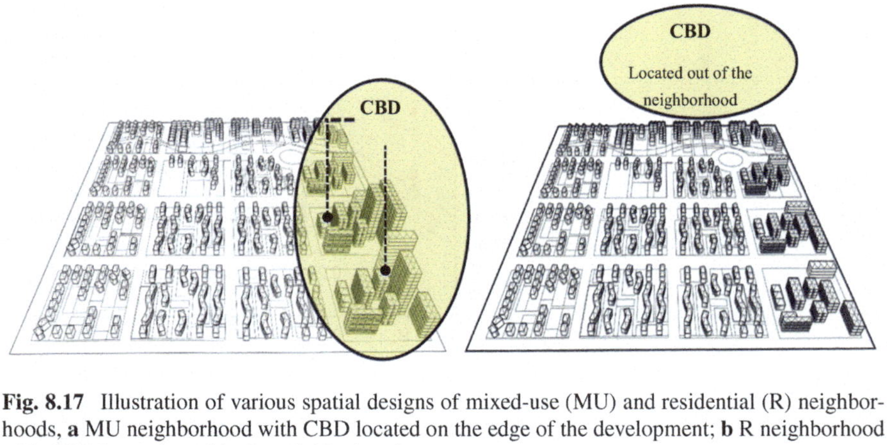

Fig. 8.17 Illustration of various spatial designs of mixed-use (MU) and residential (R) neighborhoods, **a** MU neighborhood with CBD located on the edge of the development; **b** R neighborhood with CBD located outside the neighborhood

Street Design

The mode of transport within an urban area is directly affected by the street design including layout, type, and number of intersections per km road. This section focuses on two street design parameters—bike route and number of intersections.

Bike route length: Employing bikes as transport means depends on many factors, including cultural aspects, safety, and connectivity of the bike routes. The availability of bike lanes statistically proved to be significant predictors of vehicle ownership and transit use [43]. The availability of bike lanes in the neighborhood and their length, in terms of ratio of total bike routes to roadway lengths, is analyzed to determine the impact on automobile use.

The study summarized in this section considers bike route ratios ranging from 5% to 100% of the overall major streets (100% indicates that all streets provide bike lanes).

Number of street intersections: The number of intersections per road-kilometer constitutes a potential indicator of impact of road layout on transport mode. In general, motor vehicle-oriented developments tend to reduce the intersections along main streets, so as to enhance the overall vehicular movement. The summary presented below is based on an investigation of the effect of changing the number of intersections in a street network, on automobile daily personal vehicle and public transit passenger km traveled. The number of intersections in the studied neighborhood (Fig. 8.17) is varied from 3 (25% of the 12 default intersections) to 24 (200%) of the base number of 12 intersections, by 25% (3 intersections) increments. It should be borne in mind that the number of intersections in the street network of a neighborhood affects the whole neighborhood design including the accessibility to buildings, and thus cannot be arbitrarily determined. Such considerations are not accounted for in the present study, which focuses mainly on the impact of street design on transportation modes and associated GHG emissions, employing statistical data [48, 49].

Main Observations

The main observations related to the interaction between neighborhood design and energy and transport performance are summarized below [48].

- The analysis of the effect of density (number of residential units per acre) indicates that the effect of this parameter on energy use for transport is not significant on per capita basis. Notwithstanding the substantial attention given to density effect in research conducted on transportation in urban areas, results of this study indicate that density per se, isolated from other factors, does not have a significant effect on travel mode. It is critical therefore to distinguish between the impact of density as an isolated factor and the cumulative effects of various additional factors of a compact development such as land use, transit accessibility, job availability, walkability, and others.
- Mixed-use communities consisting of residential, employment, and retail/services in close proximity to each other can significantly reduce the distance traveled per

day and therefore transport related energy use and associated GHG emissions. For residential neighborhoods, the effect of the distance from the central business district (CBD) is very significant. The increase in GHG emissions of a neighborhood with a CBD at a distance of 30 km can reach 40%, relative to a CBD distance of 5 km or less.

- Street design can affect transport mode. The study summarized in this section indicates that the availability of biking lanes can decrease the number of trips per vehicles and associated GHG emissions. Adopting bikes as transport mode depends on several immeasurable factors, including cultural issues. The use of individual private cars can be further reduced when designing streets with increased number of intersections.
- Assessment of energy use by transport and by building operations highlights the significance of reducing energy consumption by both sectors. Building operations have a substantial negative impact on the environment in low-energy efficiency neighborhoods. Improving energy performance of this sector is within reach as shown in the sections above. Although net-zero energy status of buildings in a neighborhood is achievable, transport, when taken into account, can change this balance. Improvement in the transport sector needs to be considered as a high priority objective in planning of sustainable neighborhoods and urban areas.

Although the summary presented above gives an insight into the effect of various design parameters on energy consumption by buildings and transportation, and associated GHG emissions, some socio-geographic parameters need to be addressed, in conjunction with these design parameters. For instance, employment opportunities within a specific distance (5 km, as specified in [49]) are a key component affecting the distance traveled, and therefore, require more investigation. Another important factor is the average distance from home or work to nearest public transport stations. This factor also represents a significant impact on the mode of transport.

8.5.2 *Impact of Mixed-Use Design on Resilience*

The neighborhood design, and more specifically the layout of streets, can affect the resilience and functionality of the neighborhood during disasters or other types of emergency interruptions. A main component of the neighborhood resilience is to maintain a functional connectivity between areas of the neighborhood under all circumstances.

This section presents a summary of an assessment of the vulnerability of various street layouts. The vulnerability of the street network to various interruptions is evaluated employing the primal graph, consisting of a network of nodes and links representing the streets and their intersections [50]. The analysis entails the study of a number of indicators of connectivity and centrality [31, 51]. These measures are based on a number of parameters related to graph theory network analysis.

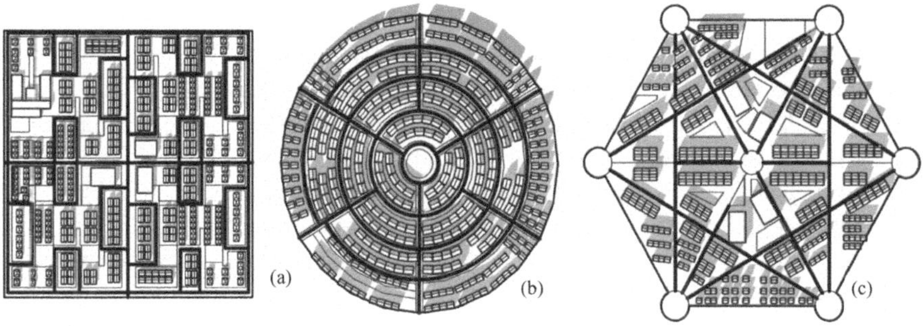

Fig. 8.18 Illustration of 3 neighborhoods associated with different street layouts, **a** rectilinear, **b** radial, **c** hexagonal

Summary of the Investigation

Studied Layouts

The neighborhood employed for this study forms a part (one quarter) of the larger mixed-use neighborhood presented above (Fig. 8.13). The buildings are rearranged to present all the characteristics of the mixed development, contained within a small area (40 acres) [31].

Assessment of the street network resilience is conducted through the analysis of functionality of the network in case of disruption, attributed to extreme events including natural disasters. The neighborhood layouts considered include rectilinear street network based on the fused grid design [6], a radial (circular) layout, and a hexagonal layout, with all nodes linked by straight roads. The neighborhoods are high density developments (average of 25 residential units per acre), featuring the same number of buildings having the same geometrical and thermal characteristics. Figure 8.18 presents an illustration of the 3 studied layouts.

Parameters

Centrality Parameters

Several centrality parameters are analyzed to assess the impact of various nodes and links in the neighborhood's street network, employing the primal graph theory [52]. These parameters are employed as well to determine the accessibility within the street network, which is a key factor during emergencies [51]. The centrality parameters are detailed below [31].

1. *Degree centrality*—The degree centrality expresses the number of direct paths between a node in a specific street network to other nodes in the primal graph. Higher number of connections of a node implies a higher degree centrality. The functionality and reliability of the street network can be significantly affected by any disruption in nodes/links with high degree centrality [53].
2. *Closeness centrality*—This indicator represents a measure of the closeness of a specific node to other nodes in the street network along the shortest path. It is inversely proportional to the sum of shortest distances between this specific node and all other nodes [50]. It is strategically beneficial to place

evacuation areas and emergency services in near proximity to nodes with high closeness centrality values. Such design considerations will allow to improve the accessibility to critical services in time of emergency [54].

3. *Straightness centrality*—The straightness centrality reports the deviation of connections between a specific node and various other nodes, from imaginary straight lines. This indicator is used to estimate the degree of straightness of shortest paths from a node to all other nodes in the street network.
4. *Betweenness centrality*—The betweenness centrality indicates the extent to which a node is located between other nodes. A lower value of betweenness centrality is desirable to increase the resilience of a neighborhood, allowing to reduce potential disruptions during emergency.
5. *Street network efficiency*—This metric indicates the directness of links between different nodes of the street network. The street network efficiency integrates the straightness centralities of all nodes.
6. *Information centrality*—The information centrality indicates the impact of eliminating a specific node and its links from the street network, on the overall functionality of this network.

Connectivity Parameters
The resilience of the street network of a neighborhood can be further analyzed employing a number of connectivity parameters. These parameters are usually utilized to assess the functionality of street networks under disruptions and unusual circumstances [31, 51]. Connectivity related parameters are summarized below.

1. *Characteristic path length*—The characteristic path length is employed to estimate the average distance of the shortest routes among different pairs of nodes in the street network. A lower value of path length is desirable to enhance the resilience of a neighborhood's street network.
2. *Cyclomatic number*—This indicator represents a measure of the number of all potential connections between a pair of nodes in a street network. A higher value of Cyclomatic number corresponds to larger number of alternative routes available in case of disruptions.
3. *Alpha index*—The alpha index (or *meshedness coefficient*) is the ratio of the Cyclomatic number to the maximum number of loops in the street network, while keeping the same number of nodes [55].
4. *Beta index*—The beta index is a measure of the frequency of paths between the nodes. It is the ratio of the number of links to the number of nodes in the street network.
5. *Gamma index*—Similar to the Beta index, the gamma index is the ratio of the total number of links in the network to the maximum possible number of links.

Main Observations
The analysis of the three site layouts presented above—rectilinear, radial, and hexagonal—indicates that a higher degree of resilience is achieved by the hexagonal layout [31]. The majority of indicators of network centrality, efficiency, and connectivity

are more favorable in the hexagonal layout as compared to the other two layouts. The circular street network performance is basically comparable with the hexagonal layout. Some of the main observations are discussed below.

- The performance of the hexagonal layout is critical in three of the centrality indicators (out of 11 overall indicators)—degree centrality, betweenness, and information centralities. This is largely due to the high centrality of the central node. A number of design modifications in the street network can be implemented to address issues related to high centrality of specific nodes. It should be noted, however, that a strong correlation exists between various parameters, and consequently attempting to improve some of these parameters need to be carefully analyzed to not reduce the efficiency of other parameters. Regarding the overall layout efficiency, the hexagonal layout displays a significantly high efficiency of the street network system (94%), indicating a high degree of directness of links between the neighborhood nodes.
- The circular layout is analogous to the hexagonal layout in many aspects of the street network performance. The information centrality of the circular network is better than the hexagonal layout, indicating superior performance of the neighborhood when all links to a specific node are disrupted. The overall efficiency of the neighborhood is about 85%.
- The square neighborhood, encompassing four quadrants designed according to the fused grids [6], needs to be reconsidered, in terms of resilience and functionality of the streets and intersections, in case of major interruptions. Although this rectilinear street system is superior for some of the indicators, such as degree centrality, betweenness, and information centralities, it is not for other indicators. The overall efficiency of this rectilinear street network is significantly lower than the other two layouts (62%). A number of the issues identified in this neighborhood street network are inherent to the fused grid design. For instance, the length of the path that needs to be taken from one node to the other is high, even within the quadrants themselves. There is need therefore to rethink the design principles of such grid-based developments, taking the resilience to potential interruptions into account.
- Some of the centrality indicators such as the degree centrality, betweenness, and information centralities are the most favorable for the square street network. The hexagonal street layout performs better than the two other studied layouts in terms of closeness centrality and straightness centrality. Considering the connectivity indicators, the performance of the hexagonal layout is superior to the two other layouts for all indicators, followed by the circular neighborhood pattern.

The study summarized above presents a unique insight into the impact of neighborhood layout on the vulnerability of the street network within a neighborhood to various disruptions. The study brings into attention the role that various design variables of a neighborhood play in the resilience and performance of a neighborhood. A holistic design approach is needed to achieve a high-performance resilient neighborhood. Such approach should consider how the neighborhood layout will function in case of interruptions while allowing a near optimal orientations to enable passive and active solar design, enhancing thus the overall efficiency of the neighborhood.

References

1. UNEP SBCI (2009) Buildings and climate change summary for decision-makers. http://www.unep.org/sbci/pdfs/SBCI-BCCSummary.pdf
2. Steemers K (2003) Energy and the city: density, buildings and transport. Energy Build 35:3–14
3. Khan Ribeiro S et al (2007) Transport and its infrastructure. In climate change 2007: mitigation. Intergovernmental panel on climate change. Cambridge University Press, Cambridge and New York, NY
4. Song Y, Knaap GJ (2004) Measuring the effects of mixed land uses on housing values. Reg Sci Urban Econ 34:663–680
5. Jacobs J (2016) The death and life of great American cities. Vintage, New York
6. Song Y, Merlin L, Rodriguez D (2013) Comparing measures of urban land use mix. Comput Environ Urban Syst 42:1–13
7. Duany A, Plater-Zyberk E, Speck J (2010) Suburban Nation: the rise of Sprawl and the decline of the American dream. Farrar, Straus and Giroux, New York
8. Kelbaugh DS (2002) Three urbanisms: new, everyday and post
9. Badland H, Schofield G (2005) Transport, urban design, and physical activity: an evidence-based update. Transp Res Part D Transp Environ 10(3):177–196
10. Grant J (2002) Mixed use in theory and practice: Canadian experience with implementing a planning principle. J Am Plan Assoc 68(1):71–84
11. Atash F (1994) Redesigning suburbia for walking and transit: emerging concepts. J Urban Plan Dev 120(1):48–57
12. Curtis C (2008) Evolution of the transit-oriented development model for low-density cities: a case study of Perth's new railway corridor. Plan Pract Res 23(3):285–302
13. Cervero R, Dai D (2014) BRT TOD: leveraging transit oriented development with bus rapid transit investments. Transp Policy 36:127–138
14. Cervero R, Sullivan C (2011) Green TODs: marrying transit-oriented development and green urbanism. Int J Sustain Dev World Ecol 18(3):210–218
15. Dittmar H, Ohland G (eds) (2012) *The new transit town: best practices in transit-oriented development*. Island Press
16. Musterd S, Andersson R (2005) Housing mix, social mix, and social opportunities. Urban Aff Rev 40(6):761–790
17. Atkinson R (2005) Occasional paper 1: neighbourhoods and the impacts of social mix: crime, tenure diversification and assisted mobility. Housing and Community Research Unit/ESRC Centre for Neighbourhood Research
18. Barton H (2013) Conflicting perceptions of neighbourhood. In: Sustainable communities. Routledge, pp 25–40
19. Morello E, Ratti C (2009) Sunscapes: "solar envelopes" and the analysis of urban DEMs. Comput Environ Urban Syst 33(1):26–34
20. Conceição António CA, Monteiro JB, Afonso CF (2014) Optimal topology of urban buildings for maximization of annual solar irradiation availability using a genetic algorithm. Appl Therm Eng 73(1):424–437
21. Sanaieian H, Tenpierik M, van den Linden K, Seraj FM, Shemrani SMM (2014) Review of the impact of urban block form on thermal performance, solar access and ventilation. Renew Sustain Energy Rev 38:551–560
22. Steemers K (2003) Energy and the city: density, buildings and transport. Energy Build 35(1):3–14

23. Van Esch MME, Looman RHJ, de Bruin-Hordijk GJ (2012) The effects of urban and building design parameters on solar access to the urban canyon and the potential for direct passive solar heating strategies. Energy Build 47:189–200
24. Strømann-Andersen J, Sattrup PA (2011) The urban canyon and building energy use: urban density versus daylight and passive solar gains. Energy Build 43(8)
25. Chatzipoulka C, Nikolopoulou M, Watkins R (2015) The impact of urban geometry on the radiant environment in outdoor spaces. In: 9th international conference on urban climate
26. Jabareen YR (2006) Sustainable urban forms: their typologies, models, and concepts. J Plan Educ Res 26(1):38–52
27. Kristl Ž, Krainer A (2001) Energy evaluation of urban structure and dimensioning of building site using iso-shadow method. Sol Energy 70(1):23–34
28. Lee KS, Lee JW, Lee JS (2016) Feasibility study on the relation between housing density and solar accessibility and potential uses. Renew Energy 85:749–758
29. Hachem C, Athienitis A, Fazio P (2014) Energy performance enhancement in multistory residential buildings. Appl Energy 116:9–19
30. Hachem-Vermette C (2015) Integrated design considerations for solar communities. J Green Build 10(2):134–156
31. Hachem-Vermette C, Singh K (2019) Mixed-use neighborhoods layout patterns: impact on solar access and resilience. Sustain Cities Soc 51:101771
32. Lund H, Werner S, Wiltshire R, Svendsen S, Thorsen JE, Hvelplund F, Mathiesen BV (2014) 4th generation district heating (4GDH): integrating smart thermal grids into future sustainable energy systems. Energy 68:1–11
33. De Winter F (1991) Solar collectors, energy storage, and materials. The MIT press, Cambridge
34. Zalba B, Marin JM, Cabeza LF, Mehling H (2003) Review on thermal energy storage with phase change: materials, heat transfer analysis and applications. Appl Therm Eng 23:251–283
35. Sharma A, Tyagi VV, Chen CR, Buddhi D (2009) Review on thermal energy storage with phase change materials and applications. Renew Sustain Energy Rev 13:318–345
36. Ellabban O, Abu-Rub H, Blaabjerg F (2014) Renewable energy resources: current status, future prospects and their enabling technology. Renew Sustain Energy Rev 39:748–764
37. Zhai XQ, Qu M, Yu X, Yang Y, Wang RZ (2011) A review for the applications and integrated approaches of ground-coupled heat pump systems. Renew Sustain Energy Rev 15(6):3133–3140
38. Sibbit B, McClenahan D, Djebbar R, Paget K (2015) Groundbreaking solar, high performance buildings magazine, p 1–12. www.hpbmagazine.org
39. Scognamiglio A, Garde F (2016) Photovoltaics' architectural and landscape design options for Net Zero Energy Buildings, towards Net Zero Energy Communities: spatial features and outdoor thermal comfort related considerations. Prog Photovolt Res Appl 24(4):477–495
40. ASHRAE (2004) Ventilation for acceptable indoor air quality. ANSI/ASHRAE Standard 62.1-2004. American Society of Heating, Refrigerating and Air-Conditioning Engineers, Atlanta, GA
41. ASHRAE (2004) Energy standard for buildings except low-rise residential buildings. ANSI/ ASHRAE/IESNA Standard 90.1-2004. American Society of Heating, Refrigerating and Air-Conditioning Engineers, Atlanta, GA
42. Hachem-Vermette C, Guarino F, La Rocca V, Cellura M (2019) Towards achieving net-zero energy communities: investigation of design strategies and seasonal solar collection and storage net-zero. Sol Energy 192:169–185
43. Hachem-Vermette C, Singh K (2019) Optimization of the mixture of building types in a neighborhood and their energy and environmental performance. Energy Build 204:109499
44. Singh K, Hachem-Vermette C (2019) Impact of commercial land mixture on energy and environmental performance of mixed use neighbourhoods. Build Environ 154:182–199
45. Hachem-Vermette C, Grewal KS (2019) Investigation of the impact of residential mixture on energy and environmental performance of mixed use neighborhoods. Appl Energy 241:362–379
46. Deb K (2001) Multi-objective optimization using evolutionary algorithms. Wiley, New York

47. Singh K, Das R (2016) An experimental and multi-objective optimization study of a forced draft cooling tower with different fills. Energy Convers Manage 111. https://doi.org/10.1016/j.enconman.2015.12.080
48. Hachem C (2016) Impact of neighborhood design on energy performance and GHG emissions. Appl Energy 177:422–434
49. IBI Group (2000) Greenhouse gas emissions from urban travel: tool for evaluating neighbourhood sustainability
50. Porta S, Crucitti P, Latora V (2006) The network analysis of urban streets: a primal approach. Environ Plan B Plan Des 33(5):705–725
51. Sharifi A (2019) Resilient urban forms: a review of literature on streets and street networks. Build Environ 147(July 2018):171–187
52. Caprì S, Ignaccolo M, Inturri G, Le Pira M (2016) Green walking networks for climate change adaptation. Transp Res Part D Transp Environ 45:84–95
53. Batty M (2013) Resilient cities, networks, and disruption. Environ Plan B Plan Des 40:571–573
54. Novak DC, Sullivan JL (2014) A link-focused methodology for evaluating accessibility to emergency services. Decis Support Syst 57:309–319
55. Buhl J, Gautrais J, Reeves N, Solé RV, Valverde S, Kuntz P, Theraulaz G (2006) Topological patterns in street networks of self-organized urban settlements. Eur Phys J B 49(4):513–522

The manufacturer's authorised representative in the EU is Springer Nature Customer Service Centre GmbH, Europaplatz 3, 69115 Heidelberg, Germany. If you have any concerns regarding our products, please contact ProductSafety@springernature.com

Printed and bound by CPI Group (UK) Ltd, Croydon, CR0 4YY
30/09/2026
02225648-0008